Zoophysiology Volume 39

Editors:
S.D. Bradshaw   W. Burggren
H.C. Heller   S. Ishii   H. Langer
G. Neuweiler   D.J. Randall

Springer
Berlin
Heidelberg
New York
Barcelona
Hong Kong
London
Milan
Paris
Tokyo

Zoophysiology

Volumes already published in the series:

P.J. Bentley

Endocrines and Osmoregulation

A Comparative Account in Vertebrates

Second Edition of Volume 1

With 30 Figures and 23 Tables

 Springer

Prof. Dr. Peter J. Bentley
The University of Western Australia
Dept. of Physiology
Nedlands, Western Australia 6907

QP
90
.6
.B45
2002

ISBN 3-540-42683-3 Springer-Verlag Berlin Heidelberg New York
ISSN 0720-1842

The first edition was published 1971 as Volume 1 of the series.

Library of Congress Cataloging-in-Publication Data

Bentley, P.J.
   Endocrines and osmoregulation : a comparative account in vertebrates / by P.J. Bentley. – 2nd ed.
     p. cm.
   Includes bibliographical references and index.
   ISBN 3540426833
   1. Osmoregulation. 2. Endocrinology. 3. Vertebrates – Physiology. I. Title.
QP90.6 .B45 2002
573.4'16 – dc21

                        2001054976

Springer-Verlag Berlin Heidelberg New York
a member of BertelsmannSpringer Science+Business Media GmbH

http://www.springer.de

The use of general descriptive names, registered names, trademarks, etc. in this publication does not imply, even in the absence of a specific statement, that such names are exempt from the relevant protective laws and regulations and therefore free for general use.

Production: PRO EDIT GmbH, Heidelberg, Germany
Cover design: design & production GmbH, Heidelberg, Germany
Typesetting: SNP Best-set Typesetter Ltd., Hong Kong

Printed on acid-free paper   SPIN 10760822   31/3130/Di   5 4 3 2 1 0

Dedication

To Sebastian Dicker For Friendship

Preface to the Second Edition

The opportunity to prepare a second edition of a book that was originally written over 30 years ago has provided me with both a challenge and a source of pleasure; the former as it needed to be spatially constrained within its original limits. Nevertheless, over 1000 references have been added. I must apologize to the many biologists whose contributions could not be included. I have attempted to keep the original format and historical perspective. The information has been principally described within the context of each phyletic group of the vertebrates and their habitats. Each chapter is reasonably self-contained, but appreciation of material in later chapters, as often indicated, can be amplified by reference to Chapters 1 and 2. Information that was provided in tables in the first edition has now often been summarized in the text. Reviewing the work of earlier contributions to this field of study has evoked many pleasant memories of friends and acquaintances, some deceased, events and occasions. It has been a particular pleasure to perceive the consequences of such observations and know some of the answers to the questions that they raised. A new generation of such questions has now emerged, which is one of the reasons for preparing this summary.

I would like to thank Professor Don Bradshaw for suggesting that this book may be welcome and Springer-Verlag for making it possible. The Physiology Department and the Biological Sciences Library of the University of Western Australia have provided me with the facilities and succour that I needed to complete this book. Dr. Norman Goodchild provided invaluable help in preparing some of the figures for the publisher. Thank you all.

Peter Bentley

The University of Western Australia
December 2001

Preface to the First Edition

It is an old and trite saying that a preface is written last, placed first and read least. It may also be an explanation or justification for what follows, especially if the author feels that the reader may not agree with him. This is, thus, something of an apologia.

When I first agreed to write this book, I felt that it was to be directed principally towards the endocrine system of vertebrates, especially as there were available some excellent accounts of their osmoregulation. However, it was soon apparent to me that in order to appreciate the role of the endocrines, one must strongly emphasize non-endocrine functions involved in the regulation of the animals' water and salt balance. In addition, several years have elapsed since a full account of vertebrate osmoregulation has been given. I have thus felt free to take an overall look at the animals' water and salt metabolism, but have especially emphasized more recent contributions. Possibly the book should now be called *Osmoregulation and the Endocrines*.

A rather conventional format has been retained for, despite many claims to the contrary, I do not feel that all things new, experimental and innovative are necessarily desirable. In the first two chapters I have attempted to introduce, and summarize, at a general level, the two major topics, osmoregulation and endocrines, while emphasizing the diversity of such processes in the vertebrates. In the second chapter I have also included a section on the *Criteria, Methods and Difficulties* of experimental comparative endocrinology. I felt that this information may be useful for the reader who desires to make a critical appreciation and evaluation of many of the experiments that are described. It may, hopefully, also be useful to the younger workers interested in comparative endocrinology and help them to glean something about the sort of work that they may expect to do. The vertebrate groups have been described within the framework of their customary classification, a decision arrived at primarily because this also makes a little evolutionary sense. Some have questioned the wisdom of placing the mammals first, and travelling down, rather than up, the phyletic scale. However, each chapter is reasonably complete in itself so that those who disagree can resort to the subterfuge of reading the book backwards.

I have attempted to relate the physiology of the various beasts to that of the environments that they normally occupy. At the same time I have not hesitated to descend to the more currently fashionable "molecular level". I hope that this may help both to remind some of the more "mature" readers and to emphasize to the younger ones the very broad scope of interest that animals have to offer. It is principally for the younger group that this book has been written.

New York
1970

P.J. Bentley

Acknowledgements (First Edition)

The observations of many people, only some of whom are quoted, made this book possible. I would like to extend my thanks to all who have contributed to these subjects and only wish that it would have been possible to acknowledge the work of all.

The latter part of this book was written while I was a guest of the Commissariat à L'Energie Atomique. I would like to thank them for their hospitality, particularly Professeur J. Coursaget, Chef du Departement de Biologie, and Dr. Jean Maetz and his Groupe de Biologie. The help and guidance of Jean Maetz made the latter part of this book possible. Dr. Erik Skadhauge kindly gave me some advice on avian matters.

The Mount Sinai School of Medicine of the City University of New York kindly granted me leave so that I could join the Commissariat à L'Energie Atomique. A number of nice ladies have helped me during the various stages in preparation of this manuscript; special thanks are due to Mrs. Jean Homola, Mrs. Norma Beasley, Miss Marguerite Murphy and Mrs. Wolina Shapiro.

My wife showed considerable forbearance and gave much-needed help in the final preparation of the proofs. The editorial help and advice of Dr. D. S. Farner is gratefully acknowledged as well as the cooperation of the staff of Springer-Verlag in Heidelberg and Mr. B. Grossmann in New York.

Contents

Chapter 1

Osmotic Problems of Vertebrates

General Introduction

Water provides the vehicle and molecular framework in which life on earth is maintained and perpetuated. It is the host and mentor of the macromolecules of proteins and nucleic acids that are intrinsic to this process. Under prevailing cosmic conditions water exhibits many chemical and physical properties that contribute to the stability, tertiary structure and replication of such proteins and nucleic acids, and the normal functioning of the assembled organism. Molecules of water in liquid form are electrically dipolar and readily form hydrogen bonds with each other and neighbouring compounds. They can also dissociate to provide essential hydrogen and hydroxyl ions. The physical properties of water include an appropriate diffusible and bulk mobility, tensile strength and a relatively high heat capacity, boiling point, melting point and heat of vaporization. Water acts as a solvent for crystalline salts such as sodium and potassium chlorides, which can then form electrically charged ions (Na^+, K^+, Cl^-). Such ions can participate in processes involving the transmission of information and the transformation of energy. Many other molecules, including essential nutrients and carbon dioxide, can dissolve in water. The properties and behaviour of water may be changed by the presence of such solutes, which can disturb hydrogen bonding, change the solubility and tertiary structures of macromolecules and decrease its mobility. The marriage of water and solutes is essential for the perpetuation of life, but can sometimes lead to disharmony. Osmoregulation is the process that attempts to forestall and correct such problems.

The vertebrates originated about 500 million years ago, probably evolving from a marine protochordate. They have since dispersed to occupy many diverse habitats. This colonization of the planet is exemplified by the great variety of extant species. A synoptic classification of such vertebrates is provided in Table 1.1. It is principally based on their anatomy and physiology and reflects, and illustrates, their progressive evolution. It is intended to provide a useful framework for the subsequent discussion of their osmoregulation.

1 Geographical and Ecological Distribution of the Vertebrates (Osmotic Distribution)

While it is uncertain if the vertebrates originated in the sea or fresh water, subsequently both environments, as well as the dry land, have been occupied by rep-

Table 1.1. A classification of the vertebrates

Phylum Chordata
  Subphylum Urochordata (tunicates)
  Subphylum Cephalochordata (amphioxus)
  Subphylum Vertebrata
Superclass Agnatha (jawless fishes, cyclostomes)
  Class Myxini
    Order Myxiniformes (hagfish)
    Order Pteraspidomorpha (includes extinct ostracoderms)
  Class Cephalaspidopmorpha
    Order Osteostraci (extinct ostradoderms)
    Order Petromyzontiformes (lampreys)
Superclass Gnathostomata (jawed vertebrates)
  Class Placodermi (extinct heavily armoured jawed fish)
  Class Chondrichthyes (cartilaginous fish, placoid scales, no swim bladder)
    Subclass Elasmobranchii (uncovered gill slits)
      Order Selachimorpha (Squaliformes) (sharks)
      Order Batoidimorpha (Rajiformes) (skates, rays)
    Subclass Holocephali (gill slits covered, chimaeras, ratfish, rabbitfish)
Class Osteichthyes (bony fish, gill slits covered, bony scales, usually a swim bladder)
  Subclass Actinopterygii (ray-finned fishes, includes most extant bony fishes)
    Superorder Chondrostei (partly cartilaginous skeleton, mostly extinct)
      Order Acipenseriformes (sturgeons, paddlefish)
      Order Polypteriformes (African freshwater bichirs and reedfish, also classified In the
        Subclass Brachyioptergii)
    Infraclass Holostei
    Order Lepisosteiformes (freshwater gars)
    Order Amiiformes (freshwater bowfin)
    Infraclass Teleostei (most extant bony fishes)
    (The Holostei and Teleostei are also classified as being in the Superorder Neopterygii)
  Subclass Sarcopterygii (Choanichthyes) (lobe-finned fishes)
    Superorder Crossopterygii (most extinct)
      Division Rhipidistia (extinct, possible ancestors of the amphibians)
      Division Coelacanthiformes (one extant marine species, *Latimeria chalumae*, the
        coelacanth)
    Superorder Dipnoi (lungfish, three extant freshwater genera)
Tetrapoda (an informal name for vertebrates with four legs: amphibians, reptiles, birds and
  mammals)
Class Amphibia (phyleticaly highest anamniotes, mostly terrestrial, aquatic larvae, soft
  permeable skin)
  Subclass Lissamphibia (includes all extant amphibians)
    Order Apoda (Gymnophiona) (caecilians; limbless, tropical and subtropical, burrowing
      or aquatic)
    Order Urodela (Caudata) (tailed amphibians, newts and salamanders)
    Order Anura (Salientia) (frogs and toads)
Amniotes; an informal grouping of reptiles, birds and mammals, extraembryonic membranes
  form a developmental sac enabling eggs to be laid on dry land.
Class Reptilia (formerly predominant land vertebrates, ectodermal scales, metanephric
  kidney)
  Subclass Anapsida
    Order Chelonia (Testudinata) turtles, terrapins, tortoises
  Subclass Lepidosauria (a diapsid skull)
    Order Rhynchocephalia (Sphenodonta) (lizard-like reptiles from the Triassic, one extant,
      Spenodon punctatus)
    Order Squamata (include majority of extant reptiles)
      Suborder Ophidia (Serpentes) (snakes)

Table 1.1. *Continued*

Suborder Lacertilia (Sauria) (lizards)
Subclass Archosauria ("ruling reptiles", include the dinosaurs and bird stock)
  Order Crocodilia (crocodile, alligators, caimans, gavials)
Class Aves (birds, feathered vertebrates)
  Subclass Neornithes
    Superorder Palaeognathae (ratite running flightless birds, ostriches, emus, rheas. No keel on sternum)
    Superorder Neognathae (carinate birds, with a keel on the sternum, which usually fly. About 20 extant orders, the Passeriformes (perching birds) comprise about half of all bird species.)
Class Mammalia (vertebrates with hair and which suckle their young)
  Suclass Prototheria (egg-laying mammals)
    Infraclass Monotremata (monotremes, echidna, platypus)
  Subclass Theria
    Infraclass Metatheria (yolk sac forms placenta, relatively undeveloped at birth)
      Order Marsupialia (marsupials, young usually suuckled in an pouch)
        Cohort Ameridelphia (N. and S. American marsupials, includes opossum)
        Cohort Australidelphia (Mostly live in Australia and New Guinea, one order In S. America. Kangaroos, wallabies, possums, koala, wombats etc.)
    Infraclass Eutheria (placentals, a "true" allantoic placenta)

resentatives of most of the major groups. The geographical location(s) for the transition to a vertebrate is also unknown, but subsequent dispersal has been widespread. The major groups of fishes, the Agnatha, Chondrichthyes and Osteichthyes, have representatives in the sea and fresh water. The tetrapods are primarily terrestrial, but the Amphibia, reptiles, birds and mammals, all have species that normally live in marine or freshwater environments. Osmoregulatory processes play a major role in such dispersal, and it is not surprising to observe that the physiological mechanisms by which such adaptations take place, often differ greatly. Even among the fishes, an apparently novel occupation of a certain medium may represent a secondary, tertiary or even further removed alteration of the habitat that was primary in the group. Darlington (1957) has traced the probable historic origins of the endemic catfishes of Australia: teleost fish originated in the sea from whence a fish related to the order Isospondyli moved into fresh water to give rise to the order Ostariophysi. Some of this group probably re-entered the sea and returned subsequently to fresh water in Australia. Among the tetrapods at least one amphibian, the crab-eating frog, *Rana cancrivora*, has returned to a marine environment, and even more numerous examples of such a transition can be seen among the reptiles, that have marine and freshwater as well as terrestrial representatives. The biological adaptations accompanying such transitions are sometimes very similar in different groups, providing examples of convergent evolution. In other instances, more novel and unique mechanisms have evolved. Thus, both the marine frog, *Rana cancrivora*, and the sharks and rays maintain their body fluids hyperosmotic to sea-water by accumulating urea. On the other hand, marine reptiles like the loggerhead turtle, *Caretta caretta*, have body fluids similar in concentration to their terrestrial relatives which is hyposmotic to sea-water. However, they can excrete salt as a highly concentrated solu-

3

tion from a modified orbital gland. Thus, *Rana cancrivora* has evolved a mechanism parallel to that of the sharks and rays, while the reptiles have utilized a novel mechanism not seen in their phyletic forbears. Both similarities and diversities in osmoregulatory mechanisms are found within, and between, the major phyletic groups of vertebrates.

The vertebrates have occupied most of the earth's geographic areas, being sparse in the cold terrestrial polar regions, and not as numerous in hot dry deserts as in the wetter tropical zones. The desert regions, where the supply of water may be limiting to life, make up about one third (50 million square kilometers) of the land surface of the earth (Schmidt-Nielsen 1964a). Despite the potential osmoregulatory problems, vertebrates do live in even the extremely dry parts of such desert areas, and exhibit interesting physiological and behavioural patterns consistent with their life there. The seas have a diverse vertebrate fauna of fish, reptiles, birds and mammals. Major geographic limitations exist for the various vertebrate groups and will be discussed later, but in the instance of the Amphibia and reptiles, they are dictated mainly by temperature rather than by water.

Geographic dispersal, followed by genetic isolation, has played an important role in the process of evolution. Both the ability and inability to osmoregulate in different situations have played a part in breaking and maintaining the physical barriers involved. Deserts can retard the dispersal of animals (Darlington 1957), as shown by the distribution of the Amphibia in Australia. Darwin (1839) considered the sea to be the major physical barrier to dispersal, a view that is still current (Darlington 1957). The sea constitutes the most effective barrier to the movement of the freshwater fishes, and Darwin noted the absence of Amphibia and terrestrial mammals from most oceanic islands, where reptiles are frequent inhabitants. This probably reflects the osmoregulatory powers of the different groups. Reptiles, with their less permeable integument, considerable tolerance to internal osmotic change, extrarenal salt excretion and marked independence of environmental temperature, would appear to be better suited to prolonged oceanic voyages than freshwater and terrestrial Amphibia and mammals. At Duke University I kept a freshwater turtle, *Pseudemys scripta*, for 30 days in sea-water. At the end of this time it was sluggish and had a plasma sodium concentration of nearly $300\,mEq\,l^{-1}$, but on being returned to fresh water it resumed its normal life. Prolonged oceanic voyages are thus conceivable, even by contempory species of freshwater-terrestrial reptiles, a tribute to their osmoregulatory capabilities.

2 The Osmotic Anatomy of the Vertebrates

Water is the predominant constituent of living organisms. In vertebrates it normally makes up about 70% of the total body weight and there are no consistent differences between the phyletic groups. However, tolerance to dehydration can vary. Humans can only tolerate a loss of water equivalent to about 12% of their body weight (Adolph 1947) and this value appears to be similar to that in other mammals. Some frogs, such as the Australian desert species *Cyclorana platy-*

cephala can survive, following rehydration, a loss by evaporation of 50% of their body weight (Main and Bentley 1964). However, tree frogs (*Litoria*, formerly *Hyla*), also from Australia, die after losing half this amount of water. Mortality due to water loss appears to reflect several events, including the effects of the increased osmotic and solute concentrations on tissue life processes and a failure of the blood circulatory system. The total body water is sequestered in separate regions (spaces or compartments) that have different chemical compositions. Their size can be estimated by measuring the volume of distribution of chemical markers such as inulin and Evans blue in the body fluids. Fluid equivalent to about 45 to 55% of a vertebrate's body weight is present inside cells (intracellular compartment) where it is contained by the cell membranes. The extracellular fluid (the inulin space) amounts to 15 to 25% of the total body weight (it varies from about 16% in some fish to 25% in other vertebrates) and is divided into the blood plasma (measured with Evans blue) and the intercellular (or interstitial) fluids. The two latter fluid compartments are separated by the capillaries. The plasma contains higher concentrations of proteins than the intercellular fluid. The plasma volume may be as low as 2% of the body weight in some fish but it is usually about 5%. The physiological maintenance of these different fluid compartments primarily depends on the functioning of the cell membranes and the cardiovascular-capillary systems. Secondarily, it depends on the regulation of the solute levels in the extracellular fluid. This process results from the activities of the kidneys and gut, and phyletically diverse salt-secreting tissues such as salt glands in some fish, reptiles and birds, and the gills in many fish.

The osmotic concentrations in the different fluid compartments in the body are similar, though their particular solute constituents differ. The composition of the extracellular and intracellular fluids differs; the latter has relatively high concentrations of K^+ while in the former the Na^+ and Cl^- are higher (Tables 1.2, 1.3). (In humans the total body content of osmotically active K^+ is 46 mmol kg^{-1} body weight while that of Na^+ is 42 mmol kg^{-1}: Thorn 1960.) The intracellular concentrations of K^+ and Na^+ are remarkably similar in different vertebrates (Table 1.3). Such uniformity appears to reflect requirements for the stability and activity of their contained macromolecules, especially those of enzymes (see Yancey et al. 1982). However, the total intracellular osmotic concentrations of solutes can vary in different species. In some marine fish, such as hagfish (Agnatha), sharks and rays (Chondrichthyes) and the coelacanth, *Latimeria* (Crossoptergygii), it is similar to that of the sea-water in which they live. Nevertheless, the concentrations of salts in the intracellular fluids of such fish remain similar to those in other vertebrates (Table 1.3). The differences between the ionic and total osmotic concentrations are mainly made up by the presence of organic solutes (osmolytes) such as amino acids, urea and methylamine compounds including trimethylamine oxide (TMAO) and betaine. The extracellular fluids are not as fastidious with respect to their salt content, as the intracellular fluid and the salt concentration may be increased to achieve an osmotic balance in some species, especially hagfish (Table 1.2). In sharks and rays and the coelacanth, organic osmolytes, including urea, may also be accumulated in the extracellular fluids. Such organic osmolytes appear to have minimal perturbing effects on the activity of cell macromolecules as compared to increased levels of inorganic ions. However, in many

5

Table 1.2. Principal osmotic constituents of the blood plasma or serum of various species

	Osmotic conc. (mosmol kg^{-1} H$_2$O)	Na	K	Cl	Urea
		(mmol kg^{-1} H$_2$O or l^{-1})			
Sea-water	about 1000	470	10	550	
Fresh water	0.2–25	0.1–6	0.06–0.11	0.03–13	
Invertebrates					
Cuttlefish[a] SW		465	22	591	
Vertebrates					
Agnatha					
Hagfish[b] SW	969	487	9	510	
(*Myxine glutinosa*)					
Lamprey FW	270	120	3	96	<1
(*Lampetra fluviatilis*)					
Chondrichthyes					
Spiny dogfish[c] SW	980	296	7	276	308
(*Squalus acanthias*)					TMAO 72
Freshwater sawfish FW	540			170	130
(*Pristis microdon*)					
Amazon stingray[d] FW	308	146	–	135	<1
(*Potamotrygon* sp.)					
Osteichthyes					
Coelacanth[e] SW	957	105	53	100	290
(*Latimeria chalumnae*)					TMAO 109
African lungfish FW	238	99	8	44	500
(*Protopterus aethiopicus*)					
(aestivation)					
Flounder SW	364	194	5	166	–
FW	304	157	5	114	–
(*Platichthys flesus*)					
Amphibia					
European green toad	279	113	5	99	16
(*Bufo viridis*)					
Aestivation 3 months[f]	1320	231	–	181	900
Crab-eating frog[g] (80% SW)	830	252	14	227	350
(*Rana cancrivora*)					
Reptilia					
Bobtail lizard[h] Winter		151	5	–	
(*Tiliqua rugosa*)					
Summer	–	196	5	–	–
Birds					
Gull[i]		154	4		
(*Larus glaucescens*)					
Mammals					
Laboratory rat	324	150	6	119	7

The remainder of the information is from the summary of Bentley (1971a).
FW = fresh water; SW = sea-water.
[a] Robertson (1965).
[b] Robertson (1976).
[c] Robertson (1975).
[d] Gerst and Thorson (1977).
[e] Lutz and Robertson (1971).
[f] Degani et al. (1984).
[g] Gordon et al. (1961).
[h] Bentley (1959).
[i] Holmes et al. (1961).

Table 1.3. Intracellular solute concentrations in various species

	Na	K	Cl mmol kg⁻¹ water	Amino acids	Urea	TMAO	Betaine
Sea-water	470	10	550				
Bacteria							
Salmonella (150 mM NaCl)	131	238	<5	109			
Halobacterium salinarium (5.1 M NaCl)	400–800	7500	3610	209			
Invertebrate							
Cuttlefish (*Sepia officianalis*) SW	31	189	–	483	–	86	108
Vertebrates							
Hagfish (*Myxine glutinosa*) SW	32	142	41	291	2	87	–
Spiny dogfish (*Squalus acanthias*) SW	18	130	13	–	333	180	100
Coelacanth (*Latimeria chalumnae*) SW	30	74	35	290	442	–	
Flounder (*Platichthys flesus*) SW	10	157	30	44	–	14	
FW	15	158	42	71	–	30	
Green toad (*Bufo viridis*)	30	95	17	100	17		
Laboratory rat	16	152	5	3	7		

Based on Yancey et al. (1982) and Bentley (1971a). Bacteria from Christian and Waltho (1962).

instances some osmolytes may also exert protective "counteracting" effects with respect to possible adverse actions of high concentrations of urea. Thus, TMAO can counteract the protein-perturbing effects of urea when the ratio urea/TMAO is about 2. Other such counteracting osmolyes are glycerophosphocholine and betaine. Compatible osmolytes, where the presence of such counteracting solutes is not needed, include polyols, such as sorbitol and glycerol, simple amino acids and amino acid derivatives such as taurine. An accumulation of organic osmolytes, especially urea, has also been observed to occur in several amphibians including the crab-eating frog, *Rana cancrivora*, which lives among coastal mangroves in Southeast Asia. Some toads that live in desert conditions may experience prolonged periods of seasonal drought when water is not freely available. They may then live in burrows, such as seen in the European green toad, *Bufo viridis*. Some amphibians may also aestivate under such conditions and form protective cocoons, as observed in the Australian water-holding frog *Cyclorana platycephala*. Under such conditions these toads can store high concentrations of urea in their body fluids (Table 1.2). African lungfish, *Protopterus aethiopicus*, aestivate in mud chambers when their water habitat dries up and they also store urea in their body fluids until water again becomes available.

In 1962, Christian and Waltho described the intracellular ion concentrations in a remarkable archaea bacterium, *Halobacterium salinarum*, which lives in concentrated solutions of brine (Table 1.3). Potassium concentrations of over 4570 mmol kg⁻¹ water were observed. In mammals this concentration is about 150 mmol kg⁻¹ water. The enzymes present in these bacteria appear to have undergone considerable adaptation that involves many intramolecular amino acid substitutions. Such adaptations reflect another way of adjusting to life in solutions of high osmotic concentration that does not require a continual extra expenditure of energy for active osmoregulation or the synthesis of organic osmolytes. However, the strategic use of organic osmolytes may be a simpler evolutionary solution to such osmotic problems and reflect parsimonious "genetic simplicity" (Yancey et al. 1982).

3 Osmotic Exchanges with the Environment

Water, solutes and nutrients continually move in and out of animals in a pattern that results ideally in their "steady state" with the environment. Man of the osmotic exchanges are obligatory for the animal and not under its direct control, though their magnitude determines the extent of the regulatory processes that are ultimately necessary in order to achieve the steady state.

This obligatory osmotic exchange or turnover depends on several aspects of the animal's physiology.

3.1 Physico-Chemical Gradients with the Environment

Differences between the chemical composition of an animal and its environment will influence water and solute movements in, or out of, the beast. Thus, aquatic species in fresh water will tend to accumulate water and lose solutes, while the opposite usually occurs in fishes living in the sea. Temperature gradients will influence evaporative water losses especially for cooling in homoiotherms.

3.2 Size and Surface Area

Exchange of water and solutes is related to surface area. Large animals have a relatively smaller surface area exposed to the environment than small animals. The gill surface in fishes varies considerably and, apart from a relation to the animal's size, may reflect the normal oxygen needs of the species.

3.3 Nature of the Integument

The barriers exposed to the environment vary in their permeability to water and solutes. Thus, the skin of the Amphibia is usually more permeable to water than that of reptiles, while the gills of elasmobranch fish are less permeable to solutes than those of the teleosts.

3.4 Metabolism and Oxygen Consumption

The exchanges of oxygen and carbon dioxide in terrestrial animals are usually accompanied by water loss, due to saturation of the expired air with water vapour. Metabolism is accompanied by the production of heat and waste solutes; the first facilitates evaporative water loss, while the latter often requires body water for its excretion.

3.5 Feeding

Excess water and solutes may be taken in with nutrients. Animals consuming a succulent herbivorous diet containing 98% water will usually have a water intake in excess of their needs, while a diet of marine invertebrates will provide superfluous salt.

The magnitude of these effects varies considerably in different species. Water exchanges in a day may be equivalent to as much as 50% of the body water in the leopard frog, *Rana pipiens*, or less than 1% in a lizard like the chuckwalla, *Sauromalus obesus*. The daily sodium exchange of a marine fish, like the flounder, is more than 20 times as great as the total sodium content of the body, but in man this exchange represents only about 3% of the total present.

4 Forces Affecting Exchanges of Water and Solutes

The osmotic composition of vertebrates differs considerably from that of their external environments. As described above, there are also considerable differences between the solute concentrations in the body fluid compartments within the animal itself. The external environment tends to dissipate such concentration gradients. Animals exploit a variety of strategies to counter such forces and maintain their chemical composition at levels that are consistent with life. The primary barriers across which such physico-chemical forces act are cell membranes. They exhibit a selective permeability to the solutes and water that tend to travel across them. In some instances they may even incorporate biological mechanisms that can reverse such dissipative molecular flows.

4.1 Diffusion

Molecules in solutions, both solutes and solvent, tend to move from regions of their higher to their lower concentration. This process of diffusion occurs as a result of random molecular movements dictated by their thermal energy. Equilibration with adjoining phases may be restricted by the presence of the membranes with which the molecules may collide. The number of molecules eventually crossing such a barrier in unit time is called the *flux* and will depend on the properties of the membrane (the *permeability coefficient*), the concentration of the molecule and, if it is an ion with an electrical charge, the electrical potential difference (PD) that may exist between the two sides of the membrane. Such diffusible movements of molecules will simultaneously occur between the phases on each side of the membrane (*influx* and *efflux*). The rate will be greatest down the steepest electrochemical gradient and result in a *net flux*. Ultimately, without further biological intervention an equilibrium state will be established when no further net change will occur.

4.2 Facilitated Diffusion

Molecular equilibration across biological membranes may sometimes occur more rapidly than predicted. Such facilitation can reflect an interaction with a membrane-associated *carrier* that speeds its transmembrane diffusion. A variety of such carriers have been identified (see later). It should be emphasized that no direct input of additional energy is involved in this type of diffusion. The carrier-molecule complex moves through the membrane by diffusion but more rapidly than the molecule would alone.

4.3 Osmosis

When two solutions with different solute concentrations are separated by a semipermeable membrane (that limits or excludes movement of the solute) there will be a net movement of solvent, by diffusion, from the side with the lower to that with the higher solute concentration. This movement of solvent is called *osmosis* and commonly influences the movements of water both within the body and between the animal and the aqueous environment that it occupies. The application of an opposing hydrostatic pressure will impede the movement of the solvent. If the solution on one side is pure solvent then the pressure that must be applied to stop net water transfer is called its *osmotic pressure*. At 0 °C, a pressure of 2.26 megapascals (Mpa) is required to prevent movement of water across a semipermeable membrane into a 1-osmolal solution. (Osmotic effects are proportional to the number of molecules in solution so that a molal solution of a salt, such as sodium chloride, which ionizes into more than one constituent, will exert a proportionally higher osmotic pressure than, for instance, a molal solution of urea.) Solutions with identical osmotic concentrations are described as being iso-osmotic or isosmotic to each other, while solutions with different concentrations may be hyperosmotic or hypo-osmotic (hyposmotic) to each other. These are physico-chemical terms. Isotonic, hypertonic and hypotonic are the equivalent biological terms and refer to such solutions with reference to their potential abilities to influence the volumes of living cells and organisms.

4.4 Evaporation

The transformation of a liquid (or solid) into a vapour or gas is called vaporization or evaporation. Kinetically, it can be viewed as an escape of molecules with a greater than average kinetic energy from a liquid to a gas phase. The excess kinetic energy of the departing molecules of water will be lost from the liquid and a loss of heat will result. At a physiological temperature of about 35 °C this heat loss will be about 584 calories g^{-1} (2443 Joules). Heat lost as a result of evaporation from the skin and respiratory tract plays a vital role in thermoregulation in mammals and birds. Evaporation is also the major cause of dehydration in terrestrial vertebrates. The particular structure of the skin and its functioning as a barrier membrane are important determinants of the rate of evaporation that occurs in terrestrial vertebrates.

Evaporation of water occurs more rapidly at higher environmental and skin temperatures, at low barometric pressures, and when the vapour pressure of water in the gas phase is low. Evaporation will cease when the latter is saturated with water vapour (the saturated vapour pressure) and this value increases with increased temperature. The thickness of the adhering layer of water vapour at the interface between the skin and the gas phase is particularly important as it contributes to the length of the pathway across which the water must diffuse. If mixing with the surrounding air is slow, evaporation will decrease, as the diffusion pathway will be impeded. Thus, increased wind speed and convectional air movements will facilitate evaporation by decreasing the effective thickness of this boundary layer. The structure and arrangement of scales, hair and feathers can retard or promote evaporation by influencing the mixing of this layer with the external air. Secretion from skin glands can provide water for evaporation. The blood supply to the skin will affect the temperature at this external interface. Evaporation from the body surface can thus be influenced by a variety of factors.

For practical comparisons of rates of evaporative water loss of different animals the values can be corrected for skin surface area and expressed in terms of the *vapour pressure difference* or saturation deficit. This value is the difference, in mmHg, between the saturated water vapour pressure of the air over a free water surface at the ambient environmental temperature and the actual vapour pressure of the water that is present. However, this adjustment provides only an approximation of comparative rates of evaporation. Measurements of evaporation from the skin can be used to calculate the animals *skin resistance* (r; units are as $s\,cm^{-1}$) to water loss (Spotila and Berman 1976; Robertson and Smith 1982). Values for r can be used for interspecific comparisons of evaporation between animals. It allows for differences due to the thickness of the external boundary layer of air such as may result from the mixing effects of different rates of air flow.

4.5 Active Solute Transport

Active transport of molecules has been defined by Hans Ussing (1960) as "a transfer which cannot be accounted for by physical forces". Solute movements down the sum of gradients of chemical concentration, electrical potential, temperature, pressure and as a result of drag (frictional) forces accompanying solvent movement *do not* constitute active transport. Transport *against* the sum of such gradients across biological membranes requires the expenditure of energy (usually supplied by ATP) by the cell into or out of which the solute transfer is occurring. Currently, active transport is often classified as primary active transport and secondary active transport. The former refers to a process in which there is a direct utilization of energy by the molecular mechanism that promotes the solute transfer. For instance, the active extrusion of Na^+ from cells as a result of its interaction with Na-K-activated ATPase (Na-K-ATPase, the Na pump). On the other hand, chloride ions (Cl^-) can also be transported actively against prevailing electrochemical gradients into cells. However, this process does not involve a direct linkage of the Cl^- to an energy-consuming event. Two Cl^- are instead coupled to the transfer of K^+ and Na^+ on a carrier molecule that can diffuse across cell membranes. This carrier utilizes the diffusion gradient of Na^+ concentration created by Na-K-ATPase to cross the cell membrane. This process, which is called *cotransport*, is an example of secondary active transport (see Table 1.4). In effect, it utilizes a tandem-like linkage of solutes and the processes of both facilitated diffusion and active solute transport.

The possibility of an active transport of water has received sporadic and sometimes semantic attention, but it is generally not considered to be a widespread process. However, "molecular water pumps" are currently receiving some attention (MacAulay et al. 2001). The forces of osmosis, diffusion and hydrostatic pressure appear to account for most movements of water across biological membranes. There remain some phenomena, such as the movement of water across the mammalian intestine, that defy such an explanation and could involve secondary active transport.

5 Transporter Proteins Involved in Solute and Water Movements Across Cell Membranes

Careful experimental measurements involving the processes of solute and water movements across cell membranes resulted in the prediction of the presence of various "pumps", "carriers", and "pores" or "channels" in cells. The enzyme Na-K-ATPase, or Na pump, was originally identified and purified using classical biochemical procedures. Recombinant DNA techniques have since been utilized for the identification and cloning of a host of genes that express proteins that function as such pumps, channels, exchangers and cotransporters when expressed in cells (expression cloning) such as the oocytes of the toad *Xenopus laevis*. Thus, molecular entities involved in basic osmoregulatory processes, which were once considered to be hypothetical concepts, are now available for direct study. To the amazement of some, they have actually been shown to exist! Some of these transport proteins are summarized in Table 1.4.

Table 1.4. Some molecular transport proteins (pumps, carriers and channels) associated with epithelial membranes that participate in osmoregulation

Transport protein	Action	Principal sites
Ion pumps (primary active transport)		
Na$^+$-K$^+$-ATPase (P-type)[a]	3Na$^+$ out of cells and 2K$^+$ in	All cells, basolateral side in epithelia
H$^+$-K$^+$-ATPase (P-type, colonic isoform)[b]	K$^+$ absorption and H$^+$ secretion	Apical surface late distal renal tubule and colon
Proton (H$^+$)-ATPase (V-type)[c]	H$^+$ secretion, secondary absorption HCO$_3^-$, Na$^+$	Apical surface late distal renal tubule, frog skin, fish gills, turtle bladder
Ion channels (uniporters, diffusion pathways)		
Epithelial Na$^+$ channels (ENaC)[d]	Na$^+$ absorption	Apical surface late distal nephron, colon, frog skin and bladder etc.
Inward rectifying K$^+$ channels[e] (K$_{ir}$ gene family)		
ROMK	Maintain negative membrane PD, K$^+$ secretion	Apical surface distal nephron, thick ascending loop of Henle
K$^+_{(ATP)}$	Recycling K$^+$ across basal plasma membrane	Basolateral side kidney, intestine and (?) gills
Cl$^-$ channels[f]		
CIC family (voltage-gated)	Transepithelial Cl$^-$, cell volume regulation	Kidney, intestine, muscle
CFTR (cAMP-sensitive) (Cystic fibrosis conductance regulator)	Cl$^-$ and fluid secretion	Apical and basal side; intestine, bronchi, salt glands, (?)gills etc.
Volume-sensitive organic anion channels (VSOAC)[g]	Cell volume regulation	Ubiquitous
Ion exchangers (antiporters, secondary active transport)		
Na$^+$/H$^+$ (NHE gene family)[h]	Acid-base balance, Na$^+$ absorption, regulation cell volume	Most cells, apical side renal tubule, intestine, gills
Cl$^-$/HCO$_3^-$[i]	Acid base balance, Cl$^-$ absorption, regulation cell volume	Many cells; apical, basal side, gills, intestine, frog skin
Na$^+$/Na$^+$[j]	Limits accumulation Na$^+$	Muscle, gills
Ion cotransport (symporters, secondary active transport)		
Na$^+$-K$^+$-2Cl$^-$[k] (NKCC)	Cell volume regulation, Cl$^-$ absorption, secretion, K$^+$ secretion	Many cells, thick ascending loop of Henle, intestine, salt glands, gills
Na$^+$-Cl$^-$[k]	Absorption Na$^+$, Cl$^-$	Distal convoluted renal tubule, fish bladder, intestine

Table 1.4. *Continued*

Transport protein	Action	Principal sites
K^+-Cl^{-1} (KCC)	Cell volume regulation, extrusion K^+ and Cl^-	Many cells; renal tubule, intestine
Na^+-HCO_3^{-m} (NBC)	Acid-base regulation, HCO_3^- absorption	Proximal and thick ascending limb renal tubule, intestine
$2Na^+$-glucose[n] (SGLT)-amino acid	Nutrient, Na^+ absorption	Kidney tubule, intestine
Na^+-, Na^+-Cl^-osmolytes[o]	Cell volume regulation	Kidney cells
Urea transporters		
UT-A[p]	Maintain renal medullary concentration gradients	Renal collecting ducts
ShUT[q]	Urea conservation	Elasmobranch kidney, gills(?)
Na^+-linked urea cotransporter[q,r]	Urea conservation	Elasmobranch kidney, gills(?), mammal kidney
Aquaporins[s] (water channels)	Cell volume regulation enhance osmotic flows and conservation of water	Cell membrane, may be unilateral
AQP1		Renal tubule; proximal, thin loop Henle
AQP2	Responsive to vasopressin	Renal cortical collecting duct, amphibian bladder
AQP3		Renal medullary collecting duct, intestine
AQP4		Renal medullary collecting duct
AQP8		Colon

[a] Robinson (1995), Blanco and Mercer (1998).
[b] Wingo and Smolka (1995), Silver and Soleimani (1999), Jaisser and Beggaln (1999).
[c] Ehrenfeld et al. (1985), Harvey (1992a,b), Lin and Randall (1995).
[d] Canessa (1996), Garty and Palmer (1997), Horisberger (1998).
[e] Jan and Jan (1994), Lee and Hebert (1995), Schultz (1997), Giebisch (1998), Derst and Karschin (1998).
[f] Larsen (1991), Silva et al. (1999a), Jentsch et al. (1999), Sheppard and Welsh (1999), Evans et al. (1999).
[g] Perlman and Goldstein (1999).
[h] Wakabayashi et al. (1997), Perry (1997).
[i] Kopito and Lodish (1985); Larsen (1991); Perry (1997).
[j] Ussing (1947), Motais et al. (1966).
[k] Xu et al. (1994), Mount et al. (1998), Isenring et al. (1998), Russell (2000).
[l] Mount et al. (1998).
[m] Romero and Boron (1999).
[n] Crane (1965); Hediger et al. (1987).
[o] Burg (1995), Perlman and Goldstein (1999).
[p] Sands et al. (1997).
[q] Smith and Wright (1999).
[r] Pärt et al. (1998).
[s] Yamamoto and Sasaki (1998); Heyman and Engel (1999), Borgnia et al. (1999).

Ion Pumps. A variety of macromolecules can actively transport solutes, including ions, across cell membranes utilizing processes that depend energetically on the hydrolysis of ATP. Such pumps (Table 1.4) make basic contributions to the regulation of ion concentrations and the fluid volume of cells and the composition of the extracellular fluid. Na-K-ATPase and H-K-ATPase belong to a superfamily of such macromolecules that have been identified in a variety of organisms including bacteria, fungi, plants and animals. Comparisons of their structures indicate that they have evolved from a distant, but common, ancestor. They are called the P-type ATPases (P refers to the common mechanism of their phosphorylation). Na-K-ATPase is the prototype that was first identified in crab nerves. It appears to be present in all animal cells, where it mediates the active extrusion of Na^+ from the cytosol and the accumulation of K^+ from the extracellular fluid. This pump can thus regulate the cell's volume and maintain the inwardly directed gradient of Na^+ concentration that is necessary for the movements by diffusion of various Na^+-linked cotransporters (see below) across the cell membrane. Thus, Na-K-ATPase has a vital basic physiological role in maintaining the osmoregulatory activities of cells. It spans the cell membrane in which it may have a quite uniform symmetrical distribution or, as seen in epithelial cells, be confined to their basolateral borders. This enzyme is heteromeric and consists of two or three subunits; α (about 112 kDa), β (40 to 60 kDa) and γ (8 to 14 kDa). The precise structures are homologous, but vary in different species. Isomers are often present in a single animal and may each contribute to the special needs of particular cells. Specific binding sites for Na^+, K^+, ATP and inhibitory molecules, such as the drug ouabain, are present. The cycle of their functioning involves allosteric changes in molecular structure between E-1 and E-2 forms. Initially, three Na^+ combine from the cytosolic side with sites on the α-subunit, and this interaction results in hydrolysis of the ATP, phosphorylation of this subunit and its transformation to the E-2 form. The three Na^+ then dissociate from the molecule at the extracellular face of the cell. Two K^+, from the extracellular fluid, then combine with the α-subunit and, along with the phosphate, subsequently dissociate from its cytosolic side. The E-1 configuration is then restored. The β-subunit is necessary for the normal activity of the process of activity of the enzyme but its precise function is unknown. The role of the γ-subunit, which is sometimes present, is also obscure. Regulation of the activity of Na-K-ATPase may involve responses to ion concentration, and the actions of mediators, which in vertebrates include adrenocortical steroid hormones and catecholamines (Therien and Blostein 2000). Regulation involves quantitative increases of the enzyme as a result of genetic expression, changes in its distribution, its affinity for solutes and its rate of degradation. The H-K-ATPase exists in several isomeric forms and has been identified in the apical plasma membrane of epithelial cells in the mammalian colon and renal tubule. It contributes to the secretion of H^+ and absorption of K^+.

A variety of H^+-ATPases are present in nature, but one type appears to have special roles in ion regulation. It was first identified in the vacuolar membranes of plants and belongs to a distinct genetic group called the V-type ATPases. This enzyme is present in the apical plasma membrane of a variety of ion-transporting epithelial cells including frog skin, turtle urinary bladder, fish gills and distal regions of the mammalian renal tubule. It secretes three H^+ per ATP molecule, which results in the generation of an electrical potential difference across the cell membrane. Apart from promoting the conservation of HCO_3^- by the kidney ($H^+ + HCO_3^- \rightarrow H_2CO_3 \rightarrow CO_2 + H_2O$) it creates a favourable electrical gradient to promote the entry of Na^+ into epithelial cells. This Na^+ may then be actively extruded across the opposite basolateral surface due to the action of Na-K-ATPase.

Ion Channels (Uniporters). The cell membrane contains a diversity of proteins, often oligomeric, that incorporate channels that have selected abilities to allow the diffusion of solutes, including Na^+, K^+ and Cl^-, from one side of the membrane to the other. Such flows of ions are usually unidirectional, or subject to substantial rectification. Such properties reflect the molecular makeup of the channels which influence the dimensions, electrical charges, the presence of specific binding sites and the activities of restricting gates. Such gates can open and close in response to such stimuli as changes in membrane voltage, mechanical stretch due to cell volume changes, the local concentrations of ATP and Ca^{2+}, and the activities of phosphorylation mechanisms mediated by G-proteins and cAMP (see Chap. 2). The opening and closing of ion channels can also be promoted by neurotransmitters and hormones. The different ion channels are expressed by specific genes, and several forms may exist for each type of ion. Such variations appear to reflect their different roles and the activities of the tissues in which they are present. As summarized in Table 1.4, apart from contributing to the electrical activities of cells sodium channels, potassium channels and chloride channels are frequently involved in the osmoregulatory activities of epithelial cells such as are present in the kidney tubules, gut, salt glands and gills. Ion channels are present in all organisms. Similarities in the structures of ion channels from distantly related organisms suggest the possibility that they may have an "evolutionary kinship" (Jan and Jan 1994; Derst and Karschin 1998).

Ion Cotransporters (Symporters). The entry or exit of Na^+, K^+ and Cl^- from cells can be facilitated by a variety of specific carrier protein molecules that are present in the cell membrane. Mutually dependent linkages of ions and solutes such as Na^+—Cl^-, K^+—Cl^-, Na^+—K^+—$2Cl^-$ and Na^+-glucose activate various cotransporter carrier proteins, which then diffuse across the cell membrane. The driving force for such diffusion into the cell is the low concentration of Na^+ inside the cell, or, in the case of the exit of K^+—Cl^-, that of K^+ in the extracellular fluid. Such gradients are primarily maintained as a result of the activity of Na-K-ATPase. These processes and their cotransport carrier proteins are widespread in nature and have been identified in vertebrates, invertebrates and plants (Mount et al. 1998). They play important roles in the regulation of cell volume and the processes of secretion and absorption of ions, especially Cl^-, in epithelia such as the kidney tubules, gut, various salt glands and fish gills. The first description of such a linked mechanism involved the cotransport of Na^+ and glucose across the intestine (Crane 1965). This process also occurs in the renal tubule, where in the proximal segment it accounts for about 50% of the Na^+ absorption. The first demonstration of linked Na^+—K^+—$2Cl^-$ transport was made by Geck and his collaborators (1980) in cultured Ehrlich tumour cells. The genes for several such carriers have been cloned, including Na^+—Cl^- (originally from the urinary bladder of flounder, Gamba et al. 1993) and Na^+—K^+—$2Cl^-$ (from the shark rectal gland, Xu et al. 1994). Homologous proteins have since been identified at many other sites in humans and other species. Comparisons of the structures of Na^+—Cl^-, Na^+—K^+—$2Cl^-$ and K^+—Cl^- cotransporter proteins indicate that they belong to a common gene family which share structural motifs with such proteins in invertebrates and plants.

The transport of HCO_3^- across epithelial tissues such as the proximal renal tubule and gut can occur in association with Na^+ and involves a cotransport protein that has been dubbed NBC (Table 1.4). Bicarbonate transport can also takes place in a process that involves the activity of a Cl^-/HCO_3^- exchange protein. The NBC and Cl^-/HCO_3^- membrane proteins have a 30 to 35% identity in their amino acid sequences and thus appear to belong to a common bicarbonate ion transporter family. The stoichiometry of Na^+ and HCO_3^- transport in NBC varies from $1:1$, which is electrically neutral, to $1:3$, which is electrogenic.

Ion Exchangers (Antiporters). There are a number of ion transport mechanisms in cell membranes which utilize electrically neutral exchanges of two cations or two anions. Such exchanges can be homogeneous, such as Na^+/Na^+, or heterogeneous, such as Cl^-/HCO_3^- and Na^+/H^+ (Table 1.4). One of the ions moves down its electrochemical gradient while the other may move against it. There is no direct input of new energy into such processes, which are described as countertransport.

An early example of the involvement of such processes in the osmoregulatory activities of epithelia was discovered by A. Krogh in 1937. He demonstrated a Cl^-/HCO_3^- and Na^+/NH_4^+ exchange that could account for the active uptake of NaCl across the skin of frogs maintained in solutions with a low salt concentration. Respiratory physiologists have long been aware of the process of Cl^-/HCO_3^- exchange which occurs across the erythrocyte cell membrane when it unloads accumulations of HCO_3^- for subsequent excretion as carbon dioxide. This process, called the chloride shift, assures that the process is electrically neutral and that no change in cell volume occurs. It involves the interaction of Cl^- and HCO_3^- with a specific cell membrane protein called the band 3 protein. Structural relatives of this anion exchange protein have been identified from other tissue sites and have been cloned (Alper et al. 1988). They occur in the kidney tubules, gut and the gills of fish, where they contribute to the regulation of pH, cell volume and transepithelial Cl^- transport.

Several Na^+/H^+ ion exchange transport proteins have been cloned from human tissues (NHE proteins) (Sardet et al. 1989). Their presence in cells is ubiquitous, where they contribute to the regulation of pH and cell volume (NHE1). Others, for instance NHE3, are present in the apical plasma membranes of epithelial cells in kidney tubules, gut and fish gills, where they are involved in the uptake and transepithelial transport of Na^+. The activities of these proteins appear to be directly regulated by cell pH but they also can respond to external stimuli such as growth factors and hormones that may act by a process involving their phosphorylation.

Urea Transporters. Urea, as described earlier, can function as an osmolyte in cells and extracellular fluids especially in elasmobranch fish that live in the sea. It also has an important role in maintaining local hyperosmotic gradients in the medullary tissue of the kidneys of mammals. At this site it subserves their ability to form a hyperosmotic urine. Urea has a high solubility in water but a low one in lipids, yet it readily crosses lipoidal cell membranes. Such anomalous behaviour suggested that it may utilize special pathways for entry into cells. Physiological observations on its urinary excretion suggest that in some species special mechanisms, including active transport, may exist in the renal tubule which fosters its conservation. Vasopressin, the antidiuretic hormone that promotes osmotic absorption of water from the distal regions of the renal tubule, has also been observed to increase the permeability of epithelia to urea, as observed in frog skin and toad urinary bladder in vitro. This response is also seen in the renal tubules of mammals, amphibians and elasmobranch fish (Sands et al. 1997; Smith and Wright 1999). These effects of vasopressin on urea are specifically blocked by the drug phlorizin. There is thus a considerable amount of evidence suggesting the presence of specific urea carriers in cell membranes. The use of gene cloning and expression techniques has resulted in the identification of several urea transporter proteins (Table 1.4). They are present in a variety of tissues including erythrocytes, brain and kidney, and include the products, UT-A and UT-B, of at least two genes in mammals. An isoform of UT-A (UT-A2) from the rabbit kidney was the first to be cloned (You et al. 1993). It is a glycoprotein (about 43kDa). UT-A1, cloned from rat kidney, can be activated by vasopressin to transport urea. The reaction appears to involve its phosphorylation as a result of the action of protein kinase A. A urea transporter has also been cloned from the kidney of the dogfish shark, *Squalus acanthias*. It has been called shUT and is structurally similar to UT-A2, suggesting that they may belong to a common protein family. These urea transporters facilitate the diffusion of urea down concentration gradients across cell membranes.

Proteins that possibly mediate an active transport of urea have not yet been directly identified. However, as described earlier, experimental observations suggest that they may exist in the kidneys of mammals, amphibians and elasmobranchs. They could be mediating a Na^+-linked secondary active transport of urea (an Na^+-urea cotransporter). Such a linkage has been observed during the activity of the Na^+-glucose cotransporter (Leung et al. 2000).

Aquaporins. The lipoidal nature of the cell membrane is expected to considerably restrict the transport of water between the cell and its bathing solutions. Nevertheless, substantial movements of water do occur, which appear, at least partly, to involve diffusion between the interstices of hydrocarbon chains in the lipids. However, in many cells, such as erythrocytes, water moves much more rapidly than expected and exhibits low activation energy, about $5 \, kcal \, mol^{-1}$, which is similar to that of the diffusion of water in water. The water thus appears to be moving through an aqueous phase across cell membranes, suggesting that water-filled protein channels or pores may be present. Such a rapid transfer of water may be necessary for adequate adjustments of cell volume, the secretion of fluids and the massive reabsorption of water that takes place across the epithelia of the renal tubules and gut.

The frog and toad urinary bladder (in vitro) provide models for studying the transepithelial water permeability of the distal regions of the renal tubule and its response to vasopressin. (Vasopressin is the hormone that decreases urinary water excretion in mammals.) In 1974 Chevalier et al., using a freeze-fracture etching technique, were able to visualize, by electron microscopy, "membrane-associated particles" in the apical cell membrane of frog bladder epithelial cells. The presence of these MAPs, was found to be associated with increases in osmotic water transfer, which occurred in response to the action of a vasopressin-like hormone (oxytocin). The particles were thought be proteins and provided a biological reality to formerly hypothetical water channels.

In 1984, a protein was identified in the cell membranes of ocular lens fibre cells Gorin et al. 1984). As it makes up about 60% of the total protein present at this site it was called major integral protein or MIP. At the time of its identification its function was unknown, but it is now thought to contribute to the loss of water by the lens fibre cells. Its absence in mice results in lens opacity (cataracts) and blindness. In 1992 the aqueous water channel in erythrocyte cell membranes was cloned by Agre and his collaborators (Preston et al. 1992). When expressed in oocytes of the toad Xenopus, it was found to behave as a water channel, which was subsequently called aquaporin 1. MIP (now called aquaporin 0) was found to be a similar protein that belongs to a large family (aquaporin or MIP family) of proteins. These molecules contain water channels or, in some, glycerol channels. About 160 such related proteins have been identified in species that include bacteria, archaea, plants and animals. So far, ten such aquaporins have been identified in mammals. They are homotetramers, with each monomer (about 30 kDa) containing a membrane-spanning water-filled pore which excludes ions but may have a limited permeability to molecules such as glycerol. The tissue sites and functions of some of these proteins are shown in Table 1.4. Aquaporin 2 is present in the renal collecting duct (Fushimi et al. 1993) and can be inserted into the apical cell membrane in response to the presence of vasopressin. It appears to be synonymous with the MAPs observed in 1974 by Bourguet and his collaborators (Chevalier et al. 1974) in the frog urinary bladder.

Other molecular mechanisms exist which may facilitate the movements of water across cell membranes, but their quantitative contributions to water transfer do not appear to be as great as those of the aquaporins. However, they may have special roles. Several molecular transporter proteins that are primarily involved in solute transport have been shown to also mediate water transport. Their effects are not directly related to osmotic changes induced as a result of possible changes in cell solute concentrations. They have been called molecular water pumps. The most closely studied are the K^+—Cl^-, and the $2Na^+$-coupled glucose (SGLT1) (Table 1.4) and the Na^+-coupled glutamate cotransporters (EAAT1). Allosteric changes associated with the activation of such transporter proteins possibly induce the opening of water channels which are a part of the molecule or closely associated with its functioning (Fischbarg et al. 1990; Zeuthen 1991; Loo et al. 1999a; MacAulay et al. 2001). The SGLT1 protein couples the transport of $2Na^+$, 1 glucose and 200–260 water molecules. The activation energy for the water transport is about $5 kcal mol^{-1}$, which is similar to that observed for aquaporins. The physiological and quantitative contributions of such molecular water pumps to the large movements of water that occur across the kidney tubules and intestine is contentious (Loo et al. 1999b; Spring 1999). Clearly, a major part of the fluid movement across the epithelial membranes is linked to Na^+ transport, but the precise mechanism is uncertain. In the intestine, Spring tentatively favours the ingenious Curran three-compartment model (Curran 1965). This theory involves a coupling of primary active Na^+ transport with secondarily generated osmotic and hydrostatic forces in three neighbouring compartments in the tissue. As a result of the generation of these forces, water movement is predicted to occur from the lumen of the gut to the blood side. The morphological presence and functioning of all these compartments remains to be demonstrated to the satisfaction of all. The precise physiological roles of molecular water pumps is also uncertain.

6 Biological Structures Participating in Fluid Exchange

The fluids of the body are separated from each other by the cell membrane and capillary, and from the external environment by the skin, by the epithelia of the respiratory organs (gills and lungs) and the gut. The integrity of the intracellular fluid is dependent on the cell itself and its surrounding membrane, as well as the composition of its bathing (extracellular) fluids. The regulation of the components of the extracellular fluid is due principally to the activities of kidneys, but also to the functioning gills, various "salt" glands, the gut and even the skin in some species.

6.1 The Cell Membrane

The outer physical barrier of the cell is called the cell membrane or plasma membrane. It separates the inner cytoplasm from the extracellular fluid, and restricts, monitors and regulates the molecular exchanges which occur between these two fluids. It regulates the volume and the concentrations of solutes in the cell. The cell membrane also participates in the transmission of information to the cell and contributes to its adhesion to neighbouring cells. It is 5 to 10 nm thick and consists principally of lipids and proteins that make up 80 to 90% of its weight. The ratio of the weight of lipid to protein depends on the type of cell and the species, but in animals is usually about 1:1. There are also some carbohydrates and water present. The precise architectural arrangement of these constituents has been the subject of speculation and argument for over a century. The lipids consist mainly of phosphatidyl ethanolamine and phosphatidyl choline (lecithin), with smaller amounts of phosphatidyl inositol and phosphatidyl serine. The lipid sterol cholesterol is also present in animal cells. The phospholipids are amphipathic and

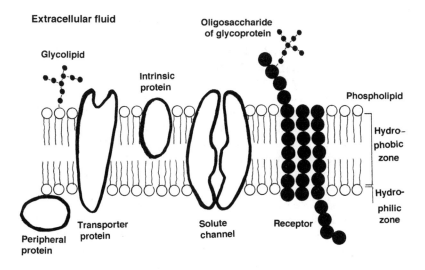

Fig. 1.1. Depiction of the Singer-Nicolson "fluid mosaic" model of the cell (plasma) membrane. Various protein components are distributed in a fluid phospholipid matrix. For further details see the text

possess a hydrophilic phosphate "head" and a hydrocarbon "tail". In the presence of water, the phospholipids aggregate to form bilayers with the heads facing the water in an outwards direction while the opposing tails face inwards (Fig. 1.1). A barrier lipid membrane is thus constituted which has a high permeability to other lipids, a restricted permeability to water and similar-sized solutes, such as urea and NH_3, and which is virtually impermeable to large water-soluble molecules, carbohydrates and ions. It has a natural lateral fluidity. Cholesterol increases the density of the packing of the hydrocarbon chains and reduces fluidity and permeability.

The arrangement and functions of the proteins in the cell membrane were uncertain until about 30 years ago. Contemporary opinion then favoured the presence of bilateral protein layers that were thought to form a "sandwich" on either side of the lipid. In 1972, Singer and Nicolson proposed that such proteins had specialized functions and were distributed heterogeneously in the lipid matrix to form a mosaic pattern. They were considered to be amphipathic, with their hydrophobic face buried in the lipid and the hydrophilic one projecting into the aqueous phase. In such situations they may span the membrane (transmembrane proteins) or occupy peripheral sites. This fluid mosaic model of the cell membrane is still current, but has been somewhat embellished. The specialized functions of many of the constituent proteins have been uncovered. They include the various solute transporters (channels, pumps and cotransporters) and water channels (aquaporins), which have been described above. Also included are specialized receptor molecules and their associated transduction proteins and enzymes that are involved in cell-to-cell communication following interactions with neurotransmitters and hormones. The cell membrane maintains functional connections with the underlying microfilaments and microtubules of the cell cytoskeleton network which helps to mediate such responses. It can also fuse with vesicles formed from the endoplasmic reticulum and Golgi apparatus (exocytosis). Such events occur in response to neural and hormonal signals.

The carbohydrates in cell membranes are usually present in oligosaccharide chains that are part of glycoprotein and glycolipid molecules. They often carry an electrical charge and project into the outer aqueous fluid. Substances such as mucins are glycoproteins which contribute to the outer cell coat (glycocalyx).

The cell membrane is involved in the process of the adhesion to neighbouring cells, from which they are separated by the intercellular or paracellular space. This expanse contains the extracellular matrix consisting of macromolecules such as polysaccharides and proteins, some of which are directly involved in cell adhesion. The paracellular spaces also contain extracellular fluid and in barrier epithelial membranes, such as the intestine and renal tubules, provide pathways through which solutes and water may diffuse. Such molecular movements are greater in "leaky" epithelia, such as the small intestine, than "tight" ones, such as the colon and urinary bladder. The magnitude of such leaks depends on the presence and integrity of tight junctions, which span and occlude part of the intercellular space. They form part of the network of cell-cell adhesion processes, including the belt desmosomes, which contain the protein actin. In epithelia, the tight junctions are near the apical surface of the cells. Together with the adherins junction (of the belt desmosomes) it forms the apical junction complex which contains the transmembrane protein occludin (Mitic and Anderson 1998). Synthesis of the proteins making up the tight junction is a process that is regulated by the cell.

6.2 Capillaries

The capillaries are the small blood vessels that perfuse all tissues and constitute the permeable barrier separating the fluids of the blood plasma and the intercellular space. Solute, water and respiratory gas exchanges occur across the capillaries. They lie at the terminus of the arterioles and are contiguous with the venules. They form a simple diffusion barrier lacking muscle attachments and consist of a single layer of vascular endothelial cells lying on a supporting basal lamina. Capillaries vary in their size and permeability characteristic depending on the tissue where they are present. They are 0.1 to 1 mm long, 0.2 to 1 µm thick with a diameter of 6 to 10 µm. (Erythrocytes, which are flexible, have a diameter of about 7 µm.) The permeability of the capillaries in the renal glomerulus is about 1000 times greater than those in skeletal muscle. Such differences are reflected in their structures. In muscle and skin the endothelial cells form a continuous layer, while in the glomerulus and intestine interruptions (fenestrations) occur which are 20 to 100 nm wide. These gaps allow for greater molecular exchanges. In liver and bone marrow the endothelial cell layer may be discontinuous and form sinusoids through which unrestricted movements of large molecules, including proteins, may take place. Molecular exchanges across capillary walls occur principally by simple diffusion with a smaller, but important component due to bulk flow of water and its contained solutes (usually excluding plasma proteins). The latter process occurs under the influence of the hydrostatic pressure generated by the heart. (As it separates the larger protein molecules from the remainder the process is referred to as ultrafiltration.) Diffusion of lipid-soluble molecules occurs across the entire capillary surface, including the cell membranes of the endothelial cells. Some water and ions may also follow this route, but it appears to occur mainly through the fenestrations and pores in the endothelial intercellular matrix. Bulk fluid flow due to hydrostatic forces occurs through pores and fenestrations, which usually exclude the proteins. It is this process that principally initiates and maintains the separate integrity of the volumes of the blood plasma and intercellular fluids. It is called the Starling-filtration-reabsorption process. It involves forces that are the algebraic sum of the hydrostatic pressure and the osmotic pressure of the plasma proteins. In the proximal region of the capillary, the hydrostatic pressure exceeds the osmotic pressure, and filtration from the plasma to the intercellular fluid occurs. Distally, following a decline in the hydrostatic pressure to levels less than that of the osmotic pressure of the plasma proteins, fluid returns to the blood plasma. The balance of these forces is normally well maintained, and even substantial changes in the blood pressure have little effect. However, if there is a drop in plasma protein concentration, or if a backpressure develops in the distal parts of the capillary, or permeability of the capillaries change, excess fluid (oedema) may accumulate in the intercellular spaces. Such changes can occur in such conditions as malnutrition, heart and kidney disease, and capillary inflammation. There is little evidence to indicate the presence of a general direct regulation of such processes, such as may be mediated by nerves or hormones. However, in the kidneys of some species vascular changes may influence filtration across the glomerulus and the formation of urine.

6.3 Skin

The outer integument of vertebrates consists of the skin and in fishes and larval amphibians also the gills. The skin forms a durable, relatively impermeable barrier that protects the animal from adverse effects of the external environment, including water and salt exchanges by diffusion and evaporation (Lillywhite and Maderson 1982; Bereiter-Hahn et al. 1984; Toledo and Jared 1993). It is composed of several layers of tissue including epithelial epidermal cells that are underlain by a basal lamina and a layer of connective tissue called the dermis or corium. The epidermis is in a continual and progressive state of development arising from its basal stratum germinativum. Epidermal cells contain the fibrous protein keratin, which eventually usually nearly fills the cells. They then die and become the outer cornified layer of the skin, which is periodically replaced, along with appendages such as scales, hair and feathers during shedding or moulting. The low permeability of the epidermis principally depends on the presence of a lipid-containing extracellular matrix. This structure seals the junctions between the cells and reduces transcellular permeability.

The skins of different species of vertebrates have different permeabilities that are usually consistent with their environmental lifestyles. In the sea and fresh water the skin restricts the osmotic movements of water and diffusion of salts, while on land evaporation is reduced. However, in amphibians the skin is generally much more permeable than in other vertebrates, which appears to reflect its role in respiration.

The skin of different species of vertebrates contains various appendages, which can modify the exchanges of water and salts. Glandular cells in the skin of fishes and amphibians secrete mucins. Some frogs and toads that live in dry environments also secrete lipids from skin glands. They wipe this secretion over the skin and thus reduce evaporation (Blaylock et al. 1976; Lillywhite et al. 1997). Most birds possess uropygial (preening) glands near the base of the tail region. They spread its secretion over the feathers to keep them flexible and waterproof. Many mammals secrete a weak salt solution from sweat glands. It is evaporated to foster cooling. The skin of some fishes contains chloride cells from which salt can be secreted as an aid to their osmoregulation in sea-water (Marshall et al. 1997; Yokota et al. 1997). As demonstrated by A. Krogh in 1937, the skin of frogs can actively take up salt from fresh water. Amphibians also utilize their permeable skin to aid their rehydration (cutaneous drinking) from pools of water or damp soil.

Reptiles, birds and mammals usually possess, respectively, keratinous, and often coloured, scales, feathers and hair, which can temper and reduce evaporative water losses and heat gains from solar radiation. African reed frogs, *Hyperolius viridiflavis*, which can live for extended periods of time in exposed hot dry environments, deposit small reflective plates made of guanine and hypoxanthine in dermal iridophore (pigment) cells (Kobelt and Linsenmair 1992). They reflect solar radiation and thus contribute to a restriction of the frog's heat gain and evaporative water loss. Other frogs that live in arid environments may withdraw to burrows during periods of drought and form enveloping cocoons composed of sheets of epithelial cells which have been disposed of during moulting (Lee and Mercer 1967; Withers 1995). Aestivation cocoons are also formed by African lungfish when the waters in which they normally live dry up (Smith 1930a).

6.4 Water Loss from the Tetrapod Respiratory Tract

Water loss from the lungs, bronchi, nasal passages and mouth takes place as a result of gaseous exchanges that support metabolism and, in mammals and birds, also a need for thermoregulatory cooling. Expired air is saturated with water vapour due to evaporation from the pulmonary surfaces. If the inspired air has a lower water vapour content, a net loss of water from the body will occur as a result of such respiration. The water vapour concentration in the expired air is related to the body temperature; at higher temperatures more water will be present. Some small mammals and birds can reduce the temperature of the expired air and thus save water. They have "cold noses" (Jackson and Schmidt-Nielsen 1964; Schmidt-Nielsen et al. 1970). Depending on the temperature and water content of the inspired air, the relatively long narrow nasal passages of such animals can be cooled due to heat loss to the incoming air and by local evaporation. The air passing out over the colder nasal surface will then be cooled and lose water vapour by condensation. The functioning of this counter-current heat exchanger depends on the presence of long narrow nasal passages such as are often present in small animals. In larger beasts, including humans, such an effect is small or absent. Mammals and birds normally extract about 5 ml of oxygen from every 100 ml of inspired air prior to expiration, and this value does not normally vary significantly. However, some reptiles, such as aquatic diving species and desert lizards, like the chuckwalla, *Sauromalus obesus*, have an irregular, discontinuous, pattern of breathing. During such breath-holding, as much as 15 ml of oxygen may be extracted from 100 ml of inspired air (Schmidt-Nielsen et al. 1966a; Thompson and Withers 1997). A considerable reduction in respiratory water loss may then occur.

Many animals hibernate in winter, aestivate in hot dry summer periods and periodically may go into states of torpor. Their body temperatures and rates of oxygen consumption decline in such conditions (Storey and Storey 1990; Guppy and Withers 1999). Such metabolic depression, which may amount to as much as 50% of the normal resting level, results in a decrease in pulmonary water loss. Some small desert rodents appear to have a metabolic rate habitually lower than predicted, which reduces their respiratory water loss (MacMillen and Lee 1970; Rubal et al. 1995).

Evaporative cooling in response to increases in body temperature can occur from the skin and respiratory tract. A rapid forced increase in pulmonary ventilation occurs in many species. It may be a shallow panting or in some birds a gular flutter which results from rapid movements of the floor of the mouth and throat. However, the need for such cooling can sometimes be delayed by allowing the body temperature to rise by 2 to 7 °C, which may then approach lethal values. Such controlled hyperthermia or heat storage has been observed in camels and desert species of rodents and birds (Schmidt-Nielsen 1964a,b; Tieleman and Williams 1999). This accumulated heat may subsequently be dissipated during the night or by periodic visits to cool burrows. A considerable reduction of the evaporative loss of water can result from such behaviour.

6.5 Gills

Gills are primarily aquatic respiratory organs that are present in fishes and larval amphibians. In fish they are the major interface with the fresh water or sea-water in which they live. They make up about 90 to 95% of the total surface area of the integument (Parry 1966). The gills are permeable not only to respiratory gases but also to water and solutes. The gills of fish are also the sites of H^+ and HCO_3^- exchanges, which aid acid-base balance, the excretion of NH_3 and NH_4^+, which contributes to their nitrogen metabolism, and for the exchanges of water, Na^+ and

Cl⁻ that are involved in their osmoregulation. Fish gain water by osmosis across their gills in fresh water, while in sea-water, depending on the osmotic concentrations of their body fluids, they either lose water in this manner, as seen in most bony fish, or they may gain small amounts of water, as seen in cartilaginous fish and hagfish. The gills are the site of a loss of salts by diffusion in fresh water and a gain in sea-water. Fish in fresh water take up large amounts of water in this way. The skin of fish has a relatively low permeability to water and salts and in fresh water little drinking occurs (Lahlou et al. 1969a; Motais et al. 1969). Most of the water uptake in freshwater fish occurs across the gills (Motais et al. 1969). The gills are major sites of salt exchanges of fish in sea-water. Flounder, *Platichthys flesus*, in sea-water can accumulate Na^+ equivalent to about 40% of their total exchangeable body Na^+ each hour. About 75% of this uptake occurs across the gills, the remainder being due to drinking (Motais and Maetz 1965). In fresh water, flounder exchange less than 2% of their body Na^+ each hour across their gills. While the gills are the principal avenue for water and salt exchanges in fish, they also provide a mechanism for compensating for water loss and salt gain in sea-water and salt losses in fresh water. Apart from often being able to selectively limit such water and salt exchanges that occur by diffusion, they are the site of mechanisms for the active uptake of salt in fresh water and its secretion in sea-water. Such processes involve the activity of epithelial chloride cells in the gills. These cells appear to be better developed in bony than in cartilaginous fish.

The morphology of the gills (Perry 1997; Evans et al. 1999) has been studied most closely in teleost fish, but the general structural pattern appears to be similar in other species. The respiratory and osmoregulatory epithelial cells are supported on the gill arches (or bars) which bear the gill filaments (also called primary lamellae). The filaments bear numerous small parallel gill lamellae (also called the secondary lamellae) that are the respiratory surfaces. The filaments and lamellae are covered by epithelial cells, which are principally composed of flat pavement cells (also called squamous cells) that make up about 85 to 97% of the total cells present. The remainder are principally mitochondria-rich cells, which are the major site for the active ion exchanges and are thus also called ionocytes, chloride-secreting cells or, more commonly, *chloride cells*. Other types of cells that are present include accessory cells, which abut the chloride cells, but whose function is unknown, and mucous cells. The predominant pavement cells are principally involved in respiratory functions, but they may also contribute to exchanges of H^+ and HCO_3^- and be sites of Na^+ uptake (Goss et al. 1998). While some chloride cells may be present on the lamellae, most are situated among the epithelial cells that cover the interlamellar surfaces of the filaments. The intercellular spaces of the paracellular pathways between the gill epithelial cells are usually closed by tight junctions. However, in euryhaline fish adapted to sea-water they expand and become leaky (Sardet et al. 1979). This change provides an avenue for increases in the diffusion of solutes, including Na^+. Mitochondria-rich cells, which are thought to be functional chloride cells, have also been identified in the gills of elasmobranch fish (Garcia Romeu and Masoni 1970; Laurent 1984), lampreys (Morris and Pickering 1976; Peek and Youson 1979) and hagfish (Bartels 1985).

Morphological changes occur in gill epithelia of fish during their transitional adaptation from life in fresh water to sea-water (Conte and Lin 1967). Such changes have been observed to involve the chloride cells. Apart from proliferation, there may be increases in their size, changes in the structure of their apical cell membranes, increases in the concentrations of Na-K-ATPase on their basolateral surfaces and modification of the intracellular tubular system. Two subtypes of chloride cells (α and β) have been identified in euryhaline fish (Pisam et al. 1987). The β-type are present only in fresh water. The α-type, which has a distinctive apical plasma membrane and contains novel cytoplasmic inclusions, is present on the gill filament near the base of the lamellae. It is thought to transform into the marine type of chloride cell during adaptation to sea-water.

The gills have a complex vasculature (Laurent 1984; Nilsson and Sundin 1998) which is consistent with their various functions. The lamellae receive blood from the afferent branchial filamentous artery. The blood, on leaving through the efferent filamentous branchial artery, may follow either of two pathways. It may go directly to the systemic circulation or be "shunted" through an arteriovenous pathway, called the filamentous nutritional vasculature, before entering the central venous sinus. This shunt pathway principally supplies the filamentous epithelium including the chloride cells. Control of the passage of blood through the gills is under neural and hormonal control.

Movements of water and non-electrolytes by diffusion across the gills of fish are generally quite restricted compared to those in other vertebrate epithelia (Isaia 1984). For instance, in fresh water the permeability of the gills of rainbow trout to water is less than 10% of that in the urinary bladder of a toad (The latter is considered to be a "tight" epithelium!). The gills of elasmobranch fish have an extremely low permeability to urea. It is 70 to 80 times less than that in rainbow trout and eels, and 30 times less than that of the toad urinary bladder (Pärt et al. 1998). This property of the gills of elasmobranchs is consistent with their need to retain urea for use as an osmolyte. The permeability of the gills of rainbow trout and eels to water declines by about 50% following their transfer from sea-water to fresh water. The mechanism for this effect is uncertain, but it could involve morphological and vascular changes and, possibly, an alteration in the lipid composition of the cell membranes (Isaia 1984).

The gills of fish play a major role in maintaining the composition of their body fluids in both fresh water and sea-water. Such activities, which have been most closely examined in teleost fish, are interrelated to their role in acid-base balance. In fresh water fish the gills can actively accumulate sodium chloride from the external medium, while in sea-water they secrete the excess salt which is gained by diffusion and as a result of drinking the medium. Such activities can be regulated by processes that often involve the actions of hormones and growth factors. The molecular mechanisms involved utilize many of the transporter proteins which are summarized in Table 1.4. The include Na-K-ATPase, H^+-ATPase, the cotransporter Na^+-K^+-$2Cl^-$, ion exchangers (Na^+/H^+, Cl^-/HCO_3^- and Na^+/Na^+), ion channels (Cl^-, Na^+ and K^+) and in elasmobranch fish gills, possibly a urea transporter. Aquaporins could probably be utilized to advantage in fish gills but they do not appear to have been investigated in this tissue. The sites and effects of

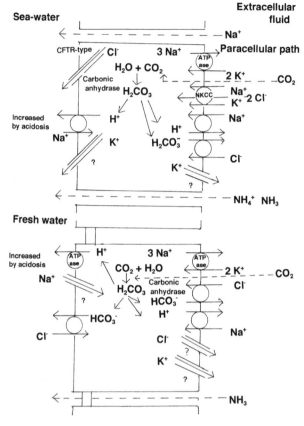

Fig. 1.2. Summary of the processes of ion transport that occurs across the gills of teleost fish in fresh water and sea-water. For further details see Table 1.4 and the text in Chapters 1 and 7

protein transporters in the gills of fish in fresh water and sea-water are summarized in Fig. 1.2.

6.6 Gut

The gut, or alimentary canal, apart from being the site for digestion and absorption of nutrients, is also the principal avenue for the exchanges of water and solutes in most vertebrates. The exceptions are fish and amphibians, which are aquatic. The absorption of fluid from the gut involves not only that which is imbibed but also a recycling of the considerable volumes that are secreted into the lumen during the process of digestion. The gut contributes to the regulation of the composition of the body fluids in several ways, which involve thirst, salt appetite and the selective absorption of salts and water. These processes mainly occur across the small and large intestine (colon), and in marine teleosts, which drink sea-water, the oesophagus. Additional such exchanges can occur across intestinal caecae and the cloaca of reptiles and birds. Herbivorous species usually have larger and longer guts than carnivores and can retain considerable volumes of digesta and fluids in alimentary sacs, such as the rumen in cattle, sheep and goats, the colon in horses and the caecum in rabbits. Such fluids are sometimes considered to be part of the animal's extracellular fluids and may provide reservoirs that can buffer the process of dehydration.

The gut is lined by an absorptive and secretory tissue called the mucosa (Lacy 1991; Madara 1991; Chang and Rao 1994). It typically consists of a single layer of columnar epithelial cells underlain by a basal lamina and a fibrous lamino propria. The latter contains blood vessels and nerve fibres. The absorptive surface of the intestine is increased by the presence of folds and villi, and microvilli (the brush border) on the apical surface of its lining epithelial cells. In some species of fish, including sharks, rays, chimaeras, sturgeons and lungfish, the intestinal surface area is increased by the presence of the partitions that form a spiral valve. The absorptive epithelial cells or enterocytes, cover the villi and new ones are produced in the crypts, or wells, at their base. The crypts contain undifferentiated epithelial cells, goblet cells, enteroendocrine cells and digestive Paneth cells. The undifferentiated epithelial cells can function as secretory cells producing fluids containing K^+, Cl^- and HCO_3^-. The epithelia can absorb fluids containing Na^+, K^+, Cl^- and HCO_3^-. Such processes involve the functional interactions of the cells and the paracellular pathways which lie between them. Such a relationship may provide the basis of a mechanism for the absorption of the considerable amounts of fluid that occur across the intestine (Curran 1965; Spring 1999). The absorptive and secretory processes that take place in the gut involve the activities of a variety of transporter proteins and are summarized in Fig. 1.3.

Drinking can be regulated by the sensation of thirst, which usually reflects the physiological need for water. A thirst centre, as well as a sodium appetite centre, is present in the anterior diencephalon region of the brain. The hormone angiotensin II (see Chap. 2) has a special role in initiating thirst and salt appetite (Fitzsimons 1998). Drinking in response to the need for water occurs in all groups of tetrapod vertebrates, with the notable exception of the Amphibia. The latter instead take up water across their skin. Freshwater fish, which accumulate water by osmosis across their gills, drink little. However, marine teleosts must drink sea-water in order to maintain their bodily hydration. They imbibe fluid equivalent to as much as 12% of their body weight each day (Motais and Maetz 1965). Elasmobranch fish apparently do not normally drink and, indeed, have little such need, as their body fluids are maintained at concentrations which are slightly hyperosmotic to sea-water. However, when injected with angiotensin II, drinking can sometimes be induced in such fish (Hazon et al. 1989). The drinking of seawater by teleosts involves a considerable adjustment by the fish, as this solution is about three times as concentrated as their body fluids. Ultimately, as described above, the excess salt is rapidly excreted as a result of the secretory activities of the chloride cells in the gills. The absorptive process for the imbibed sea-water involves an uptake of the sodium chloride and accompanying water, mainly across the wall of the oesophagus. The process involves diffusion and an active coupled

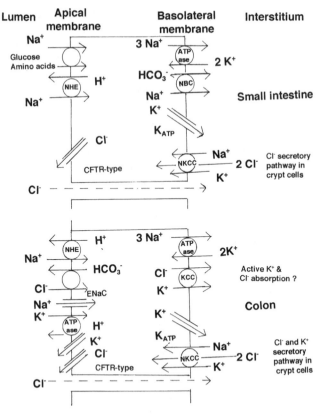

Fig. 1.3. Summary of the various transport mechanisms involved in the absorption and secretion of solutes across the small intestine and colon. Most of the information has been derived from observations in mammals, but such processes have also been identified in the other phyletic groups. For further details and references see Table 1.4 and the text

absorption of the Na$^+$ and Cl$^-$ (Hirano and Mayer-Gostan 1976; Nagashima and Ando 1994).

The colon, or large intestine, has a special role in regulating water, Na$^+$ and K$^+$ excretion in the faeces. There is a selective absorption of fluids and Na$^+$, and a secretion of K$^+$. The absorption of Na$^+$ involves the activity of special epithelial sodium channels (ENaC, see Table 1.4), which are also found in kidney tubules, amphibian skin and urinary bladder, but not usually in other regions of the gut. The ENaC can be specifically inhibited by the drug amiloride and have been identified in this way in the colons of amphibians, reptiles, birds and mammals, but apparently not in fishes. The intestinal contents can undergo considerable dehydration in the colon, where the water content may decline from about 95% in the small intestine to 60 to 70% in the faeces. It has been proposed that this fluid absorption occurs from the crypts at the base of the villi. A local osmotic gradient appears to be established across the mucosal membrane at this site as a result of active Na^+ transport (Naftalin and Pedley 1999; Naftalin et al. 1999). This concentration gradient is maintained due to the presence of a pericryptal sheath, which prevents dissipation of the Na^+ into the submucosal tissue. The osmotic pressure across the cryptal mucosal membrane is transduced into a suction hydrostatic pressure in the crypts. This pressure appears to be sufficient to remove fluid from the colonic contents. The pericryptal membrane has been functionally and microscopically identified in the rat distal colon.

The alimentary tract in non-mammals usually terminates in a cloaca, which it shares with the ureters and reproductive ducts. The cloaca can be quite large and the site of active Na^+ absorption in birds and reptiles (Bentley and Schmidt-Nielsen 1965; Skadhauge 1967). In birds, a reflux of urine into the upper regions of the colon has been demonstrated, providing an opportunity for further reabsorption of urinary water (Akester et al. 1967; Nechay et al. 1968). Schmidt-Nielsen and his collaborators (Schmidt-Nielsen et al. 1963) suggested that in some reptiles and birds, salt reabsorbed from the cloaca may be subsequently excreted by the nasal salt glands (see later). As the salt concentrations in the nasal gland secretions are much higher than those in the urine, a reduced urinary water loss may thus be accomplished. The physiological reality of this ingenious suggestion remains to be proven to the satisfaction of all (Thomas 1997).

6.7 Urinary Bladder

Fluid stored in the urinary bladder can make important contributions to the osmoregulation of some vertebrates. Urinary bladders are present in many, but not all, species. In fish they are usually quite small and are derived embryonically from mesoderm. They are expansions of the mesonephric ducts. In tetrapods the urinary bladder originates as an endodermal evagination of the cloaca. These urinary bladders can be quite large and in some amphibians can hold fluid equivalent to about 50% of their body weight. Tortoises also may have a capacious urinary bladder holding fluid equal to 20 to 30% of their body weight. Birds, crocodiles, alligators, snakes and most lizards lack such a urinary bladder. The urinary bladder can be quite permeable to water and is often the site of an active reabsorption of Na^+ and/or Cl^-. Its permeability and ion transport properties reflect the presence of a lining of epithelial cells, which in fish consists principally of mitochondria-rich cells. These cells are also present in other species but may be interspersed with other types of cells such as "granular" cells in amphibians. The mammalian urinary bladder, which is often exposed to urine with a hyperosmotic concentration, is relatively impermeable and has a multilayered epithelial lining. In amphibians and tortoises, water stored in the urinary bladder can be reabsorbed or provide a storage site for solutes during dehydration (Bentley 1966; Jørgenson 1998). Fluid can also be conserved by its reabsorption from the bladders of marine teleost fish (Howe and Gutknecht 1978). An active reabsorption of Na^+ and Cl^- may result in a physiologically significant conservation of salt in freshwater fish (Curtis and Wood 1991) and probably also amphibians (Bentley 1966) and chelonian reptiles (Brodsky and Schilb 1960). Ion transport and the control of water permeability have been studied in urinary bladders under in vitro conditions. In teleost fish a linked Na^+-Cl^- absorptive process has been identified (Lahlou 1967; Lahlou and Fossat 1971; Renfro 1975). The studies on winter flounder by Renfro provided the first genetic material for the cloning (Gamba et al. 1993) of the Na^+-Cl^- cotransport protein (Table 1.4). This protein also mediates ion transport in the distal convoluted renal tubule of mammals. In frogs and toads the urinary bladder has been widely used to study the process of active transepithelial Na^+ transport (Leaf et al. 1958), and hormonally regulated osmotic water permeability and Na^+ transport (Bentley 1958; Crabbé 1961a,b). It also provided a preparation for the early studies of amiloride-inhibitable Na^+ transport (mediated by ENaC) in epithelia (Bentley 1968).

6.8 Salt Glands

Many birds, reptiles and elasmobranch fish possess salt-secreting glands that can often produce hyperosmotic solutions, which principally contain either sodium chloride or potassium chloride. The concentrations of these solutions are much greater than can be achieved in the urine of such animals (Table 1.5). The salt glands are used to excrete excesses of ions, which are obtained in the diet, as a result of drinking sea-water or by diffusion across the gills. The relative development of these tissues usually reflects the animal's exposure to such regimens and conditions.

Table 1.5. Sodium and potassium concentrations in extra-renal salt glands in vertebrates

	$mEq\,l^{-1}$	
	Na^+	K^+
Chondrichthyes—rectal gland		
Spiny dogfish[a] (*Squalus acanthias*)	540	7
Reptilia		
Chelonia—supraorbital "tear" glands		
Loggerhead turtle[b] (*Carretta carretta*)	732–878	18–31
Lacertilia—lateral nasal gland		
Desert iguana[c] (*Dipsosaurus dorsalis*)	494	1387
Marine iguana[d] (*Amblyrhynchus crisatus*)	1434	235
Ophidia—sublingual gland		
Sea snake[e] (*Aipysurus laevis*)	798	28
Crocodilia—lingual gland		
Estuarine crocodile[f] (*Crocodylus porosus*)	386–740	10–16
Birds—supraorbital nasal gland		
Herring gull[g] (*Larus argentatus*)	718	24
Leach's petrel[g] (*Oceanodroma leucorhea*)	900–1100	–
Mallard duck[g] (*Anas playrhynchos*)	550	–

[a] Burger and Hess (1960). [b] Holmes and McBean (1964). [c] Schmidt-Nielsen et al. (1963). [d] Dunson (1969). [e] Dunson and Dunson (1974). [f] Taplin (1988). [g] Schmidt-Nielsen (1960).

Salt glands are derived embryonically from a variety of glands including tear glands in turtles, nasal glands in birds and lizards and salivary glands in some snakes. Morphologically they are similar and consist of a system of branched secretory tubules that open into a series of ducts. In elasmobranchs a central duct empties into the rectal region of the gut. (Hence this gland is called the rectal gland.) In birds and lizards this duct opens into the nasal cavity, while in sea snakes and crocodiles it opens into the mouth. All such salt glands appear to share a common mechanism for forming their secretions, which is similar to that of the chloride cells in fish (Fig. 1.2). Chloride is secreted actively across the apical side of the tubular cells following its entry, utilizing the Na^+-K^+-$2Cl^-$ cotransporter, across the basolateral surface of the cell. The initial genetic cloning of this latter transport protein was accomplished using tissue from the rectal gland of the spiny dogfish (Xu et al. 1994). Sodium ions appear to follow the Cl^- by diffusion, along paracellular pathways down the established transepithelial electrochemical gradient. The secretory cells are rich in mitochondria and have high concentrations of Na-K-ATPase on their basolateral surfaces (Shuttleworth and Hildebrant 1999; Silva et al. 1999a).

The secretory activity of the salt glands of sea birds and turtles was discovered by Knut Schmidt-Nielsen and his collaborators in 1958 (Schmidt-Nielsen and Fänge 1958; Schmidt-Nielsen et al. 1958; Schmidt-Nielsen 1960). In 1960, Burger and Hess described the functioning of the rectal gland in the spiny dogfish. Shortly afterwards, the salt-secreting function of nasal glands was described in several species of lizards (Schmidt-Nielsen et al. 1963; Templeton 1964). The secretions formed by the lizard's salt glands differed from those of the other species as they often contained high concentrations of potassium chloride instead of sodium chloride. The ability to secrete a concentrated potassium chloride solution was also observed in the marine iguana from the Galapagos Islands (Dunson 1969). The secretion of K^+ apparently occurs in response to the needs of reptiles that consume herbivorous diets, which contain large amounts of potassium. If potassium chloride solutions are given to ducks, the composition of their salt gland secretion is unchanged and the excess K^+ is then excreted in the urine (Simon and Gray 1991). The mechanism of this special ability of some reptile salt glands to secrete K^+-rich solutions is unknown.

Salt glands can secrete relatively large volumes of fluid. In herring gulls this volume can be ten times as great per gram of tissue as that occurring in the human kidney (Schmidt-Nielsen 1960). In birds the secretory response is initiated by a rise in the plasma osmotic pressure and increases in the plasma volume. It is mediated by neural stimuli arising from the parasympathetic nervous system. Acetylcholine is the principal neurotransmitter. Rectal gland secretion occurs in response to increases in plasma volume and may involve local nerve network, but hormonal stimulation (by a natriuretic peptide hormone) appears to be the predominant mediator (Valentich et al. 1995). Potassium secretion by the nasal salt glands in lizards can be initiated by the injection of potassium chloride (Shuttleworth et al. 1987). Both nerves and hormones could be involved in this process, but the details, including those of the secretory process in the cells, are unknown. It is possible that K^+ diffuses down electrochemical gradients via activated potassium channels in the apical plasma membrane.

The use of salt glands to secrete excess electrolytes appears to be an ancient process in vertebrates. Evidence for the presence of nasal salt glands has been found in fossil dinosaurs belonging to the order Ornithischia (Whybron 1981). These reptiles were apparently related to birds (they had a bird-like pelvis) and were herbivorous. Possibly these salt glands secreted a K^+-rich fluid, though that would not have been bird-like!

6.9 The Kidneys

Kidneys play a major role in the osmoregulation of vertebrates. However, their particular functions differ depending on the animal's phylogeny and whether it habitually lives in fresh water, the sea or terrestrial environments, such as deserts or more temperate regions. The particular physiological roles of kidneys are also related to the presence of other tissues and organs, which can contribute to osmoregulation, including gills, salt glands, the skin of amphibians and a reab-sorptive urinary bladder. General references include Dantzler (1992); Jamison and Gehrig (1992); Vander (1995); Braun and Dantzler (1997) and Reilly and Ellison (2000).

6.9.1 Functions

The primary role of the kidneys is to maintain the constancy of the composition of the extracellular fluids. This function includes the volume, concentrations of mineral ions (Na^+, K^+, Cl^-, Ca^{2+} and Mg^{2+}) and acid-base balance (H^+ and HCO_3^-). The kidney also contributes to the excretion of endogenous metabolites, such as result from nitrogen metabolism (NH_3, urea and uric acid). The processes of the excretion of excesses of water and solutes also involves a concurrent conservation and retention of essential constituents of the body fluids. The kidney also functions as an endocrine organ (see Chapter 2) and can secrete hormones and growth factors, including renin, erythropoietin and 1,25-dihydroxyvitamin D_3.

6.9.2 Structure

The intimate structure of the kidney varies in different species and is related to such factors as their phylogeny, body size and habitual physiological needs. The basic functional tissue unit in the kidney is the nephron. Each kidney contains many thousands, even millions, of such units. The numbers generally reflect the animal's size and metabolic rate. Nephrons are arranged in orderly arrays in relation to each other and the blood vessels that supply them. The latter arise from the renal artery or often in non-mammals from a venous renal portal system. Groups of nephrons may be arranged in the kidney in distinct lobules and zones, which may be situated centrally or peripherally. In mammals and birds they are present in an outer cortex and an inner medulla and papilla.

The basic architecture of the nephron includes a capillary bed called the glomerulus, which is surrounded by the Bowman's capsule. The latter connects with a morphologically and functionally segmented tubule lined by epithelial cells. The principal segments that are present vary in different species but often consist of a "neck" segment followed by the proximal tubule, intermediate segment, distal tubule, connecting duct and collecting ducts. Various smaller segments have also often been defined, especially in mammals. The morphological and physiological characteristics of the lining epithelial cells usually differ in the various segments. They are the sites of various arrays of transport proteins (Table 1.4; Fig. 1.4). The glomeruli are supplied with arterial blood which passes in through an afferent and out through an efferent glomerular arteriole. The latter can then break up into capillaries, which supply the tubules. Alternatively, in mammals and birds the efferent arterioles from some nephrons (juxtamedullary nephrons) send long capillary vessels deep into the renal medulla and papilla (the vasae recta) where they form hairpin bends, or loops, before returning to the renal vein. The venous renal portal system from the tail region of non-mammals supplies only the renal tubules. The afferent glomerular arterioles are the site of two local regulatory mechanisms: baroreceptors and the macula densa. The latter consists of a group of columnar epithelial cells, which in mammals adhere to part of the adjacent distal renal tubule. There are many interspecific variations of the basic structure of the nephron. Not all vertebrates have glomeruli at all. They can be reduced in number or even completely absent (aglomerular) in many species of marine fish and even in some snakes. On the other hand, hagfish (class Myxini, primitive jawless fish) lack nephrons (anephric) but possess large glomeruli that open directly into the archinephric duct. A distal tubule (the diluting segment) is absent in most marine teleost fishes but is present in most such freshwater and euryhaline species. Marine, but not freshwater, elasmobranch fish possess very long nephrons that are arranged to form two hairpin loops with each other and associated blood vessels. A countercurrent flow system appears to be formed which is comparable to that present in mammals and birds (Friedman and Hebert 1990). It has been suggested that this system may contribute to the kidney's ability to conserve urea in these fish (Boylan 1972). In birds and mammals the proximal and distal tubules are joined by an elongated thin intermediate tubule which also forms a hairpin bend, or loop, which may extend deep into the renal medulla and papilla. The entire looped structure of the nephron is called the loop of Henle and includes part of the proximal and distal tubules. This region of the latter is called the thick ascending limb of the loop of Henle. The loop of Henle lies parallel to the blood vessels of the vasa rectae and provides the countercurrent flow system that allows mammals and birds to secrete a hyperosmotic urine (Table 1.6). The countercurrent flows of blood and glomerular filtrate preserve the high concentrations of osmolytes which accumulate in the renal medullary tissue and provide the concentration gradients necessary for the formation of a hyperosmotic urine. This system is generally not as well developed in birds as in mammals.

Table 1.6. Maximum urine concentrations of various species of mammals and birds

	mosmol l^{-1}
Mammals	
Humans[a]	1450
Mountain beaver[b] (*Aplodontia rufa*)	500
Seal[c] (*Phoca vitulina*)	2200
Sand rat[d] (*Psammomys obesus*)	6300
Spinifex hopping mouse[e] (*Notomys alexis*)	9400
Quokka wallaby[f] (*Setonix brachyurus*)	2200
Echidna[g] (*Tachyglossus aculeatus*)	2300
Birds	
Domestic fowl[h] (*Gallus domesticus*)	520
Emu[i] (*Dromaius novae-hollandiae*)	485
Ostrich[j] (*Struthio camelus*)	800
Budgerygah[k] (*Melopsittacus undulates*)	850
Zebra finch[i] (*Taeniopygia castanotis*)	1005

[a] Smith (1951). [b] Dicker and Eggleton (1964). [c] Smith (1936). [d] Schmidt-Nielsen (1964a). [e] MacMillen and Lee (1969). [f] Bentley (1955). [g] Bentley and Schmidt-Nielsen (1967). [h] Skadhauge (1981). [i] Skadhauge (1974). [j] Withers (1983). [k] Krag and Skadhauge (1972).

6.9.3 Physiology

The formation of urine begins with the ultrafiltration of plasma across the capillaries of the glomerulus. This ultrafiltrate contains electrolytes and small solutes with a similar concentration to that of the plasma, but little or no protein. In humans about 120 ml of plasma are filtered in this way each minute. It is called the glomerular filtration rate or GFR. The urine flow in humans is only about 1 ml each minute so that about 99% of this filtrate is reabsorbed as it passes along the renal tubule. The GFR varies widely in different species reflecting their body size, metabolic rate and their environmental circumstances. In humans the GFR (ml kg^{-1} h^{-1}) is about 100, while in rats it is 200 and in domestic fowl 75. In terrestrial reptiles it varies from about 4 to 25 ml kg^{-1} h^{-1} and it is reduced when the animals are dehydrated (Dantzler 1992). In fresh water the GFR is about 34 ml kg^{-1} h^{-1} in frogs and 9 ml kg^{-1} h^{-1} in rainbow trout. When these trout are adapted to sea-water, the GFR decreases to about 1 ml kg^{-1} h^{-1}.

A reabsorption of filtered water and solutes occurs progressively as the filtrate passes along the nephron. The process involves the activities of the renal tubular epithelial cells and a variety of their transporter proteins (Fig. 1.4). Other solutes may be added as a result of the secretory activities of the renal tubular epithelium. Such substances that are secreted include K^+, H^+, NH_3 and uric acid. In aglomerular fish, sodium chloride and water are secreted across the proximal tubule.

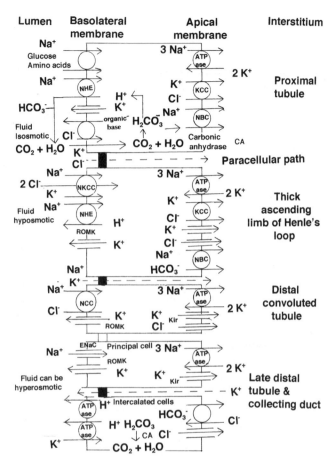

Fig. 1.4. Summary of the various transport mechanisms that are involved in the formation of urine by the kidney. Most of the information refers to the mammalian kidney, but homologous mechanisms have been identified in other vertebrates. For further details and references see Table 1.4 and the text

In mammals, about 65% of the glomerular filtrate, including all the glucose and most of the HCO_3^- are usually reabsorbed in the proximal tubule. This fluid remains isosmotic to the plasma due to its osmotic equilibration, with the aid of aquaporins. The distal tubule is referred to as the diluting segment since, following further reabsorption of Na^+ and Cl^-, the luminal fluid becomes hyposmotic due to its low permeability to water. In mammals the diluting segment is usually considered to be part of the thick ascending loop of Henle and the distal convoluted tubule. About 25% of the Na^+ and Cl^-, and the remainder of the K^+ in the filtrate are reabsorbed in this part of the nephron. The process involves the activities of the Na^+-K^+-$2Cl^-$ and the Na^+-Cl^- cotransporter proteins. The late distal tubule and the collecting ducts are the sites of the final regulated reabsorption of Na^+. This process utilizes specific epithelial sodium channels (ENaC). A secretion of K$^+$ also occurs in the distal tubules, as well as adjustments of the pH involving a secretion of H$^+$ and NH$_3$. In mammals, and possibly some other tetrapods, the transport of Na$^+$ and K$^+$ in this region of the nephron can be regulated by the hormone aldosterone. This region of the nephron is usually quite impermeable to water. However, under the influence of antidiuretic hormone (ADH) and aquaporin 2 its permeability can increase, allowing an osmotic equilibration of the filtrate with the hyperosmotic fluids which, in mammals and birds, are present in the renal medullary tissue. A hyperosmotic urine can thus be formed in such species.

6.9.4 Regulation

The regulation of the secretion of urine involves hormones and nerves and autoregulation, which arises from within the kidney itself. Such control mechanisms can influence the GFR and absorption and secretion across the renal tubule. The hormones usually are produced by the endocrine glands but "local" hormones produced by the kidney tissue can also contribute to regulation. The neural mechanisms are generally mediated by a sympathetic nerve supply to the glomeruli and renal tubules (DiBona and Kopp 1997; DiBona 2000). Local hormones include the products of arachidonic acid metabolism, such as prostaglandins, nitric oxide, dopamine and kinins. Hormones produced by endocrine glands that exert direct effects on kidney function include vasopressin (antidiuretic hormone, ADH), vasotocin, aldosterone, natriuretic peptides and angiotensin II. These hormones will be described in succeeding chapters. Glomerular function is usually well controlled by the vascular activity of the afferent and, sometimes, the efferent glomerular arterioles. The rate of filtration across single, individual, glomeruli normally changes little. However, large changes in the GFR frequently occur in non-mammalian vertebrates, and this process usually involves increases or decreases in the numbers of the functioning glomeruli (glomerular intermittency or recruitment). In individual nephrons an equilibrium usually exists between the GFR and rates of tubular reabsorption. This relationship is sometimes referred to as glomerular-tubular balance. The rate of delivery of water and solutes to the tubule is related to the GFR and their concentrations in the plasma. If, for instance, delivery is increased, the normal reabsorptive processes may tend to reject the excess, which will then be excreted in the urine. Such an effect is seen dramatically in the disease of diabetes mellitus, when an increased delivery of plasma glucose to the renal tubule results in its appearance in the urine. The excretion of Na$^+$ and urea can be similarly influenced locally in the kidney due to a situation that has been called glomerular tubular imbalance. Conditions in the renal tubule can also influence the GFR. This process of tubuloglomerular feedback results from the ability of the macula densa cells to monitor the concentrations of sodium chloride present in the filtrate and convey this information back to the afferent glomerular arteriole. Small, appropriate, changes in the GFR can then be induced. Such autoregulatory processes are clearly important for the regulation of kidney function. Nevertheless, an absence or excess of hormonal regulation can have disastrous effects on kidney functioning, as seen in several human and animal diseases.

7 Relationships of Nitrogen Metabolism to Osmoregulation

Proteins can be converted to carbohydrates and lipids, and also provide precursors for nucleic acids. This process involves breakdown to their constituent amino acids, which undergo oxidative deamination resulting in the release of α-amino nitrogen and the formation of ammonia (NH_3) (Fig. 1.5). This NH_3 is highly toxic, especially to the nervous system. In rabbits a concentration of 0.03 mmoll^{-1} is fatal. The NH_3 is readily diffusible in water so that vertebrates that live in aquatic environments can excrete it directly into their bathing solutions. Ammonia is readily converted to ammonium (NH_4^+) at a physiological pH, and much of it usually exists in this form. Ammonium is also water-soluble but it is less lipid-soluble than NH_3 and hence is less toxic. Excretion occurs across the gills of bony and agnathan fish and the skin of larval forms of amphibians. As this process is the predominant manner in which they excrete waste nitrogen, they are said to be *ammonotelic* (also called ammoniotelic). Ammonia can be formed from glutamate in the distal renal tubule, where its secretion results in the neutralization

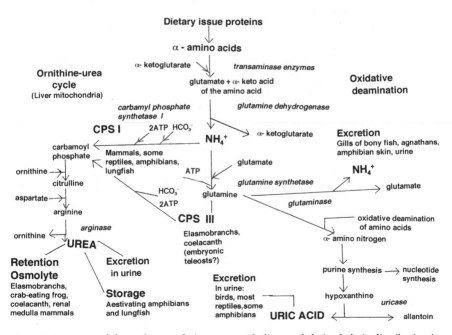

Fig. 1.5. Summary of the pathways of nitrogen metabolism and their phyletic distribution in vertebrates. General references are: Anderson (1995); Wright (1995); Campbell et al. (1987); Walsh (1997); Withers (1998b)

39

of H^+ to form NH_4^+, which is excreted in the urine. This process helps maintain the pH of the urine at physiologically acceptable levels. The NH_4^+ also can substitute for Na^+, which is reabsorbed, and would otherwise be excreted in excessive amounts. Some tetrapod vertebrates with a surfeit of available water excrete large amounts of NH_4^+ in their urine and have been described as exhibiting facultative ammonotelism and include crocodilians and, remarkably, humming birds (Preest and Beuchat 1997). However, tetrapod vertebrates, as well as some fish, especially elasmobranchs, resort to other strategies to detoxify NH_3. Some convert it to urea and others to uric acid. Such species are said to be, respectively, *ureotelic* and *uricotelic*. There are no clear phyletic distinctions as to which groups of vertebrates utilize a particular process for nitrogen excretion nor are they mutually exclusive. Ammonotelic fish often also excrete small amounts of urea, while crocodiles excrete both ammonia and uric acid. Turtles and tortoises can excrete both urea and uric acid. Hence the prefix "mainly" is often used. Mammals and amphibians are then described as being mainly ureotelic while birds and most reptiles are mainly uricotelic. Elasmobranch fish, the coelacanth, *Latimeria*, and the crab-eating frog, *Rana cancrivora*, retain urea and utilize it as a retention osmolye to maintain the osmotic concentrations of their body fluids similar to that of their environmental sea-water. They are described as being ureosmotic, though they appear to be also ureotelic.

Urea is very water-soluble and appears to principally cross cell membranes by passing along paracellular pathways and by utilizing special urea transporter proteins in cell membranes (see Table 1.4). It has a low toxicity and in some ureosmotic species possible toxic effects may be reduced, as described above, by the presence of counteracting solutes such as methylamine compounds. Urea can be readily excreted in the urine, though a retention mechanism exists in the kidneys of ureosmotic species and in the renal medullary tissue of mammals. Uric acid contains four nitrogen atoms compared to two in urea. Uric acid is quite insoluble but remains in solution in the renal tubule. However, in birds and reptiles it is subsequently precipitated in the cloaca, where a reabsorption of urinary water occurs. The excretion of nitrogen as uric acid in the urine requires less water than when urea is utilized as the end product of such metabolism. It has been estimated that 1 g of nitrogen requires 10 ml of water for its excretion in uricotelic animals as compared to 10 to 50 ml in ureotelic species and 500 ml in ammonotelic ones (Smith 1951). It has been suggested that the savings of water that occur as a result of uricotelism were a major factor in the success of the archosaurian "ruling reptiles", which included the dinosaurs and ancestors of the birds (Campbell et al. 1987). The ornithine-urea cycle appears to have been a requisite for the movement of the Amphibia onto dry land, and they may have inherited this ability from crossopterygian relatives. Extant ancestors of the latter are the coelacanth and lungfish, which can synthesize urea.

A number of vertebrates, which live under osmotically stressful conditions, have been observed to modify the habitude of their nitrogen metabolism. African lungfish are normally ammonotelic but when they experience periods of drought and aestivate they synthesize urea. This metabolite is accumulated in their body fluids at concentrations that may be as high as 500 mmol l^{-1} (Smith 1930a). A similar storage of urea, at concentrations in the body fluids up to 900 mmol l^{-1}, has been observed in some frogs and toads that live in deserts (McClanahan 1967; Degnani et al. 1984; Withers and Guppy 1996). Even more remarkable was the discovery of several species of tree frogs from Africa and South America (genera *Chiromantis* and *Phyllomedusa*) which are uricotelic (Loveridge 1970; Shoemaker et al. 1972a; Drewes et al. 1977). African reed frogs, *Hyperolius viridiflavus*, which live in dry exposed situations, are ureotelic (Schmück et al. 1988). As described above, they can deposit platelets that reflect sunlight in their dermal iridophores. These platelets are composed of the purine compounds guanine and hypoxanthine derived from the metabolic nucleic acid-uric acid pathway. Formation of these platelets continues in dry periods, when it apparently provides a "sink" for waste nitrogen, thus reducing the need to synthesize urea.

The biosynthetic pathways that mediate the synthesis of NH_4^+ are summarized in Fig. 1.5. The synthesis of urea involves the formation of carbamoyl phosphate which enters the ornithine-urea cycle. This synthesis is mediated by the enzyme carbamoyl phosphate synthetase (CPS). Elasmobranch fish and the coelacanth possess a variant of this enzyme called CPS III, while that in mammals, amphibians and chelonian reptiles and lungfish is CPS I. Some embryonic teleost fish possess CPS III, but few adult teleosts utilize the ornithine-urea cycle. However, it appears that the requisite genes may still be present in these fish. Uric acid is also a waste product of nucleic acid metabolism, but in most ureotelic species it is converted to allantoin by uricase. However, this enzyme is not present in uricotelic species or higher primates such as humans. Nucleic acid synthesis utilizes α-amino nitrogen to form purine and pyrimidine nucleotides. The purine pathway has apparently been coopted by uricotelic species for more general use in disposing of α-amino nitrogen as uric acid. Urea can also be formed from uric acid, and this process can occur in some teleost fish.

8 Osmoregulation and the Evolution of Vertebrates

The nature of the processes of osmoregulation in extant vertebrates has provided the basis for some stimulating speculation about their origins and evolution. Extrapolations to the physiological nature of osmoregulation in early vertebrates may be relevant to questions such as their possible origins in fresh water or the sea, and if tetrapods arose from freshwater or marine ancestors. Fossils from the Palaeozoic era have provided morphological evidence regarding the distribution of the various phyletic groups in geological time and the nature of their environments. However, such information is often difficult to interpret. It is generally agreed that the vertebrates evolved from a protochordate, which may have been a cephalochordate (Chen et al. 1995) related to amphioxus (*Branchiostoma*). All known protochordates live in the sea. The earliest vertebrates were jawless fish (superclass Agnatha, the cyclostomes) the fossils of which (ostracoderms) first appeared in the Ordovician period about 500 million years ago. Contemporary agnathans, hagfish and lampreys, are, respectively, exclusively marine and euryhaline. Like most other fish, their kidney function involves the presence of glomeruli. As these structures are well suited to the excretion of excess water, it has been suggested that their presence reflects the evolution of such fish from a freshwater ancestor (Marshall and Smith 1930; Romer 1955). However, this explanation has not been accepted by all (Robertson 1957; Halstead 1985; Griffith 1987). It is now generally agreed that the geological sediments where the first fossil agnathan remains were discovered were marine. Furthermore, glomerular-like vascular structures also function as excretory organs in marine invertebrates that have no freshwater lineage. Their roles in marine vertebrates, including hagfish, appear to be related to the regulation of ion, rather than water, concentrations in the body. The presence of glomeruli and their use for the excretion of excess water could be considered to have arisen as a preadaptation, which was subsequently utilized by a freshwater ancestor.

Hagfish (superclass Agnatha; class Myxini; order Myxiniformes) are considered to be extant relatives of fossil agnathans that are collectively described as ostracoderms (class Pteraspidomorphi; order Heterostraci and others). They are exclusively marine and the composition of their extracellular body fluids, except for divalent ions, is remarkably similar to that of sea-water (Table 1.2). The divalent ions are present in higher concentrations than in sea-water and are, apparently, mainly excreted by the kidney. However, intracellular concentrations of ions in hagfish are similar to those in other vertebrates and are much lower than those in the extracellular fluid (Table 1.3). Osmotic balance is achieved by utilizing amino acids as intracellular osmolytes. Hagfish are now often considered to provide an osmotic prototype for the first vertebrates. (The Agnatha are considered to be a diphyletic group with the lampreys, class Petromyzontiformes, displaying the clearest affinities to contemporary vertebrates (Bardack and Zangerl 1968).) The ancestral chondrichthyean fish appear to have been marine, like most extant species, and presumably utilized urea as an osmolyte to allow for a decrease in the concentrations of ions in the body fluids. Such ion concentrations may have been advantageous with respect to the stability and functioning of cell macromolecules. The maintenance of isosmoticity between the body fluids and the bathing sea-water may also have been energetically advantageous at the low prevailing levels of atmospheric oxygen prior to the Carboniferous period (Ballantyne et al. 1987). However, it is notable that the relative concentrations of the ions in sea-water and the body fluids of most vertebrates are still similar (Epstein 1999). The utilization of urea as an osmolyte by the Chondrichthyes presumably was made possible following the acquisition of the ornithine-urea cycle. Some osteichthyean fish, especially crossopterygians like the extant coelacanth, probably also developed such an osmotic strategy based on the use of urea as an osmolyte.

The fossil record identifies many freshwater, as well marine, species of ostracoderms in the Devonian. The transition to fresh water from the sea may have occurred at the Silurian-Devonian boundary (Halstead 1985). It may have involved an anadromous breeding migration by bony fish that were incipiently euryhaline (Griffith 1987). Apart from an ability to excrete accumulated water, such fish may have developed an ability to limit loss of, or even accumulate, salts from across their gills. Extant freshwater fish can utilize branchial Na^+ / H^+ and Cl^- / HCO_3^- exchange mechanisms to take up sodium chloride. These exchange processes are also present in marine hagfish, where they are utilized for maintaining acid-base balance (Evans 1984). Thus, an ability to accumulate sodium chloride across the gills could have been present in fishes before they entered fresh water.

The transition of fish to tetrapod vertebrates and a terrestrial life appears to have occurred in the Devonian and to have involved a crossopterygian (lobe-finned) fish (order Rhipidistia) (Thompson 1980; Bray 1985). This momentous evolutionary step to an amphibian could have been taken by a marine or fresh-water species of fish. It could have been prompted by the drying up of a fresh-water river or lake or the pollution of a coastal marine environment as a result of flooding. A marine origin for the tetrapods is usually favoured by the pundits. Such a fish, like the extant coelacanth, could have utilized urea as an osmolyte. This strategy is also utilized by a contemporary estuarine amphibian, the crab-eating frog. The presence of the ornithine-urea cycle and ureotelism would appear to have been an essential prerequisite for a terrestrial life. (Uricotelism did not, it seems, appear until much later.) Amphibians living on dry land have special problems due to evaporation from their permeable skin. This limitation was apparently not overcome until the evolution of the reptiles in the Carboniferous period.

Cotylosaurian stem reptiles appear to have evolved from salamander-like labyrinthodontian amphibians. Reptiles have been very successful in adapting to dry terrestrial conditions. The stem cotylosaurians are thought, like extant turtles and tortoises, to have been both ureotelic and uricotelic (Campbell et al. 1987). (It will be recalled that some extant amphibians that live in dry exposed con-ditions have also developed uricotelism.) Contemporary offspring of the Lepidosauria, the lizards and snakes, and the Archosauria, the crocodiles and birds, are uricotelic, which can result in a considerable reduction of their urinary water loss. It seems likely that the dinosaurs, which were archosaurians, were also uricotelic. Fish, amphibians and reptiles cannot form hyperosmotic urine and in birds this ability is quite limited compared to that in mammals. Urates are relatively insoluble and the danger of their precipitation in the renal tubule may have limited such renal concentrating ability. However, the ability to excrete excess salt as a hyperosmotic solution can have considerable benefits in water-saving. Various secretory glands have been utilized for such salt excretion in reptiles and birds. They have even been identified in fossil dinosaurs (Osmolska 1979; Whybron 1981). However, they do not appear to have evolved in mammals, which have, instead, a special capacity to form a hyperosmotic urine. This ability is linked to the presence of the ornithine-urea cycle, which is necessary for the establishment of high renal medullary tissue concentration gradients. They may have acquired this ureotelism from the cotylosaurian reptiles.

Chapter 2

The Vertebrate Endocrine System

Osmoregulation is the process of physiological adjustment of the volume of water and the concentrations of solutes in the cells and body fluids of organisms, including the vertebrates. Such responses are usually necessitated by changes that can occur as a result of the animal's interactions with its external environment but also can be due to metabolic changes and disturbances, including disease. Changes in the animal's fluid content and composition occur continually in potentially hostile osmotic environments such as the sea, freshwater rivers and lakes, and terrestrial habitats, which can even include hot dry deserts. Several types of physiological responses may be involved in maintaining homeostasis of the body fluids. They are mediated, depending on the species, by such organs and tissues as the kidneys, urinary bladder, gut, salt-secreting glands, skin, gills and the blood vascular system. The nature and duration of their physiological responses, coordination with each other and the environmental events, necessitates the transmission of chemical messages that are provided by neighbouring cells, nerves and the endocrine glands. Osmoregulation can involve all of these coordinating processes. However, the hormones secreted by the endocrine glands have special roles that probably reflect their more persistent and ubiquitous effects.

1 Endocrine Function

Endocrine glands secrete hormones directly into the blood, which disseminates them to distant organs and tissues. Such a release of hormones usually occurs in response to specific stimuli. The glands may detect changes in the composition and volume of the blood plasma or be stimulated to secrete by tropic hormones and nerves. Hormones are widely distributed in the body but only certain tissues and organs respond. Tissues identify hormones by the presence of "receptor" molecules that can only bind and respond to specific hormones. Receptors may be present in the cell membrane or the cytosol or be associated with genes in the cell nucleus. Hormones have constitutive roles in regulating processes such as growth and differentiation, reproduction, the metabolism of fats, proteins and carbohydrates, and the volumes and composition of the body fluids. Hormones can also contribute to rapid responses, such as may be precipitated by stressful situations, when they usually act in conjunction with nerve impulses and local hormones such as cytokines. Such physiological coordination occurs in all vertebrates. However, it may differ somewhat, depending on the animal's natural environmental circumstances and phylogeny. Thus, although all vertebrates possess kidneys, only fish and larval amphibians have gills. The occurrence of salt glands is phyletically sporadic and an ability to form hyperosmotic urine is present only in birds and mammals. Morphologically homologous endocrine glands have been identified in most of the main groups of vertebrates, but there are exceptions, such as the lack of the parathyroid glands in fish. Hormones with similar, homologous, chemical structures have a widespread distribution in vertebrates. However, their precise structures may differ and even display evidence of an evolution. The physiological roles of hormones may also display phyletic variation, and like their structures, also have evolved. This area of study is known as comparative endocrinology. Comprehensive reviews of this subject are available (Henderson 1997; Bentley 1998). There are many examples of such phyletic differences and evolution in the processes of osmoregulation of vertebrates.

1.1 Overview

The principal endocrine glands, their secreted hormones and main functions are shown in Table 2.1. It can be seen that their roles and functions encompass a wide range of metabolic and physiological processes, and adaptive responses to changes in the internal and external environments. They include the regulation of the water and salt content of the body which, among various species, can directly involve the actions of many different hormones. In tetrapod vertebrates these include vasopressin (ADH), vasotocin, aldosterone, angiotensin II, adrenaline, natriuretic peptides and the guanylins. Additional hormones with special such roles in the fishes include growth hormone, IGF-I, prolactin, cortisol and, possibly, the urotensins. Other hormones, such as adrenocorticotropin (ACTH) and the hypothalamic hormones may be involved at a secondary, and even tertiary, level, as they can influence the synthesis and release of more directly involved hormones. Other hormones have indirect effects on osmoregulation, such as those that mediate reproduction with its associated increased need for water and salts. Thyroid hormone can increase the basal metabolic rate, and hence evaporation, and its morphogenetic effects may contribute to the differentiation of tissues, such as amphibian skin and fish gills, that influence osmoregulation. Other endocrine glands are necessary for the sustenance of many tissues. These hormones include insulin, glucagon, cortisol, leptin, parathyroid hormone, calcitonin, 1,25-dihydroxyvitamin D_3, adrenaline and angiotensin II. Hormones clearly have multifacetted effects on living processes. The present account is mainly concerned with those that have primary and secondary effects on osmoregulation.

1.2 Architecture of Hormones and Receptors

Chemically, hormones may be peptides, (vasopressin and angiotensin), polypeptides, (ACTH and natriuretic peptides), proteins, (growth hormone and

Table 2.1. Summary of the principal endocrine glands in vertebrates: their secreted hormones and main roles

Endocrine glands	Principal hormones	Main roles
Pituitary		
Adenohypophysis	ACTH, TSH, FSH, LH, GH, PRL	Tropic actions on endocrines Growth and differentiation
Neurohypophysis	Vasopressin, vasotocin, oxytocin-like peptides	Water metabolism, vascular contractility Reproduction
Pars intermedia	MSH	Pigmentation (skin)
Hypothalamus	CRH, TRH, Gn-RH, GH-RH GH-R-IH, dopamine etc.	Regulate hormones of adenohypophysis
Adrenal		
Cortex	Aldosterone, corticosterone, cortisol	Na and K metabolism, Intermediary metabolism
Medulla	Adrenaline, noradrenaline	"Stress" responses
Thyroid	Triiodothyronine, thyroxine	Basal energy metabolism morphogenesis
Parafollicular "C" cells (of thyroid) (Ultimobranchial bodies)	Calcitonin	Calcium metabolism
Parathyroid glands Gonads	Parathyroid hormone	Calcium metabolism
Ovaries	Estradiol-17β, progesterone	Female reproductive system
Testes	Testosterone, 5α-dihydrotestosterone	Male reproductive system
Islets of Langerhans	Insulin (B-cells)	Metabolism, proteins, carbohydrates, fats
	Glucagon (A-cells)	As above, generally opposes insulin
Pineal	Melatonin	Biological rhythms, especially reproduction
Liver	Angiotensinogen (prohormone), Angiotensin II	Adrenal cortex, thirst, blood pressure
	IGF-I (induced by GH)	Growth, differentiation
Kidney	Renin	Angiotensin I from angiotensinogen
Heart	Natriuretic peptides	Regulation extracellular volume
Adipose tissue	Leptin	Appetite regulation, proreproductive
Skin	Vitamin D_3 (prohormone)	Calcium metabolism
Gastrointestinal tract	Gastrin, cholecystokinin secretin, guanylins etc.	Digestion, absorption, feeding
Corpuscles of Stannius (some fish)	Stanniocalcin	Calcium metabolism
Urophysis (some fish)	Urotensin I and II	Possible roles in osmoregulation

prolactin), amino acid derivatives (adrenaline and thyroid hormone) and steroids (aldosterone and cortisol). Some hormones, such as the amino acid derivatives, are quite small molecules, but the proteins can have a molecular mass over 30 kDa. Receptors for hormones are large proteins and glycoproteins. The chemical structures of hormones and receptors are prescribed by the animal's genes. The structures of hormones and receptors determine their abilities to interact with each other and initiate a specific response. For a hormone this configuration is called its structure-activity relationship (SAR). In nature, and as a result of the manipulations in the chemist's laboratory, hormones may display variations (polymorphism), especially in different species of animals. Structural variants of a particular hormone may also exist in a single species or even an individual animal. Receptors also display such structural variability. Variations in the structures of a particular hormone and its receptors may have little effect on the qualitative nature of the response, though quantitative differences may result. However, more critical changes in structure can precipitate in the abolition of an ability to initiate a particular response and even presage the evolution of a novel hormone. Chemically homologous hormones and receptors can often be classified into families and superfamilies, the individual members of which may, nevertheless, mediate quite different types of responses. At least 14 different peptides that are related to mammalian vasopressin have been identified in different species of vertebrates. Most animals have at least two of them, which can react with at least two different types of receptors and initiate a wide variety of effects. The responses, in different species, include increases in the water permeability of kidney tubules and amphibian skin and urinary bladder, a contraction of blood vessels, a release of ACTH, contraction of the uterus or oviduct, and the release of milk from mammary glands.

Different parts of hormones and receptors contribute to various aspects of their functioning. In polypeptides and proteins these regions are often described as domains which contain specific sequences of amino acids. Recognition and binding domains are concerned with the interactions between hormones and their receptors. Receptors may also have transmembrane domains, which anchor them in the cell membrane, dimerization domains, which are involved in their abilities to associate with each other, and transactivation domains, which are concerned with initiation of the response. The structures of hormones can also incorporate other properties such as may contribute to their abilities to bind to plasma proteins and their metabolic destruction and clearance from the blood.

1.3 Morphology of the Endocrine Glands

The endocrine glands usually occupy comparable positions in the bodies of all vertebrates and exhibit characteristic morphological structures. They generally have a plentiful blood supply into which they directly secrete one or more hormones. A nerve supply may contribute to such a release of hormones and regulate the blood flow. There are some phyletic differences in the occurrence of the endocrine glands. Parathyroid glands are not present in fish. Hagfish, crocodiles and whales lack a pineal gland. Some fish possess special endocrine glands, for instance the corpuscles of Stannius and the urophysis, which are absent in tetrapods. The various cells that aggregate to form endocrine glands may have quite different embryonic origins. Thus, the neurohypophysis and adrenal medullary tissue are derived from neural tissue, the adenohypophysis from ectoderm and the adrenal cortex from mesoderm. Endocrine glands may be divided into distinct lobes, as seen in the pituitary, a cortex and medulla, as in the mammalian adrenal, and the zones of the adenohypophysis and mammalian adrenal cortex. Such morphological divisions are usually related to the synthesis and secretion of different hormones. Not all endocrine cells aggregate into distinct glands. There are, for instance, many types of hormone-secreting cells that are individually dispersed in the mucosal lining of the gut. Various tissues and visceral organs may also function as endocrine glands and secrete hormones. The heart secretes natriuretic peptides, fat cells produce leptin, the liver IGF-I and angiotensinogen and the kidney renin and 1,25-dihydroxyvitamin D_3. On specific occasions, a variety of other tissues can secrete hormones, including the placenta, mammary glands and some tumours.

1.4 Synthesis of Hormones

Endocrine cells sometimes synthesize hormones in situ from either amino acids or cholesterol. This process often occurs in response to, and with the aid of tropic hormones from other endocrine glands. Examples include the actions of ACTH and angiotensin II on the adrenal cortex, thyrotropin on the thyroid gland and gonadotropins on the gonads. Some hormones are stored following their synthesis in granules, vesicles and tissue follicles. Others, especially the steroidal hormones, are promptly released into the circulation. The chemical design of hormones and their biosynthesis are directed by their genes. Peptide, polypeptide and protein hormones are made from precursor proteins (preprohormones and prehormones) that are produced by genetic transcription and translation. On the other hand, adrenaline and thyroid hormones are made from tyrosine, melatonin from tryptophan and steroid hormones from cholesterol. The chemical substitutions involved in the conversions of these substrates to the hormones are the results of sequences of chemical reactions that are controlled by enzymes. They usually occur in the endocrine cell, but an "activation" of a hormone may sometimes occur in peripheral tissues.

Following their formation on the ribosomes, preprohormones and prehormones are directed by the N-terminal signal sequence of amino acids to the endoplasmic reticulum (ER). This signal peptide is then removed, which results in the formation of a prohormone or a hormone. The latter occurs in the synthesis of growth hormone and prolactin. A prohormone or newly formed hormone may undergo further posttranslational processing in the cisternae of the ER, where molecular folding, cross-linking, acetylation and glycosylation can occur. On passing to the Golgi apparatus, the prohormone may undergo selective cleavages of its amino acid sequence to produce a single or even several hormones. Multiple copies of a single hormone or several different hormones may be produced in this way from a single prohormone molecule. Thus, the prohormone proopiomelanocortin (POMC), which is produced by the pituitary, can provide ACTH, α-MSH, β-MSH and several other biologically active molecules, including Met-enkephalin and β-endorphin. Such cleavage of the prohormone occurs at predetermined sites in the molecule, usually involving basic arginine and lysine residues. This process results from the actions of enzymes called subtilisin-related convertases (SPCs). The posttranslational processing of prohormones may vary in different endocrine glands. Thus, ACTH and β-endorphin may be produced from POMC in the adenohypophysis and α- and β-MSH, and β-endorphin in the pars intermedia. (A single gene may also give rise to different hormones as a result of differences in posttranscriptional processing, alternate-splicing, of its precursor RNA.)

The synthesis of steroid hormones from cholesterol involves the actions of a succession of enzymes that are associated with the ER and mitochondria in steroidogenic cells. The early steps in this synthesis are shared by the adrenocortical and gonadal steroid hormones. However, the subsequent biosynthetic changes are more specific. Thus, aromatization enzymes produce oestrogens from androgen substrates, 11β-hydroxylase produces corticosterone from 11-deoxycorticosterone and cortisol from 11-deoxycortisol, and aldosterone synthetase converts corticosterone to aldosterone.

1.5 Release of Hormones

A release of hormones from an endocrine gland into the blood may occur in response to stimuli which signal a physiological imbalance, changes in environmental conditions, pending events, such as reproduction and migration, and physical and emotional stress. Such hormone release may be constitutive, and occur continuously, sporadically, such as in response to an acute need, or rhythmically. Hormones may be released in small pulsatile bursts or tonically in a steady stream. Rhythmical release may be related to regular events, such as day and night, the seasons and the activities of endogenous biological clocks. A variety of receptor mechanisms may be involved in triggering the release of a hormone. Endocrine cells may respond directly to concentrations of ions, nutrients and the osmotic and hydrostatic pressures of its perfusing blood. Receptors for tropic hormones may be present in the cell membrane. Responses of endocrine glands to neural stimuli may be mediated by receptors and ion channels. Hormones are secreted from endocrine cells by diffusion or exocytosis following the fusion of storage granules and vesicles with the cell membrane. Such an event may be linked to electrical depolarization or transduced by elevations of cell Ca^{2+} concentrations and the activities of the adenylate cyclase-cAMP and phospholipase C-phosphatidylinositol signalling systems (see later).

1.6 Hormones in the Blood

Physiologically effective concentrations of hormones in the blood plasma range from about 10^{-7} to 10^{-11} M. They may be in aqueous solution or bound to plasma proteins. The lipid-soluble steroid hormones and thyroid hormone can achieve higher concentrations by such binding. The transport of hormones to their effector target sites will also be facilitated. Other hormones, including growth hormone, prolactin and IGF I, may also be bound to proteins in the plasma. For a response to occur, the hormones must enter the aqueous phase and be "free" in solution so that they can combine with their receptors. In some instances, as much as 99% of a hormone may be bound while only 1% is free in solution. However, this proportion varies with the different hormones. Some hormones may bind to plasma albumin in a relatively non-specific manner (low-affinity binding) though in relatively large amounts (high capacity). However, some proteins may be present in the plasma, at lower concentrations, which can bind hormones more specifically (high-affinity but low-capacity). They include corticosteroid-binding globulin, (CBG), which binds 80 to 90% of the cortisol and corticosterone that is present in the plasma, though only 20% of the aldosterone. The binding of hormones to proteins in the plasma can have a number of consequences on their actions, apart from facilitating their transport. They may also protect tissues from possible toxic effects of high aqueous concentrations of a hormone. Binding of hormones may impede or facilitate the transport of a hormone across barriers in capillary beds, such as those in the kidneys and brain. The clearance of a hormone from the plasma may also be reduced so that its action will be more prolonged.

1.7 Peripheral Metabolism of Hormones

Hormones, following their release into the blood, undergo metabolic changes which can either increase (activation) or decrease their effectiveness. Such processes occur in peripheral tissues, especially the liver and kidneys, and also their target organs. These chemical changes include the cleavage of polypeptides and proteins by peptidase enzymes. The enzymes may act at internal sites or the amino- or carboxy-terminals of such molecules. They are then called, respectively, endopeptidases and amino- or carboxy-peptidases. The initial formation, activation and final demise of the octapeptide hormone angiotensin II involves the actions of all such types of these enzymes. Hormones can be inactivated by a reduction of cross-linking disulphide bonds. Oxidation reactions, such as the conversion of hydroxyl to keto groups, are often involved in the inactivation of steroid hormones. Conversely, a reduction of a keto to a hydroxyl group can occur and can, for instance, activate cortisone, to form cortisol. Vitamin D_3 is activated to 1,25-dihydroxyvitamin D_3 by two successive hydroxylations, which occur first in the liver and then the kidney. The inactivation of adrenaline involves the processes of methylation of its aromatic ring and the deamination of its side chain. Thyroxine (T_4) contains four atoms of iodine. Removal of one of these, at the 5^l position, results in the formation of triiodothyronine (T_3), which is much more active. Further such deiodination, by deiodinase enzymes, however, results in its inactivation. Steroid and thyroid hormones can undergo conjugation reactions to form glucuronide and sulphate compounds. This process abolishes their biological activity and also increases their water solubility so that they can be excreted more readily in the urine and bile.

1.8 Mechanisms of Hormone Action

Osmoregulatory responses to hormones involve such processes as the absorption and secretion of water and electrolytes, especially Na^+, K^+ and Cl^-, across epithelial membranes in the kidneys, gut, urinary bladder, gills, skin and salt glands. The structure and metabolic integrity of such tissues and organs can be modulated by hormones. Such processes include cell growth and differentiation, and the functioning and replication of their transporter proteins. The cardiovascular system can influence osmoregulation. It not only provides sustenance but also mediates the exchanges of fluids between the plasma and the intercellular compartment, and the formation of the glomerular filtrate by the kidneys. Hormones can have special roles in regulating such processes by influencing the contractility of vascular smooth muscle in small blood vessels. The effects of a hormone commence with its binding to its receptors and conclude with the response of its target tissue. The molecular processes that link these events are called signal transduction (Hunter 2000). It involves a translation and amplification of the hormone's message to induce a morphological or physiological change in the effector. The mechanism usually involves a complex cascade of biochemical events in the cell. An activation of a variety of proteins, especially enzymes, usually takes place. The latter include many protein kinases (Hunter 2000), which, by utilizing ATP as a substrate, can phosphorylate tyrosine, serine and threonine residues on proteins and so change their inherent activities. Such events usually involve the induction of conformational changes. These proteins may be other enzymes, transcription factors and structural components of cells.

Receptors for hormones, as described earlier, may be present in the cell membrane or the cytoplasm, or be associated with the nucleus of the cell. They are proteins or glycoproteins, which include a number of functional domains. Such sequences of amino acids are involved with ligand (hormone) binding (the recognition site), anchorage in the cell membrane (transmembrane domain), binding to a nuclear-response element and the formation of receptor dimers. Membrane-associated receptors are elongated molecules which lie in the plasma membrane. Their N-terminus projects into the extracellular fluid and their C-terminus into the cytoplasm. The recognition site of these receptors is usually associated with the extracellular portion of the molecule or, sometimes, the transmembrane domain. The cytoplasmic portion of the receptor often, but not always, has enzymatic activity and contains a catalytic site. The membrane receptor may cross the cell membrane once or recross its several times in a serpentine-like manner. Such receptors have a capacity to move laterally and some may then associate with others to form dimers. Such combinations may be necessary for their functioning. Similarities in the structures of receptors often allow them to be classified into families and superfamilies which appear to reflect a common primeval ancestry.

The binding of a hormone to its receptor is thought to result in a conformational change in its structure, which provides the initiating transduction signal. It may result in the formation of a dimer with a neighbouring receptor of the same, or even different ilk, the dissociation of a contained inhibitory component, migration to the cell nucleus or the activation of an endogenous enzymatic activity. Such events constitute the first period of the transduction process.

G-Protein-Linked Receptor Mechanisms. Many membrane receptors are associated with heterotrimeric G-proteins that transduce the response to juxtaposed enzymes, such as adenylate cyclase and phospholipase C. The activity of G-proteins is influenced by their binding to the guanosine nucleotides GDP and GTP. Hormone receptors which utilize this mechanism include vasopressin, adrenaline, ACTH and angiotensin II. The transmembrane domains of these receptors characteristically cross and recross the cell membrane seven times.

The Actions of Vasopressin V_2 Receptors, Adrenaline β-Receptors and ACTH Receptors. These are linked to adenylate cyclase in the cell membrane (Fig. 2.1).

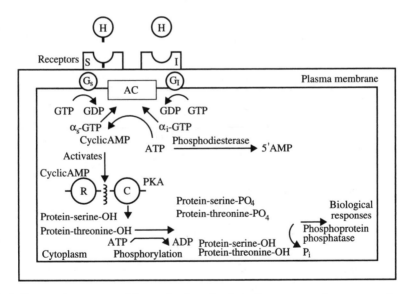

Fig. 2.1. A diagramatic summary of the mechanism of action of hormones, such as vasopressin and vasotocin (V_1 type receptors) and adrenaline (β-adrenergic receptors), that utilize the adenylate cyclase-cyclic AMP and G protein systems. *AC* Adenylate cyclase; G_s and G_i G proteins with stimulatory (*S*) and inhibitory (*I*) effects; α_i and α_s activated inhibitory and stimulatory subunits, respectively, of G proteins; *H* hormone; *I* inhibition; *PKA* Proteinkinase A; *C* catalytic unit of PKA; *R* regulatory subunit of PKA; *S* stimulation. For further information see the text. (Bentley 1998)

The G-proteins consist of α, β and γ subunits. The α-subunit bind is bound to GDP. Following the activation of the receptor by the hormone, the GDP on this subunit is replaced by GTP. A dissociation of α-GTP subunit from the G-protein then follows and results in an activation of adenylate cyclase. The α-GTP possesses intrinsic GTPase activity and quickly reverts to α-GDP, which then recombines with the β-γ subunit. The α subunit can exist in two forms, α_s, which stimulates adenylate cyclase activity, and the other, α_i, which inhibits it.

The activated adenylate cyclase forms cAMP from ATP. This nucleotide, which is called a second messenger (the hormone being the first messenger), combines with the regulatory subunit of serine / threonine protein kinase A (PKA), releasing its active catalytic unit. The PKA can phosphorylate a variety of proteins that are involved in the response of the effector. Such proteins include the transcription factor CREB that binds to cAMP-response elements (CRE) in the cell nucleus.

G-proteins also link hormonal responses that involve the activation of phospholipase C (PLC) (Fig. 2.2). This membrane enzyme is linked to α_1-adrenergic receptors, V_{1a} vasopressin receptors and angiotensin AT_1 receptors. The G-protein subunit involved is α_q. The membrane lipid phosphatidylinositol 4,5-bisphosphate (PIP_2) is its substrate. Two second messengers result: diacylglycerol (DAG) and inositol-1,4,5-trisphosphate (IP_3). The DAG activates protein kinase C (PKC) while IP_3 mobilizes Ca^{2+} from the endoplasmic reticulum. The Ca^{2+} can combine

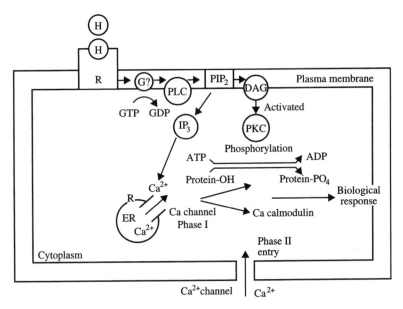

Fig. 2.2. A diagramatic summary of the mechanism of action of hormones, such as vasopressin and vasotocin (V_2-type receptors), angiotensin II and adrenaline (α-adrenergic receptors) that utilize the phospholipase C-phospatidylinositol system. *DAG* Diacylglycerol; *ER* endoplasmic reticulum; G protein G_q; *H* hormone; IP_3 inositol-1,4,5-trisphosphate; *PLC* phospholipase C; *R* receptor. For further information see the text. (Bentley 1998)

with the intracellular protein calmodulin and initiate various responses, including activation of protein kinases and phosphodiesterase. The latter inactivates cAMP.

The Insulin Receptor Protein Tyrosine Kinase-MAP Kinase Pathway. The insulin receptor superfamily includes membrane receptors for IGF-I, which mediate many effects of growth hormone, including the adaptation of some teleost fish to life in the sea. Insulin and IGF-I have also been observed (in vitro) to increase Na^+ transport across some epithelial membranes, such as the amphibian urinary bladder and skin (Blazer-Yost et al. 1989). Insulin and IGF-I receptors have a homology in their amino acid sequences of about 50% (Prager and Melmed 1993). A homologous insulin-IGF-I type receptor has been described in a protochordate (amphioxus, *Branchiostoma*) (Chan et al. 1992). Such receptors mediate the various effects of such hormones on metabolic processes and cell division, growth and differentiation. They typically consist of two pairs of subunits, α and β, that are linked by disulphide bonds (Kahn et al. 1993). The α subunit projects into the extracellular fluid and contains the hormone-binding site while the β subunit has a cytoplasmic domain that possesses intrinsic tyrosine kinase activity. The binding to the hormone results in a lateral movement of the receptors. They then form dimers, which results in conformational changes and the activation, by autophosphorylation, of the cytosolic tyrosine kinase domain of the receptor (Saltiel 1996; Schlessinger 2000). This event initiates the activation of a cascade of kinase enzymes. This process is initiated by the activation of a small G-protein called Ras that is associated with the cell membrane. It is linked to the receptor tyrosine kinase by an adaptor protein (Grb_2) and a guanylate nucleotide exchange factor (GEF or SOS protein). The phosphorylation kinase cascade can result in the activation of MAP-kinases (mitogen-activated proteins), which can enter the nucleus and activate transcription factors. The activated tyrosine kinase may also contribute to the functioning of other intracellular processes.

The Janus Kinase-STATS Pathway. This mediates the actions of growth hormone and prolactin. Growth hormone and prolactin contribute to the osmoregulatory adaptations of some fish during their migrations between rivers and the sea. The receptors for these hormones span the cell membrane but do not appear to possess any intrinsic enzymatic activity. They belong to the cytokine receptor superfamily that also includes receptors for several cytokines such as erythropoietin and interferons (Bazan 1990; Nakagawa et al. 1994). Such receptors have been identified in osmoregulatory tissues of teleosts (Sandra et al. 2000). They mediate processes including cell growth, development, locomotion and repair. Responses to these receptors are transduced by cytosolic tyrosine kinase pathways, which results in the activation of transcription factors (Finidori and Kelly 1995; Carter-Su et al. 1996; Horseman 1997). The receptors form dimers following their binding to the hormones. This event signals their cytosolic domains, which then each combine with a molecule of Janus kinase (JAK). A number of types of these tyrosine kinases exist (JAK_1, JAK_2 etc.) in the cytoplasm. When they are associated with the receptor they can phosphorylate and activate transcription factors called STATs (signal transducer and activator of transcription).

Following such activation, STATs form homo- and hetero-dimers and enter the cell nucleus, where they can initiate transcription. One of the proteins induced in the liver by growth hormone is IGF-I, which, as described above, contributes to the adaptation of some teleost fish to life in sea-water.

Membrane Guanylate Cyclase and the Receptors for Natriuretic Peptides and Guanylins. Guanylate cyclase hydrolyzes GTP to cGMP, which, like cAMP, can act as a second messenger and activate protein kinases. Guanylate cyclase exists in a soluble form in the cytoplasm, where it may be involved in such processes as neurotransmission and the relaxation of vascular smooth muscle. This form of guanylate cyclase can be activated by nitric oxide (NO), which is formed from arginine by nitric oxide synthase (NOS). A second form of guanylate cyclase is associated with the cell membrane, where it can be an integral part of receptors for the natriuretic peptides and guanylins. These peptides promote the secretion of sodium chloride by the kidney and, in the instance of the guanylins, also the intestine (Atlas and Maack 1992; Forte et al. 2000). The natriuretic peptide hormone receptors (GC-A, GC-B and GC-C) have a single transmembrane domain linking an extracellular ligand-binding domain with a cytosolic domain. The latter possesses intrinsic guanylate cyclase activity (Chinkers et al. 1989). Transmembrane signalling, as a result of ligand binding, results in the activation of this enzyme and the formation of cGMP. This nucleotide can activate protein kinases including cGMP-activated protein kinase II and it can also activate cAMP-dependent protein kinase. The CFTR chloride channels in intestinal mucosal cells can be activated as a result of phosphorylation by these kinases (Forte et al. 1992; Vaandrager et al. 1997). It has also been shown that cGMP can inhibit the activities of sodium channels (ENaC) in the kidney collecting ducts (Light et al. 1989). Apart from phosphorylation of proteins, cGMP may have more direct effects on ion channels such as by binding to components that modulate their gating mechanisms. (Finn et al. 1996).

Steroid Hormone-Activated Transcription Factors. The receptors for steroid hormones and thyroid hormone have many structural similarities and are considered to belong to the same protein superfamily. They are principally present in the region of the cell nucleus where they can bind to DNA. These receptors are actually transcription factors that can be activated by directly binding to hormones. Steroid hormones are lipophilic and so can cross the cell membrane by diffusion (Fig. 2.3). Some may interact with receptors in the cytoplasm, but more usually this interaction occurs in the nucleus. The steroid hormones contribute to many homeostatic mechanisms. The regulation of sodium and potassium metabolism is the province of the adrenocorticosteroids, especially aldosterone, but in teleost fish cortisol is involved. Their receptors contain domains with special functions. A hormone-binding domain is present at the C-terminus and a transactivation domain, which stimulates transcription, at the N-terminus. A DNA-binding domain and dimerization domain lie in the middle of the receptor. The amino acid sequences of such domains vary for the different steroid hormones and species, but they still display considerable homologies. The DNA-binding domain of the human glucocorticoid receptor has a 97% homology

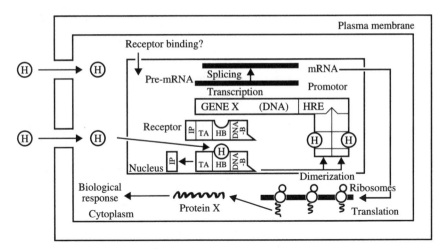

Fig. 2.3. A diagramatic summary of the mechanism of actions of steroid hormones, such as aldosterone, that utilize the process of genomic transcription. *DNA-B* DNA-binding domain; *H* hormone; *HB* hormone-binding domain; *HRE* hormone-response element; *IP* inhibitory protein; *TA* transcriptional activation domain. For further information see the text. (Bentley 1998)

in its amino acid sequence to that in the rainbow trout (Ducouret et al. 1995). The hormone-binding domains of these hormones have a 70% homology. The effects on sodium and potassium metabolism are usually mediated by mineralocorticoid receptors (MR) and those on protein, fat and carbohydrate metabolism by gluco-corticoid receptors (GR). However, teleost fish have only one adrenocorticosteroid hormone receptor that mediates both types of effects and its structure is GR-like. Interactions of the steroid hormone receptors with hormones result in a structural change, which includes the unmasking of a site that facilitates the formation of a homodimer. (Some steroid hormones receptors form heterodimers.) The activated hormone receptor-transcription factor moves to the gene and combines with a specific DNA sequence, called the hormone-response element (HRE), on the promoter. (An enhancer DNA sequence, which can increase utilization of the promoter, is sometimes activated.) A multiprotein transcription initiation complex is assembled on the promoter and guides the actions of RNA polymerase II at the initiation site. The transcription of RNA follows, resulting in the formation of specific proteins that mediate the response. In the instance of mineralocorticoids, such proteins can include Na-K-ATPase and sodium channels (ENaC).

2 Osmoregulatory Hormones

A variety of hormones secreted by endocrine glands directly contribute to the regulation of the accumulation, conservation and excretion of water and salts in the body; Table 2.2). Others indirectly influence such processes as a consequence

Table 2.2. Hormones that contribute to osmoregulation in vertebrates

Hormone	Site of origin	Chemical nature	Osmoregulatory target organs	Principal effects
Primary effects				
Vasopressin (antidiuretic hormone, ADH), vasotocin (AVT)	Neurohypophysis	Nonapeptides	Kidney, blood vessels, amphibian skin and bladder	Antidiuresis. In amphibians water uptake across skin and bladder. Vasoconstriction
Natriuretic peptides (ANP, BNP, CNP etc.)	Heart, also brain	Polypeptides	Kidney, adrenal cortex, blood vessels, gills	Regulate extracellular fluid volume; natriuresis, diuresis, vasodilatation, Inhibit synthesis of aldosterone
Angiotensin II (from angiotensinogen)	Renin-angiotensin system (kidney, plasma)	Octapeptide	Kidney, adrenal cortex, blood vessels, hypothalamic thirst-centre	Increases corticosteroids, drinking, Na-appetite. Vasoconstriction
Growth hormone	Adenohypophysis	Protein	Gills of teleost fish (Cl-cells)	Induces IGF-I, increased ability to extrude Cl⁻
Prolactin	Adenohypophysis	Protein	Gills of teleost fish (Cl-cells). Also gut, bladder, kidney?	Decreases salt loss from gills during adaptation to fresh water. Decreased permeability to water
Aldosterone	Adrenocortical tissue	Steroid	Kidney, colon, sweat glands, amphibian skin and bladder, reptile salt glands	Na$^+$ retention and absorption, K$^+$ secretion
Cortisol	Adrenocortical tissue	Steroid	Gills of teleost fish (Cl-cells) Also gut and (?) kidneys.	Promotes salt secretion in sea-water Salt uptake in fresh
Adrenaline and noradrenaline	Chromaffin tissue, adrenal medulla	Catecholamines (amino acid derivatives)	Cardiovascular system, epithelial permeability, (kidneys, gills, amphibian skin).	Vasoconstriction, vasodilatation, altered ion transport and permeability to water
Guanylins	Intestine	Peptides	Kidney and gut epithelia	Promotes Cl⁻ secretion

Secondary and Tertiary Effects

Adrenocorticotropin (ACTH)	Adenohypophysis	Polypeptide	Adrenocortical tissue	Increases synthesis of cortisol, corticosterone and can facilitate that of aldosterone
Corticotropin-releasing hormone (CRH)	Hypothalamus	Polypeptide	Adenohypophysis	Increases release of ACTH
Dopamine	Hypothalamus	Catecholamine	Adenohypophysis	Inhibits prolactin release
Growth hormone-R-H	Hypothalamus	Polypeptide	Adenohypophysis	Increases release of growth hormone
Somatostatin	Hypothalamus etc.	Peptide	Adenohypophysis	Inhibits release of growth hormone

Indirect Effects on Use of Water and Salt

Oestrogens, androgens and progesterone	Gonads	Steroids	Reproductive organs (primary and secondary)	Supports reproductive cycles, care of young, migration
Gonadotropins (LH, FSH)	Adenohypophysis	Glycoproteins	Gonads	Control reproductive cycles
Thyroid hormone (triiodothyronine, T_3)	Thyroid gland	Iodinated amino acid derivative	Most tissues	Morphogenesis (includes amphibians and fish), increased metabolic rate in homeotherms

General tissue support

Insulin, glucagon	Islets of Langerhans	Protein, polypeptide	Liver, muscle, fat cells	Regulate metabolism of proteins, fats, and carbohydrates
Cortisol, growth hormone	see above			
Parathyroid hormone	Parathyroid gland	Polypeptide	Kidney, gut, bone, gills	Regulate calcium metabolism
Calcitonin	Thyroid "C" cells	Polypeptide		
Stanniocalcin	Corpuscles of Stannius	Polypeptide		
Vitamin D_3	Skin and diet	Sterol		

of their tropic actions on these glands and effects on the morphological and metabolic integrity of the target organs involved. The particular hormones involved in the osmoregulation of vertebrates often display differences in their chemical nature, physiological effects and the particular target organs and tissues where they act. The following is a comparative description of such hormones in the vertebrates. General references are Henderson (1997) and Bentley (1998). Summaries are provided in Tables 2.1 and 2.2.

2.1 The Hypothalamus and Pituitary Gland

The hypothalamus and pituitary gland function as an integrated neuroendocrine complex that is the major interface between the nervous and endocrine systems. It makes important contributions to osmoregulation (Denton et al. 1996) and many other activities of the endocrine glands. Both tissues secrete hormones and lie adjacent to each other at the base of the forebrain (diencephalon) below the third ventricle. The hypothalamus is formed embryonically as a thickening of the floor and side of the third ventricle and runs from the optic chiasma caudally to the midbrain. The pituitary lies beneath it. They communicate by means of both neural connections and a system of blood vessels, the hypophysial-pituitary portal system, that originates in the median eminence regions of the hypothalamus.

The *hypothalamus* (see Fig. 2.4) is the site of complex neural networks in which the cell bodies form a number of aggregates called nuclei. The larger, magnocel-

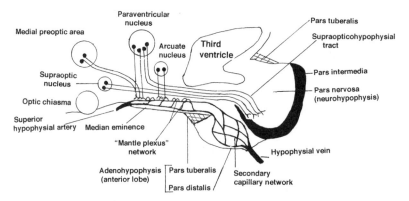

Fig. 2.4. A diagrammatic depiction of the amniote pituitary gland and the associated region of the hypothalamus. In anamniotes the neurohypophysis does not form a distinct neural lobe. In teleost fish the interactions of the hypothalamus and adenophypophysis are principally due to neural rather than vascular connections. For further details see the text

lular, neurons form the supraoptic nuclei (SON) or, in fishes and amphibians, the preoptic nucleus. These neurons have neurosecretory activity and are the sites of formation of the neurohypophysial peptide hormones. Smaller, parvicellular, neurons also form such nuclei, including the arcuate and paraventricular nuclei (PVN) which secrete several "releasing" hormones into the portal blood vessels. These hypothalamic hormones travel to the anterior lobe of the pituitary gland (the adenohypophysis or pars distalis), where they can regulate the release of its hormones. The hypothalamus has many other functions that are mostly concerned with integrating and maintaining bodily homeostasis. It contributes to the regulation of the cardiovascular system, body temperature, feeding, thirst and salt appetite. Neurons in the hypothalamus can also detect the levels of steroid hormones in the peripheral circulation and so provide feedback signals that can control the release of ACTH as well as gonadotropins. Thus, the hypothalamus has several important roles that are related to the animal's osmoregulation.

The *pituitary gland* (hypophysis) secretes at least nine different hormones, four of which have important effects on osmoregulation in various species. Embryonically, the pituitary has a dual origin, being formed by a downgrowth of neural tissue from the brain and an upgrowth of ectodermal tissue from the roof of the pharynx. The former (Fig. 2.4) becomes the *neurohypophysis* (pars nervosa) and the latter the *adenohypophysis*, which includes the pars distalis, pars tuberalis and pars intermedia. The pars distalis and pars tuberalis are sometimes called the anterior lobe of the pituitary, the pars nervosa the posterior lobe and the pars intermedia the intermediate lobe. The neurohypophysis retains neural connections to the hypothalamus through the infundibular stalk, which carries the nerve fibres of the supraopticohypophysial tract. The hormones secreted by the neurohypophysis, in mammals vasopressin and oxytocin, are formed in the cell bodies of the SON and PVN and travel down their axons to the pituitary. The adenohypophysial hormones are synthesized in specialized cells including the corticotropes (ACTH), gonadotropes (FSH and LH), thyrotropes (TSH) and somatomammotropes (GH and prolactin).

The general pattern of the morphology of the pituitary and hypothalamus is similar throughout the vertebrates, but some differences occur. The neurohypophysis is enlarged in tetrapods, as compared to the fishes, and is called the neural lobe. This change has been associated with a terrestrial manner of life and a greater quantity of stored hormones. A hypophysial-pituitary portal blood system is absent or poorly developed in teleost fish. Instead, they utilize neural connections between the two tissues. Hagfish and lampreys (Agnatha) lack both vascular and neural connections between the hypothalamus and pituitary. The hypothalamic hormones apparently depend on diffusion to gain access to the pituitary. Birds lack a pars intermedia and it is poorly developed in some mammals, including humans, elephants and whales.

Neurohormones of the Hypothalamus. At least six different hormones have been identified in the hypothalamus. However, it is suspected that several others, often referred to as factors, may exist. Usually, such a hormone or factor has been found to either increase or inhibit the release of a specific pituitary hormone.

61

[Thus, in naming such excitants the suffixes –RH (releasing) and –IH (inhibiting) are often used.] However, in some instances, it is suspected that a single such hypothalamic hormone can influence the release of more than one pituitary hormone. For instance: thyrotropin-releasing hormone (TRH) may sometimes also promote the release of growth hormone and prolactin. The release of some pituitary hormones is under dual control of hypothalamic hormones, one of which increases while the other decreases, release. All of the identified hypothalamic hormones are peptides or polypeptides with the single exception of dopamine (which is a catecholamine). Dopamine functions as a prolactin release-inhibiting hormone (P-R-IH). Growth hormone (GH) is under the dual control of GH-RH (a polypeptide containing 37 to 45 amino acid residues) and somatostatin (GH-R-IH containing 14 amino acids). The release of ACTH (corticotropin) is promoted by CRH (contains 41 amino acids) and is possibly inhibited by CRIF (Redei et al. 1995). Vasopressin is better known as a neurohypophysial hormone with an antidiuretic action, but it can also promote the release of ACTH. This response may occur under conditions where stress is a predominant stimulus for the release. Hypothalamic hormones have been identified throughout the vertebrates, but their amino acid sequences usually display differences that can be related to their phylogeny. Several of them also occur at other sites in the body such as the brain. Somatostatin can also contribute to the control of secretion of insulin and gastric juices. Corticotropin-releasing hormone is also expressed in the mammalian placenta and appears in the peripheral circulation. It may be involved in the initiation of parturition (Challis and Lye 1994).

The Peptide Hormones of the Neurohypophysis. Most vertebrates can secrete two structurally related hormones from the neurohypophysis (Fig. 2.5). In mammals they are vasopressin (ADH) and oxytocin. Other phyletic groups of vertebrates usually secrete vasotocin and an oxytocin-like peptide. Vasopressin and vasotocin regulate the excretion of water by the kidneys of tetrapods and therefore play an important role in their osmoregulation. In mammals, oxytocin contributes to the reproductive process by its ability to contract the uterus during parturition and to promote the let down of milk by the mammary glands. The role of oxytocin-like peptides in non-mammalian vertebrates is uncertain. However, about 14 such peptides have been identified in species ranging from mammals to fish. Such phyletic persistence suggests that they have physiological functions, but currently it is unknown what they may be.

The neurohypophysial hormones are nonapeptides, which, as described earlier, are synthesized in the cell bodies of the supraoptic and paraventricular nuclei in the hypothalamus of mammals, birds and reptiles. In amphibians and fish they are formed in the preoptic nucleus. Chemically, neurohypophysial peptides consist of a ring structure containing six amino acids which is closed by a disulphide bond contributed by two hemi-cystines. There is also a side chain consisting of three amino acids (Fig. 2.5). Vasopressin has phenylalanine at position 3 and either arginine or lysine at position 8. Oxytocin differs from vasopressin by two amino acid substitutions and has isoleucine at position 3 and leucine at position 8. In non-mammals, vasotocin replaces vasopressin. Like oxytocin, it has isoleucine at position 3 but retains arginine at position 8, as in vasopressin. Hence

Amino acid position	1	2	3	4	5	6	7	8	9
Common structure, variations in positions 3, 4 and 8									
Neurohypophysial hormone									
8-Arg-vasopressin Mammals	Cys	Tyr	Phe	Gln	Asn	Cys	Pro	Arg	Gly (NH$_2$)
8-Lys-vasopressin Pigs, some marsupials								Lys	
Oxytocin Mammals, Holocephali			Ile					Leu	
8-Arg-vasotocin All nonmammals			Ile					Arg	
Mesotocin Birds, reptiles, amphibians, lungfish, some marsupials			Ile					Ile	
Isotocin Teleosts and some other bony fish			Ile	Ser				Ile	
Glumitocin Skates and rays			Ile	Ser				Glu	
Valitocin Sharks			Ile	Gln				Val	

Fig. 2.5. The amino acid sequences and phyletic distribution of the major neurohypophysial hormones in vertebrates. (Based on information summarized by Acher 1996 and Acher et al. 1970, 1999)

it is called vasotocin. Its effects are less specific than those of vasopressin as it not only has antidiuretic and vasoconstrictor effects but also exerts oxytocic actions on the uterus and oviduct. When tested in mammalians it can also promote milk letdown. This hormone has been identified in agnathan fish, where it appears to be the sole such neurohypophysial peptide. Vasotocin may be the primordial neurohypophysial hormone in vertebrates. The lineage of such peptides may go back even further, as structural homologues, conopressin and cephalotocin, have been identified in molluscs (Van Kesteren et al. 1992a,b).

In tetrapods, vasopressin and vasotocin are released in response to increases in the osmotic pressure of the plasma. An osmoreceptor complex is associated with the hypothalamic neuronal nuclei. However, vasopressin and vasotocin may also be secreted in response to other stimuli, including a reduction in blood volume and stress. Several distinct receptors for the neurohypophysial hormones have been identified in mammals. There is an oxytocin receptor, a vasopressin V_{1a} receptor, which mediates the contraction of vascular muscle and a V_2 receptor, which is responsible for increases in the permeability of the renal tubule to water. The receptors in rats have been cloned and contain about 400 amino acid residues. They belong to the G-protein-coupled receptor superfamily (Birnbaumer et al. 1992; Kimura et al. 1992; Lolait et al. 1992; Morel et al. 1992). The vasotocin receptor in a teleost fish, the white sucker, has also been cloned (Mahlmann et al. 1994). It contains 435 amino acid residues, the sequence of which has a 61% identity to the rat V_{1a} receptor and a 42% identity to the V_2 receptor. It interacts not only with vasotocin but also vasopressin and even molluscan conopressin. However, it does not bind to isotocin, which is the other oxytocin-like hormone in teleosts. Responses to V_{1a} receptors are mediated by the phospholipase C-phosphatidylinositol system and those to V_2 receptors by the adenylate cyclase-CAMP mechanism.

The Growth Hormone-Prolactin Family. Growth hormone (somatotropin) and prolactin belong to the same family of protein hormones, which also includes several placental lactogens and somatolactin (Nicoll et al. 1986; Wallis 1992; Aramburg et al. 1997). The latter has been identified in the pars intermedia of several groups of bony fish. Growth hormone and prolactin contribute to the adaptation of some teleost fish to changes in the osmotic concentrations of their external environments. Growth hormone has ubiquitous effects in the regulation of growth and the metabolism of fats, carbohydrates and proteins. It can influence such processes in many tissues. Prolactin has many different effects, which involve tissue differentiation including milk production by the mammary glands, crop sac secretion in some birds, metamorphosis in larval amphibians and cutaneous changes which occur in newts during their breeding migration to fresh water. In 1956, D. C. W. Smith found that the administration of growth hormone to brown trout increased their ability to adapt and survive transfer from fresh water to sea-water. In 1959, Pickford and Phillips demonstrated that euryhaline killifish failed to survive transfer from sea-water to fresh water after they had been hypophysectomized. However, this ability was restored by the administration of prolactin. It is now known that these effects of the two hormones in fish are principally due to their actions on exchanges of sodium chloride and water across their gills. Urodele amphibians exhibit the so-called newt water-drive response to injected prolactin, and migrate to their breeding ponds (Chadwick 1940). Prolactin has been shown, in such amphibians, to decrease the permeability of their skin to water and Na^+ (Brown and Brown 1973).

Growth hormone and prolactin are synthesized in the somatomammotrope cells of the adenohypophysis. They are related proteins that are present in all vertebrates, with the apparent exception of agnathan fish. Growth hormone consists of about 190 amino acids and contains two cross-linking disulphide bonds. Prolactin usually has about 200 amino acids, though there are fewer such residues in the hormones of some fish. In most species prolactin contains three disulphide bonds but there are only two in teleost prolactin and in the bowfin (Holostei). However, in the sturgeon (Chondrostei), which is considered to be a phyletic predecessor of the teleosts and the bowfin, there are three such disulphide bonds. Prolactin in lungfish (Dipnoi) also contains three disulphide bonds. Many amino acid substitutions occur in the molecules of growth hormone and prolactin, but they each retain five common domains and have similar gene architectures. It has been suggested that they share a common ancestry and arose from a single growth hormone-like gene (Chen et al. 1994). Cattle growth hormone has a 67% homology in its amino acid sequence to that of the human hormone, but in salmon this similarity is only 34%. Cattle prolactin has a 76% homology to that in humans but when compared to salmon this value is 35%. Despite such structural differences, there is usually a considerable crossover in the abilities of such hormones to act, when injected, in other species. Chen and his collaborators (1994) have suggested that this family of hormones originated from a gene that was present in a deuterostome ancestor of the vertebrates. They also consider that the primaeval gene product may have had a role in the osmoregulation of such invertebrates, possibly contributing to the mobilization and transport of organic osmolytes, derived from their nutrients.

The synthesis and release of growth hormone in fish has been described as multifactorial (Peter and Chang 1997). However, basal levels of the hormone appear to be regulated from the hypothalamus by the stimulatory actions of GH-RH and inhibitory ones of somatostatin. Dopamine also exerts a stimulatory effect on the release of growth hormone in teleosts. The secretion of prolactin in vertebrates is predominantly under the inhibitory control of dopamine (Sharp 1997). However, peptides with a stimulating effect on the release of prolactin may also be present. Thus, TRH has been observed to stimulate the release of prolactin in amphibians. In some euryhaline teleosts the release of prolactin may result from the direct effects of decreases in the osmotic concentration of the extracellular fluid (Grau et al. 1994).

Most of the effects of growth hormone are mediated by a growth factor, which it induces in the liver and some other tissues. It is called *insulin-like growth factor-I* (IGF-I) and it belongs to the insulin hormone family. Human IGF-I contains 70 amino acids and it has a 35% homology in its amino acid sequence to insulin. Insulin-like growth factor-I has been identified in all groups of vertebrates, and its structure is highly conserved. Salmon IGF-I has an 80% homology to human IGF-I. Insulin-like growth factor-I has mitogenic effects and its many target tissues include the gills of fish. Its receptors, which are present in the cell membrane, belong to the insulin receptor tyrosine kinase family.

Cell membrane receptors for the growth hormone-prolactin family of hormones belong to the cytokine receptor superfamily (Bazan 1990; Nakagawa et al. 1994). It also includes receptors for several cytokines.

Adrenocorticotropic Hormone. The adrenocorticotropic hormone (ACTH, corticotropin) stimulates and maintains the steroidogenic activity of adrenocortical tissue. It principally influences the synthesis of cortisol and corticosterone. Its relationship to osmoregulation is generally indirect. This hormone is present in the pituitary of all groups of vertebrates. It is formed from the prohormone, proopiomelanocortin, in the corticotrope cells of the adenohypophysis. The adrenocorticotropic hormone contains 39 amino acid residues. Interspecific differences principally occur in a C-terminus "tail" region. The amino acids are invariant at 19 positions, 18 of which are present near the N-terminus of the molecule. Release occurs in response to stimulation by CRH from the hypothalamus and, in stressful situations, also vasopressin. Various cytokines, including some interleukins and tumour necrosis factor-α, can also stimulate release of ACTH. Cortisol and corticosterone exhibit negative feedback effects, both in the hypothalamus and on the corticotropes in the pars distalis, and inhibit release.

2.2 The Renin-Angiotensin System (RAS)

Angiotensin II is an octapeptide hormone with several important roles in the processes of osmoregulation (Laragh and Sealey 1992; Kobayashi and Takei 1997; Fitzsimons 1998). It is formed from a precursor in the plasma, the globulin protein angiotensinogen, as a result of the action of renin, which is secreted by the kidney. A decapeptide, angiotensin I, is initially formed, which is subsequently "activated" by angiotensin-converting enzyme (ACE) to form angiotensin II. This peptidase enzyme is mainly associated with the pulmonary blood vessels. Angiotensin II is a powerful vasoconstrictor and can increase blood pressure. It can also promote sodium retention by the kidney and gut by stimulating the secretion of aldosterone by adrenocortical tissue. The latter steroid hormone promotes the reabsorption of Na^+ from the fluids in the renal tubule and colon. Angiotensin II may also have direct effects on Na^+ reabsorption at these sites. It also can act in the hypothalamus, to induce the sensation of thirst and salt appetite and promote the release of vasopressin. Thus, the overall effect of angiotensin II on osmoregulation is to maintain the volume and the composition of the extracellular fluid.

Renin is secreted by the kidneys. It is a glycoprotein protease enzyme that is formed in the granulated juxtaglomerular cells of the afferent glomerular arteriole. These are epithelioid cells, which in mammals, and probably in birds, lie near the macula densa. The latter group of cells are present in the wall of the adjacent distal renal tubule. Together with a tuft of tissue that is associated with the glomerulus, the extraglomerular mesangium (or polkissen), these two types of cells make up the juxtaglomerular apparatus. This tissue association is apparently present only in mammals, birds and, somewhat unexpectedly, elasmobranch fish (Lacy and Reale 1990). However, the granulated juxtaglomerular cells, which secrete renin, have been identified throughout the vertebrates, with the apparent exception of agnathan fish. Isorenins occur in other tissues, including the brain, cardiovascular system and uterus.

Angiotensinogen (renin-substrate) is an $\alpha2$-globulin, which is synthesized in the liver and is present in the blood plasma. It contains 453 amino acid residues of which the first 10, at the N-terminal, are split off, as a result of the action of renin, to form angiotensin I. As described above, this peptide is converted to angiotensin II [angiotensin-(1–8)] by ACE. Other angiotensin peptides are also formed as a result of the actions of peptidase enzymes including angiotensin III [angiotensin-(2–8)] (Blair-West et al. 1971). It has a much weaker vasoconstrictor action than angiotensin II, but has a similar ability to stimulate aldosterone synthesis. Angiotensin-(1–7), under experimental conditions, can antagonize the effects of angiotensin II and *increase* urinary sodium excretion (Handa et al. 1996). Its possible physiological role has not been defined. Homologous forms of angiotensin I have been identified in many vertebrates (Fig 2.6). In most fish, except the bowfin, asparagine is present at position 1 (fish angiotensin). In the tetrapods, aspartic acid is present at this site. Many amino acid substitutions have been observed at position 9, which is the site of action of ACE. A notable and unique amino acid substitution is that of proline for valine at position 3 in the elasmobranch angiotensin I.

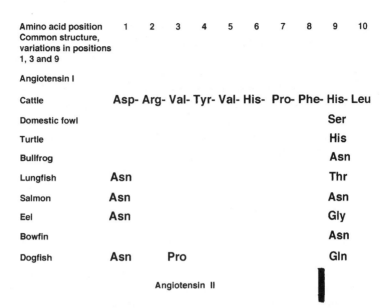

Amino acid position	1	2	3	4	5	6	7	8	9	10
Common structure, variations in positions 1, 3 and 9										
Angiotensin I										
Cattle	Asp-	Arg-	Val-	Tyr-	Val-	His-	Pro-	Phe-	His-	Leu
Domestic fowl									Ser	
Turtle									His	
Bullfrog									Asn	
Lungfish	Asn								Thr	
Salmon	Asn								Asn	
Eel	Asn								Gly	
Bowfin									Asn	
Dogfish	Asn		Pro						Gln	

Angiotensin II

Fig. 2.6. The amino acid sequences of angiotensin I from different vertebrates. (Based on information given by Takei et al. 1993, 1998 and Joss et al. 1999)

The activation of the renin-angiotensin system begins with the release of renin from the juxtaglomerular cells in the kidney. This event can occur in response to several stimuli. Baroreceptors associated with the afferent glomerular arteriole can sense a drop in hydraulic pressure and trigger release. The macula densa (Schnermann 1998) can detect changes in the sodium chloride (especially Cl⁻) concentrations in the distal renal tubular fluid and transmit this information to the adjacent juxtaglomerular cells. A decrease in the renal tubular concentration of sodium chloride results in an increased secretion of renin. Conversely, increases in sodium chloride concentration or a rise in hydraulic pressure inhibit release. The juxtaglomerular cells have sympathetic nerve supply and β-adrenergic receptors. Neural stimuli originating in the brain can trigger a release of renin in this way.

Angiotensin II can constrict local intrarenal blood vessels, including the afferent and efferent glomerular arterioles and vasae rectae. It may thus influence the GFR and renal medullary solute concentration gradients. However, the possible physiological relevance of such effects on urinary Na⁺ excretion and urine concentration is not clear. Angiotensin appears to have other physiological roles, apart from maintaining the integrity of the extracellular fluids. The presence of a renin-angiotensin system in the brain suggests that it may function as neurotransmitter, and it could be contributing to such functions as cognition and memory (Mosimann et al. 1996).

Several types of angiotensin receptors have been identified in various tissues (Wright et al. 1995; Nishimura et al. 1997). They are associated with the cell membrane. The AT_1 and AT_2 receptors have been cloned. An AT_1-like receptor has been identified in birds, amphibians and teleost fish. The AT_1 receptors mediate the responses related to the maintenance of the integrity of the extracellular fluids, including cardiovascular actions. These effects are transduced by the phospholipase C-phosphatidylinositol mechanism.

The actions of angiotensin II on the cardiovascular system, kidney function, thirst and corticosteroidogenesis have been observed in most groups of the vertebrates (Nishimura 1987; Joss et al. 1994; Hanke 1997; Kobayashi and Takei 1997). However, possible roles in the agnathan fish are in doubt.

2.3 Adrenocorticosteroids

The adrenocorticosteroid hormones, which have been identified in all groups of the vertebrates, have two major types of effects. They contribute to the regulation of the metabolism of proteins, fats and carbohydrates. These effects are called their glucocorticoid actions. They also help regulate the concentrations of sodium and potassium ion the body. Such effects are known as their mineralocorticoid actions. These steroid hormones occur throughout the vertebrates. In tetrapods they promote the retention of sodium and excretion of potassium. In marine teleost fish, salt excretion across the gills is increased.

In mammals, adrenocorticosteroids are secreted by the adrenal cortex. The adrenal gland lies near the kidneys. In most mammals it is divided into a distinct outer cortex, which secretes steroid hormones, and an inner medulla composed of chromaffin tissue, which secretes adrenaline and noradrenaline. This type of zonation does not occur in other vertebrates, though an intermingling of the two types of tissues occurs in most species. However, in some fish, including elasmobranchs and agnathans, they each occupy separate sites. In the elasmobranchs the steroidogenic tissue forms distinct aggregates that lie between the kidneys and are called the interrenal glands. In most mammals the cortical tissue is arranged in three layers; an outer zona glomerulosa, an underlying zona fasciculata and an inner zona reticularis.

Adrenocorticosteroids are so named because of their site of origin in mammals. As such hormones are also secreted by homologous (adrenocortical or interrenal) tissue in other vertebrates, this generic name has been retained. However, the shorter version, corticosteroids, is used more often. The particular chemical type of steroid hormone and its relative abundance may vary in the different groups of vertebrates (Henderson and Kime 1987). Mammals secrete aldosterone, which is the prototypical mineralocorticoid, and also cortisol and corticosterone (Fig. 2.7; Table 2.3). The latter two steroids generally function as glucocorticoids. However, these steroids in sufficient concentration can also exhibit mineralocorticoid effects. In teleost fish, cortisol acts as both a glucocorticoid and mineralocorticoid. In birds, reptiles and amphibians, the predominant corticosteroid hormones are aldosterone and corticosterone. In teleost fish, cortisol predominates though in some species small amounts of corticosterone and

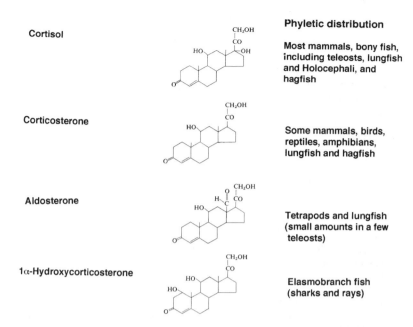

		Phyletic distribution
Cortisol		Most mammals, bony fish, including teleosts, lungfish and Holocephali, and hagfish
Corticosterone		Some mammals, birds, reptiles, amphibians, lungfish and hagfish
Aldosterone		Tetrapods and lungfish (small amounts in a few teleosts)
1α-Hydroxycorticosterone		Elasmobranch fish (sharks and rays)

Fig. 2.7. The chemical structures and phyletic distribution of adrenocorticosteroid hormones in vertebrates

even aldosterone have been identified. Elasmobranchs possess a unique such hormone, 1α-hydroxycorticosterone. This steroid is not present in their cartilaginous relatives the Holocephali, which, instead, have cortisol. Lungfish, like tetrapods, secrete relatively large amounts of aldosterone as well as cortisol and corticosterone.

The synthesis of corticosteroid hormones from cholesterol can be increased by ACTH and angiotensin II and inhibited by natriuretic peptides. Elevated concentrations of K^+ in the plasma can also directly promote such synthesis. The relative importance of the effects of the tropic hormones in the different phyletic groups of vertebrates varies. In mammals, ACTH is especially important in promoting cortisol and corticosterone synthesis. This process occurs in the zona fasciculata. Angiotensin II acts in the zona glomerulosa to increase the synthesis of aldosterone. In other groups of vertebrates, both these tropic hormones can exert more general effects and often promote the synthesis of all corticosteroids.

The target organs and tissues that mediate the mineralocorticoid effects of the corticosteroids vary in different species. In mammals, aldosterone promotes Na^+ absorption and K^+ secretion across the renal distal tubule-collecting duct system. A similar effect is seen in the colon of most tetrapods and the sweat glands and salivary glands of mammals. The kidneys of birds, reptiles and possibly amphibians are also responsive. Aldosterone promotes the absorption of Na^+ across the

Table 2.3. The predominant adrenocorticosteroid hormones in the vertebrates

	Cortisol (F)	Corticosterone (B)[a]	Aldosterone
Mammals			
Placentals	+		+
Rats, mice, rabbits		+	+
Marsupials	+	A few species	+
Monotremes			
Echidna	+		+
Platypus		+	+
Birds		+	+
Reptiles		+	+
Amphibians		+	+
Fishes			
Osteichthyes			
Dipnoi (lungfish)	+	+	+
Teleostei	+[b]		A few?
Holostei (bowfin)	+	+	
Chondrostei (sturgeons)	+	+	
Chondricthyes			
Elasmobranchii	1α-hydroxycorticosterone		
Holocephali (chimaeroids)	+		
Agnatha			
Myxiniformes			
(hagfish)	+	+	

References: Henderson and Kime (1987).
Marsupials: Oddie et al. (1976); McDonald and Martin (1989).
Monotremes: Weiss and McDonald (1965); McDonald et al. (1988).
[a] Small amounts of corticosterone, high F/B, are often present in the plasma of species where cortisol is predominant.
[b] Cortisone is also often present in the plasma.
(Androgens are secreted by the adrenal cortex of mammals.)

skin and urinary bladder of amphibians. Cortisol can promote the uptake of Na^+ across the gills of teleost fish in fresh water and stimulate salt excretion by the same route in sea-water.

Two types of receptors for corticosteroids have been identified, principally in mammals but also in some other species, especially amphibians. They are present inside cells, usually in the region of the nucleus. They have been characterized, and named, glucocorticoid receptors (GR, Type I) and mineralocorticoid receptors (MR, Type II). In teleost fish one type of receptor mediates both types of effects (Sandor et al. 1984). It has been cloned (Ducouret et al. 1995) and appears to have originated from a GR-like receptor prior to the origin of the tetrapod GRs. The corticosteroid receptors, when combined with their hormonal ligands, initiate genetic transcription and translation. Mineralocorticoid-type effects can involve such processes as the synthesis, activation and translocation of Na-K-ATPase and ion channels, especially the epithelial sodium channel or ENaC.

2.4 Adrenaline and Noradrenaline

Adrenaline (epinephrine) and noradrenaline (norepinephrine) are secreted by endocrine chromaffin tissue in all vertebrates. Embryonically, this tissue is derived from neural ectoderm. In mammals, it forms the adrenal medulla that is equivalent to a collection of postganglionic nerve cells of the sympathetic nervous system. The adrenal medulla retains the preganglionic cholinergic nerve connection to the sympathetic nervous system. Noradrenaline is the principal neurotransmitter of postganglionic sympathetic nerves and is a metabolic precursor of adrenaline. The latter is usually the predominant secretion of the adrenal medulla. Hormonal adrenaline and noradrenaline can mimic the effects of stimulation of the sympathetic nervous system. Their release constitutes part of the overall sympathetic response, such as occurs in response to stress. This effect is an adaptation to noxious physical and emotional events and can influence osmoregulation. Such adrenergic responses include vasoconstriction, vasodilatation and increases in blood pressure that can influence the functioning of such organs as the kidneys, skin and gills. A tissue mobilization of glucose (glycogenolysis) and fatty acids (lipolysis) supplies additional energizing metabolic substrates. Adrenaline and noradrenaline also can have direct modulating effects on the secretion of exocrine and endocrine glands, including sweat glands and the neurohypophysis, where a release of vasopressin occurs. In the gills of fish and skin of amphibians the transport of ions and permeability to water may be changed.

Adrenaline and noradrenaline (Fig 2.8), and also dopamine, are called catecholamines due to the presence of a catechol moiety (dihydroxybenzene) in their chemical structures. They are synthesized enzymatically from phenylalanine or tyrosine. The structures of these catecholamines are constant throughout the vertebrates and they are also present in invertebrates and even some plants.

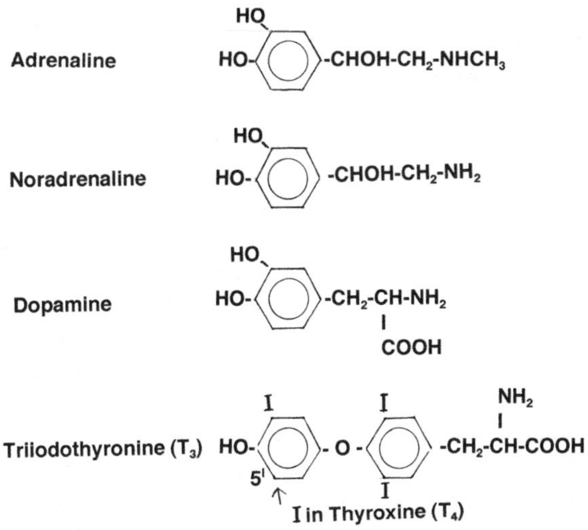

Fig. 2.8. The chemical structures of the catecholamine and thyroid hormones in vertebrates

The effects of the catecholamine neurotransmitters and hormones are mediated, and differentiated, by a variety of subtypes of receptors that are present in the cell membrane. Vasoconstriction is the result of stimulation of α_1- and α_2-adrenergic receptors and the rate and force of contraction of the heart of β_1- and β_2-receptors. The release of renin is mediated by α_1- and β_1-adrenergic receptors and that of vasopressin by β_1-receptors. Changes in the movements of ions and the osmotic permeability of the gills of fish can involve both α- and β-adrenergic receptors. The cellular mechanisms mediating such responses include the adenylate cyclase-cAMP system for stimulating effects mediated by β-receptors and inhibitory effects mediated by α_2-receptors. The phospholipase C-phosphatldylinositol system mediates responses involving α_1-adrenergic receptors.

2.5 The Natriuretic Peptide Hormones

In 1981, de Bold and his associates identified a hormonal product of mammalian atrial heart muscle that promoted the excretion of sodium by the kidney (de Bold 1985). The possible existence of a natriuretic hormone had been the subject of much earlier speculation, but its origin in atrial heart muscle was an endocrine surprise. The secretion was called, variously, atrial natriuretic peptide (ANP), atrial natriuretic factor and atriopeptin. Several other related "peptides" have been identified in the brain of many vertebrates. They also appear to be present in invertebrates, such as in the hearts of oysters and blue crabs (Poulos et al. 1995). In mammals, ANP acts on the kidney, where it increases urine flow and sodium excretion (natriuresis) (Brenner et al. 1990; Atlas and Maack 1992). It is also a powerful vasodilator. Such effects have been observed throughout the vertebrates (Takei and Balment 1993). Natriuretic peptides can also promote salt loss in marine fish by stimulating its extrusion from the gills of teleost fish and secretion by the rectal salt gland in elasmobranchs. The general effects of the natriuretic peptides appear to counteract expansion of the extracellular fluid space due to the excessive accumulation of sodium.

The chemical structures of the natriuretic peptides (Atlas and Maack 1992; Takei 1994) all include a ring configuration containing 17 amino acid residues joined by a disulphide bridge provided by two hemicystines. ANP contains 28 amino acids, 6 at the N-terminus precede the ring structure and 5 form a "tail" at the C-terminus (Fig. 2.9). The amino acid sequences in the ring structures of different species are highly conserved. However, there are considerable differences in the remainder of such molecules, depending on the species and the tissue site of its origin. The various natriuretic peptides have been classified into three main groups (Fig. 2.9). Type A, which includes mammalian ANP-28; type B (BNP), which is found in the brain as well as heart muscle; and type C (CNP), which is mainly present in the brain. A fourth type has been identified in the heart ventricular muscle of teleost fish. It has an elongated tail containing 14 amino acid residues at its C-terminus. Interspecific variations in their amino acid sequences suggest that type-B natriuretic peptides are the least conserved and type C are

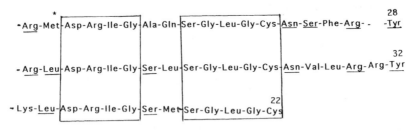

Type A (ANP-28) Ser-Leu-Arg-Arg-Ser-Ser-Cys-Phe-Gly-Gly-

Type B (BNP-32) Ser-Pro-Lys-Thr-Met-Arg-Asp-Ser-Gly-Cys-Phe-Gly-Arg-

Type C (CNP-32) Gly-Leu-Ser-Lys-Gly-Cys-Phe-Gly-Leu-

-Arg-Met-Asp-Arg-Ile-Gly-Ala-Gln-Ser-Gly-Leu-Gly-Cys-Asn-Ser-Phe-Arg- - -Tyr (28)

-Arg-Leu-Asp-Arg-Ile-Gly-Ser-Leu-Ser-Gly-Leu-Gly-Cys-Asn-Val-Leu-Arg-Arg-Tyr (32)

-Lys-Leu-Asp-Arg-Ile-Gly-Ser-Met-Ser-Gly-Leu-Gly-Cys (22)

Fig. 2.9. Amino acid sequences of the atrial natriuretic peptides (*type A*, *type B* and *type C*). The *boxed sections* indicate identical sequences within the ring structure in all three types of the hormones. The *underlined residues* indicate homologies in two of the types. * replaced by isoleucine in some mammals. Type B may have an extended N-terminus to give, for instance, BNP-32 (as shown), BNP-26 in the domestic pig and BNP-45 in rodents. (After Atlas and Maack 1992)

the most conserved. Elasmobranch brain CNP has an 85% identity of its amino acid sequence with its counterpart in mammals (Suzuki et al. 1992). A unique natriuretic peptide has also been identified in the mammalian kidney (Goetz 1991). It has been called urodilatin, but its possible physiological role is unknown. A peptide, containing 29 amino acids, has been identified in the heart of Atlantic salmon that appears to be related to the A-, B- and C-types of natriuretic peptides found in other species (Tervonen et al. 1998). However, it is distinct, and it has been suggested that it could be an ancestral type of such peptides.

Natriuretic peptides in mammalian heart muscle are released in response to expansion of the plasma volume that results in the activation of stretch receptors in the right atrium (Thibault et al. 1999). Volume expansion of the extracellular fluids also promotes such a release in teleost and elasmobranch fish. In teleosts, hyperosmotic stimuli may also increase such secretion.

The natriuretic and diuretic actions of the natriuretic peptides can involve several types of effects and mechanisms. Increases in the GFR, such as could be mediated by a dilatation of the afferent glomerular arterioles, could be involved. However, aglomerular teleost fish also exhibit such renal effects. Natriuretic peptides can inhibit the synthesis of aldosterone, which would be expected to result in a decreased reabsorption of sodium from the renal tubule. A direct blockade of sodium channels in the apical cell membrane of the renal tubule may also occur. The infusion of ANP has also been observed to increase the movement of fluid and proteins across the capillary bed (Renkin and Tucker 1996). Such an effect would be expected to decrease the plasma volume. Natriuretic peptides have been shown to decrease the release of vasopressin, which could be contributing to the observed diuretic effects.

Three types of membrane receptors have been identified that can bind natriuretic peptides. They are present in a variety of tissues including blood vessels, kidney, adrenal, brain and the gills of fish. Two of these receptors are linked to guanylate cyclase and are called GC-A and GC-B. The former interacts with ANP and BNP and the latter with CNP. Such binding results in an activation of the guanylate cyclase (Chinkers et al. 1989). The third type of receptor binds natriuretic peptides but does not appear to be linked to known responses. It has been called a clearance, or C-type, receptor. A more definitive role for it may yet emerge.

2.6 Adrenomedullin

In 1993, Kitamura and his associates (Kitamura et al. 1993a) described a polypeptide in a human adrenal medullary tumour (a phaeochromocytoma) that had a powerful vasodilatatory effect and decreased the blood pressure of rats. It was also found in normal adrenal medulla and a number of other tissues including, kidneys, heart, intestine, pituitary gland and the hypothalamus (Kitamura et al. 1995). However, its concentrations are much higher in the adrenal medulla. It is also present in blood, and it has been suggested that it may function as a hormone.

Apart from its vasodilatatory effect, adrenomedullin has a number of other actions that may influence osmoregulatory processes (Samson 1999; Samson et al. 1999). They include diuretic and natriuretic effects on the kidneys, and an inhibition of thirst and salt appetite, which is mediated in the hypothalamus. Adrenomedullin can also inhibit the release of ACTH, vasopressin and aldosterone. Its presence in the blood has been related to diseases such as congestive heart failure, that result in an expansion of the extracellular fluid volume (Cheung and Leung 1997). Such observations strongly suggest that adrenomedullin has a physiological role in the regulation of fluid volume in the body.

Human adrenomedullin is a polypeptide containing 52 amino acid residues that incorporate a six-membered ring structure closed by a disulphide bridge (Kitamura et al. 1993a,b). Its precursor, proadremomedullin, consists of 164 amino acid residues. Adrenomedullin has a structural similarity to calcitonin-gene-related peptide (CGRP, a neuropeptide) and amylin (secreted by the B-cells of the islets of Langerhans). These molecules have even been observed on occasion to interact with each other's receptors and are considered to be members of the same superfamily (Wimalawansa 1996). Interspecific differences have been observed in the structures of adrenomedullin. Thus, the rat polypeptide has only 50 amino acids and differs from that in humans by six amino acid substitutions.

Adrenomedullin-like immunoreactivity has been identified in the nervous system of an echinoderm, suggesting that it may have a long phylogenetic history (Martinez et al. 1996).

The N-terminus of proadrenomedullin includes a 20-membered peptide that also exhibits biological activity (Kitamura et al. 1993b; Samson 1998). It is called proadrenomedullin N-terminal peptide or ProAM-N20 peptide (PAMP). Like adrenomedullin, it can decrease blood pressure and inhibit the release of ACTH and aldosterone. However, its vasodilatatory action is mediated by a decrease in neural sympathetic activity, including inhibition of release of adrenaline.

The effects of adrenomedullin on the kidney are associated with an increased renal blood flow and GFR, and decreased sodium reabsorption from the distal renal tubule (Jougasaki et al. 1995). Receptors for adrenomedullin are associated with the renal distal convoluted tubules (Jensen et al. 1998). It is possible that it is exerting a local, paracrine, effect in the kidneys. Its vascular effects appear to be mediated by the formation of nitric oxide (NO) by the vascular endothelium. However, the mechanism of its action of the renal tubule is unknown. In blood platelets it has been observed to increase the production of cAMP (Kitamura et al. 1993b).

2.7 Guanylin Peptides

The guanylin peptides are a recently discovered group of putative hormones that can promote the secretion of Cl^- and HCO_3^- by the intestinal mucosa (Forte and Hamra 1996; Forte et al. 2000). They also have diuretic, natriuretic and kaliuretic effects on the mammalian kidney. Their discovery arose from the observations of M. Field and his collaborators (1978) on the causes of watery diarrhoea in humans. This disease results from infections by *Escherichia coli* bacteria. A heat-stable enterotoxin was isolated from incubates of these bacteria that increased intestinal secretion of Cl^-. This response was associated with an increase in tissue cGMP, which was apparently formed as a result of a stimulation of intestinal guanylate cyclase. The bacterial enterotoxin (ST, for stable toxin) contains 16 amino acid residues, including six cysteines that form three disulphide bonds. A search for endogenous homologues of this toxin uncovered three related peptides; guanylin, uroguanylin (originally found in the urine of the opossum) and lymphoguanylin. They contain 15 or 16 amino acids. Guanylin and uroguanylin have two disulphide bonds and lymphoguanylin one. The bacterial enterotoxin mimics the effects of the guanylins in the intestine. The homology of the amino acid sequences in uroguanylin and bacterial ST is 69%.

Guanylin peptides are produced in the intestinal mucosa by the enterochromaffin cells. Lymphoguanylin appears to be synthesized in lymphoid tissue (Forte et al. 1999). There are interspecific variations in the structures of the guanylins. Human and opossum uroguanylins have a homology of 80% in the sequences of their amino acid residues. This value is 73% for their guanylins. Uroguanylin appears to be more active than guanylin, possibly reflecting a greater resistance to enzymatic inactivation.

Guanylins have been identified in the blood and are also secreted into the intestinal lumen. The precise stimuli for their release do not appear to have been defined, but it has been suggested that secretion occurs postprandially and reflects the sodium chloride content of the food. Their excretion in the urine is increased by a high salt diet.

Two types of receptors have been identified for guanylins; in the intestinal mucosa and the proximal renal tubule of the opossum. They are present in the cell membrane, where they are associated with membrane guanylate cyclase. The responses to guanylin are mediated by cGMP and may involve an interaction with cGMP-dependent protein kinase II (PKG II) or even cAMP-dependent protein kinase II (PKA II). The subsequent protein phosphorylation may involve activation of ion channels, including the CFTR chloride channel.

2.8 The Thyroid Hormones

A thyroid gland or homologous follicular tissue is present in all vertebrates and the chemical structures of its secretions are identical in all species. They are iodine-containing derivatives of tyrosine (Fig. 2.8). Thyroxine contains four iodine atoms (hence it is referred to as T_4) and triiodothyronine three (T_3). The latter has the predominant hormonal activity, but T_4 can be deiodinated at the 5^l position in peripheral tissues, by monodeiodinase enzymes, to Form T_3. Thyroxine thus functions mainly as a prohormone. The final synthesis of thyroid hormone takes place in the follicles of the thyroid gland. The process can be regulated by adenohypophysial thyroid-stimulating hormone (TSH, thyrotropin), the release of which is partially under the control of TSH-releasing hormone (TRH) from the hypothalamus. Synthesis involves the 'prohormone' thyroglobulin, which is a large tyrosine-containing protein that accumulates in the thyroid follicles. Iodide is actively accumulated by the thyroid cells from the blood and, following its oxidation to iodine, is utilized to iodinate specific sites on the tyrosine residues.

Thyroid hormone has ubiquitous actions in the body, which usually have only indirect effects on osmoregulation (Grau 1987). In mammals and birds, cell metabolism and oxygen consumption are increased by thyroid hormone (calorigenic effect). This effect can result in increased evaporative water losses. A stimulation of oxygen consumption has been sporadically reported in cold-blooded vertebrates, but such an effect does not appear to be a general one in such species. Thyroid hormone also can promote cell proliferation and differentiation in many vertebrates (morphogenetic effect). A dramatic example of this action is the promotion of metamorphosis by amphibian tadpoles into juvenile adults. Such a metamorphosis is also seen in some fish, such as flounder.

The activity of the thyroid gland and plasma thyroid hormone concentrations increase during the transformation of freshwater salmonid parr to smolt (Hoar 1951; Prunet et al. 1989). The latter can then migrate and adapt to life in the sea. This ability is associated with the differentiation of salt-secreting chloride cells in their gills. However, as will be described later, other hormones also contribute to this process. Metamorphosis in amphibians is associated with the ability to adapt to a terrestrial life. This process results in many changes, such as involve the functioning of the skln and an ability to respond to neurohypophysial hormones (Howes 1940; Bentley and Greenwald 1970). In mammals, thyroid hormone has been shown to increase the levels of Na-K-ATPase in the kidney (Ismael-Beigi 1993).

2.9 The Urotensins

Urotensin I and II are peptides that have been identified in large spinal cord neurons (Dahlgren cells) in a wide variety of fish, including teleosts and elasmobranchs. They exhibit a number of actions, including an ability to influence the contractility of vascular and urinogenital smooth muscle (hence their names). Urotensins can change the rate of salt transport across some epithelial membranes. In teleosts, the spinal cord neurons often aggregate in the tail region to form a gland-like structure called the urophysis (Arsaky 1813). Such a structure has been identified in over 400 species of teleost fish (Fridberg and Bern 1968; Kobayashi et al. 1986). Changes in its histological appearance have been observed following transfer of such fish between fresh water and sea-water (Enami 1955). These observations resulted in the suggestion that it may be contributing to their osmoregulation (see Larson and Bern 1987). However, such a role, while considered likely, has not yet been clearly established. Secretion from the urophysis could have other functions and be contributing to the functioning of the cardiovascular system and reproduction. Urotensin I is a polypeptide containing 41 amino acid residues. Its structure is similar to that of mammalian corticotropin-releasing hormone (CRH), which is present in the hypothalamus (Lederis et al. 1994; Vaughan et al. 1995). It has been classified as a member of the CRH family of neuropeptides that have a widespread distribution in vertebrates (Lovejoy and Balment 1999). Urotensin II consists of 12 amino acid residues and exists in at least 15 different but homologous forms (Fig. 2.10). It is usually present in the

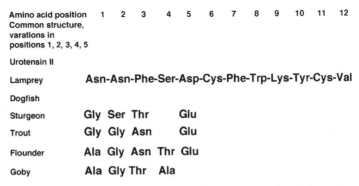

Fig. 2.10. The amino acid sequences of urotensin II from various fish. (Based on information given by Waugh and Conlon 1993 and Waugh et al. 1995)

urophyses and Dahlgren cells of fish, but it has also been identified in the brains of lampreys and frog (Conlon et al. 1992). It has even been found in the spinal cord of humans (Coulouarn et al. 1998).

3 Mechanisms of the Hormonal Regulation of Water and Salt Across Epithelial Membranes

Neurohypophysial and adrenocortical hormones have special roles in regulating water and salt transfer across epithelial membranes of tetrapod vertebrates. Such effects are particularly important in the renal tubules and colon, and in the Amphibia also occur in the skin and urinary bladder. The mechanism of action of these hormones at such sites has attracted considerable interest. Special efforts to understand their molecular nature have been fruitful. Studies on osmoregulation in non-mammmals have played seminal roles in providing experimental tissue preparations for these investigations.

3.1 Vasopressin and Vasotocin

Vasopressin and its homologue vasotocin have vital roles in limiting urinary water loss in the kidneys of tetrapod vertebrates. An analogous effect also occurs in the skin and urinary bladder of the amphibians. In mammals, this antidiuretic response to vasopressin is principally due to the hormone's ability to increase the permeability of the distal renal tubule-collecting duct system to water. A reabsorption of water occurs from the tubular fluid, down its osmotic gradient into the renal medullary interstitium. This response, as described above, is initiated by the interaction of vasopressin with V_2-type receptors situated on the basolateral borders of the renal tubular epithelial cells. An activation of the G-protein-linked adenylate cyclase system follows resulting in the formation of cAMP. This nucleotide then activates protein kinase A, which can phosphorylate proteins that initiate the final response. Utilizing the same type of receptors, vasopressin can also increase the permeability of the renal tubules to urea and promote the active absorption of Na^+. The latter two effects make a vital contribution to the ability of the mammalian kidney to form hyperosmotic urine. In conjunction with the countercurrent flow processes occurring in the loop of Henle and the vasae rectae, they help establish, and maintain, a high osmotic concentration in the interstitial fluids of the renal medulla and papilla. These three effects of vasopressin on water and solute transfer are not confined to the mammalian kidney. They also occur in response to vasotocin in non-mammalian tetrapods. Indeed, they were first demonstrated (in vitro) in the skin of frogs (Fuhrman and Ussing 1951; Koefoed-Johnsen and Ussing 1953). The urinary bladders of frogs and toads are similarly responsive to neurohypophysial peptides.

Vasopressin (and vasotocin) can also contract vascular smooth muscle, an observation that led initially to the naming of this hormone. This effect is mediated by V_{1a}-type receptors and is transduced by the phospholipase C phosphatidylinositol-DAG-IP$_3$ system (see above). An activation of protein kinase C results. In non-mammalian tetrapods, and possibly some fish, this vascular effect may be elicited on the afferent glomerular arteriole and results in a decreased GFR and even closure of glomeruli. Such a vascular effect contributes to the observed antidiuretic effects of vasotocin in such species. In mammals, it has been suggested that intrarenal vascular effects of vasopressin may influence the pathways of renal blood flow (Cowley 2000). Such an effect could contribute to the maintenance of the high osmotic concentrations of fluid in the renal medulla.

The mechanism of the actions of vasopressin and vasotocin on the permeability of the renal tubule to water, urea and Na$^+$ are of special interest (Fig. 2.11). The first direct experimental studies utilized in vitro preparations of the skin and, especially, the urinary bladders of frogs and toads. Vasopressin and related peptides increase the permeability of such epithelial membranes to water, urea and Na$^+$, just as observed in mammalian kidney tubules. Water and urea move down their concentration gradients only while an active transport of Na$^+$ is promoted. The kinetics of the process of the water movement suggested that vasopressin was creating water-filled pores or channels in the membrane (Koefoed-Johnsen and Ussing 1953). Unlike in the intestine, there was no direct linkage of the water movement to Na$^+$ transport. The presence of similar, but separate, pores was suggested to account for the movement of urea. This effect of neurohypophysial hormones on Na$^+$ transport is mediated by activation and/or mobilization, probably by phosphorylation via protein kinase A, of sodium channels in the apical plasma

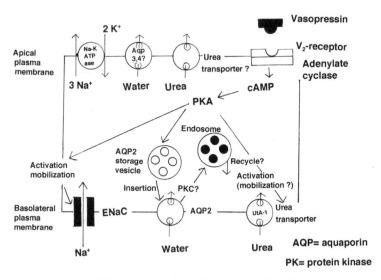

Fig. 2.11. Diagramatic summary of the various actions of neurohypophysial hormones (vasopressin and vasotocin) on the permeability and transport of water, sodium and urea across epithelial cells in the kidney tubules of tetrapods and the urinary bladder and skin of many amphibians. For further information see the text. (After Ward et al. 1999)

membrane of the cells. This response allows entry of Na$^+$ by diffusion into the cell. It is then extruded across the basolateral cell membrane by the Na-K-ATPase "pump" (Garty and Palmer 1997).

Chevalier et al. in 1974 (see also Chevalier et al. 1981) used an electron microscopy freeze-etching technique to study the apical cell membrane of the frog urinary bladder during exposure to oxytocin. (This neurohypophysial peptide is a homologue of vasopressin and vasotocin.) Clusters of particles (intramembranous aggregates) were observed within this membrane. They were found to be associated with the stimulation of water transfer by the peptide. Similar observations were subsequently made on mammalian renal tubules exposed to vasopressin. They resulted in the proposal of the particle "membrane shuttling" hypothesis to account for the action of vasopressin on osmotic water transfer (Wade et al. 1981). Thus, it was suggested that increases in permeability to water in response to vasopressin resulted from the exocytotic insertion of water channels into the apical plasma membrane. These structures were mobilized from cytoplasmic storage sites called aggrephores. The subsequent discovery of aquaporins provided a firm molecular basis for such a mechanism. In 1993, Fushima and his associates cloned such a water channel from the rat renal collecting duct. It has been named aquaporin 2 (AQP2). Aquaporin 2 contains two serine residues, in human AQP2 Ser-256 and Ser-231, which can be phosphorylated, respectively, by protein kinases A and C (Brown et al. 1998; Deen and Van Os 1998; Borgnia et al. 1999; Ward et al. 1999; Christensen et al. 2000). This water channel is stored in vesicles in the cytoplasm. Following cAMP-dependent activation of protein kinase A by vasopressin, and the resulting phosphorylation of Ser-256 in AQP2, the storage vesicles are mobilized (see Fig. 2.11). They move along the microtubules of the cell cytoskeleton towards the apical plasma membrane. This process is referred to as vesicular trafficking. It appears to involve the activity of the motor protein dynein (Marples et al. 1998). The vesicles then gain access to the membrane in a process that appears to involve an interaction with the F-actin filaments. This network lies beneath the plasma membrane. Exocytotic fusion with the plasma membrane then occurs which involves reactions between the vesicular protein vAMP-2 and plasma membrane syntaxin-4. The AQP2 is then inserted into the plasma membrane and provides channels for the water transfer. The AQP2 can leave the membrane by endocytosis and is collected in endosomes where it may possibly be reprocessed (Katsura et al. 1996). The endocytosis of AQP2 may be initiated by phosphorylation of Ser-231 by protein kinase C. A long-term regulation of the synthesis of AQP2 could involve activation of the cAMP response element (CRE) on the aquaporin gene (Marples et al. 1999).

Several urea transporters (UT-A1, UT-A2, UT-B1) have been identified that contribute to the conservation of urea in the interstitial fluid of the renal medulla (Bankir et al. 2000). The UT-A1 has been cloned and found to be present in the apical plasma membrane of the renal inner medullary collecting ducts of mammals (You et al. 1993; Nielsen et al. 1996). It facilitates the diffusion of urea from the lumen of the tubule to the medullary interstitium. Its activity is regulated by vasopressin. (The other urea transporters, UT-A2, which is present in the thin descending loop of Henle, and UT-B1 in the vasae rectae, do not respond to vasopressin.) The UT-A1 has been identified in cytoplasmic vesicles which, in contrast to those containing AQP2, do not, apparently, undergo trafficking in response to vasopressin (Inoue et al. 1999). The activity of UT-A1 appears to be regulated in situ as a result of its phosphorylation by protein kinase A.

3.2 Mineralocorticoid Hormones

Aldosterone is the most active mineralocorticoid hormone in tetrapod vertebrates. Its actions, as described above, usually commence with its binding to mineralocorticoid receptors (MR) inside the cell. Cortisol and corticosterone can also elicit mineralocorticoid-type effects but they are much less active than aldosterone. However, their plasma concentrations are usually considerably higher than those of aldosterone so that potentially they could elicit mineralocorticoid effects. Nevertheless, under normal physiological conditions their access to MR is restricted by mechanisms that include their selective inactivation at potential target sites by the enzyme 11β-hydroxysteroid dehydrogenase (11β-HSD) (Funder and Myles 1996). Most fish lack aldosterone but utilize cortisol to elicit both glucocorticoid and mineralocorticoid effects. They have a single glucocorticoid-like receptor (see above).Glucocorticoid receptors can, on occasion, apparently also mediate mineralocorticoid effects in tetrapods, as observed in a cell line (A6) derived from the kidney tubules of the toad *Xenopus laevis* (Watlington 1998).

The principal action of aldosterone is to promote the conservation of sodium in the body fluids. This effect is mediated by its ability to increase active Na$^+$ transport across a variety of epithelial membranes, including the renal tubules and colon of tetrapods. Aldosterone also stimulates the excretion of potassium, principally by its secretion across the renal tubules and colon.

The process of transepithelial Na$^+$ transport involves several events (Fig. 2.12). They include entry of the ion into the cell through specific sodium channels (ENaC) in the apical plasma membranes, followed by its extrusion across the opposite, basolateral, surface of the cell as a result of the activity of Na-K-ATPase. In 1961, Jean Crabbé found that aldosterone could stimulate such active Na$^+$ transport, across the urinary bladder of the toad *Bufo marinus* under in vitro conditions (Crabbé 1961a; Crabbé and De Weer 1964). This tissue preparation for many years provided the principal information regarding the mechanism of action of this hormone (Marver 1992). The response to aldosterone, unlike that to vasopressin, was delayed for 60 to 90 min. It could be inhibited by the presence of actinomycin D, which blocks genetic transcription, or puromycin, which blocks the translation of mRNA. These early observations were consistent with a response that involved the synthesis of an essential protein, or proteins, that mediate the response. Many subsequent studies, which have been extended to the kidney and colon, have attempted to identify such proteins and find out what precisely they do. Substantial advances have been made, but the information is still incomplete and consensus has not been reached. (Barbry and Hofman 1997; Garty and Palmer 1997; Verrey 1999; Fig. 2.12).

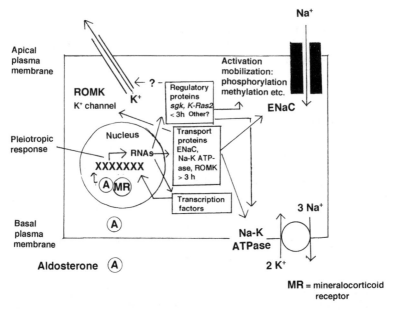

Fig. 2.12. Diagramatic summary of the mechanism of action of aldosterone on Na$^+$ and K$^+$ transport across epithelial cells, such as in the kidney and colon of many tetrapods and the urinary bladder of amphibians. The information has principally been derived from studies using the mammalian and amphibian kidney, the mammalian colon and the amphibian bladder. (After Verrey 1999 and references given in the text)

The entry of Na$^+$ into many types of epithelial cells is now known to take place through specific sodium channels (ENaC). This process can be rate-limiting for such active Na$^+$ transport. Careful kinetic studies, including the timing and magnitude of changes in intracellular Na$^+$ concentrations, suggest that aldosterone can increase the activity and possibly the numbers of ENaC. Increases in the passage of Na$^+$ through ENaC channels may be due to their activation so that their conductance for Na$^+$ increases. The possibility that they will be open could also be enhanced. It is also considered likely that inactive sodium channels, which could lie in reservoirs, such as cytoplasmic vesicles, may be inserted into the membrane. New sodium channels may also be synthesized in response to aldosterone. The ENaC channel consists of three subunits, α, β and γ. Activation of this ion channel could involve phosphorylation of a site, or sites, by a kinase. Aldosterone has been observed to promote serine / threonine phosphorylation of the β and γ but not the α subunits from such sodium channels prepared from the distal renal tubules of dogs (Shimkets et al. 1998). It has also been suggested that activation of ENaC channels could involve other processes, such as methylation, increased intracellular pH or the action of Ca-calmodulin (Garty and Palmer 1997).

The presence of aldosterone has been shown to increase the transcription of mRNA coding for various subunits of ENaC, but the results vary, depending on the particular tissue studied and the species. In rat kidney, little effect was observed, but in the colon there was an induction of the β and γ subunits (Asher et al. 1996). In the rabbit kidney, aldosterone increased mRNA for the γ-subunit (Denault et al. 1996). In the distal colon of rats, aldosterone increases the transcription of the β and γ subunits of the ENaC channels (Epple et al. 2000). The colon of chickens on a low-sodium diet, when plasma aldosterone levels are elevated, exhibits a marked increase in mRNAs for the α and β subunits (Goldstein et al. 1997). While the principal effects of aldosterone on the renal tubule appear to occur in the medullary collecting duct, a response has also been observed in the distal convoluted tubule. The entry of Na^+ across the apical side of the epithelial cells in this tubule is mediated by the Na-Cl cotransporter protein. The expression of this protein was increased nearly fourfold by aldosterone (Kim et al. 1998).

The successful transit of Na^+ out of epithelial cells depends on the activity of Na-K-ATPase on their basolateral surfaces. This enzyme can be activated by the increase in intracellular Na^+ that occurs following its entry into the cell. However, when exposure to aldosterone is prolonged for about 5 h or more, there is also an increased synthesis of this enzyme. Such an effect on Na-K-ATPase was first clearly shown in the toad urinary bladder by Geering and his collaborators (1982). This transport enzyme contains two major subunits, α and β. In cultured A6 kidney cells from the toad *Xenopus laevis*, aldosterone has been found to induce a two- to fourfold increase in the synthesis of both of these subunits (Verrey et al. 1987). Similar effects have also been observed in mammalian kidneys and colon (Ewart and Klip 1995).

The action of aldosterone has been assumed to commence with the activation of an early-response gene, resulting in the initiation of a signal-transducing cascade and the induction of a protein or proteins that contribute the final effect. The response appears to be pleiotropic and could involve the formation of further transcription factors. It has been difficult to identify such components from among the many mRNAs and proteins that are produced in the target cells. Several candidates have been investigated. The Ras proteins are small G-proteins that are expressed by the *Ras* gene family. They are involved in many signal-transduction pathways, especially those involved in mitogenic responses. Aldosterone has been observed to increase the expression of K-*Ras2* in A6 cells from the kidney of *Xenopus laevis* (Spindler and Verrey 1999). It has also been observed to increase the methylation of Ras (Al-Baldawi et al. 2000). Inhibition of such methylation inhibits the response to aldosterone. Apart from inducing the expression of K-Ras2, aldosterone could also be inducing modulatory proteins that activate an essential methyltransferase enzyme. When K-Ras2 and ENaC channels are coexpressed in *Xenopus* oocytes, Na^+ transport through the ion channels is increased (Mastroberardino et al. 1998). The Ras protein may be playing an integrative role in the aldosterone-signalling pathway. However, a more primary signal would appear to be occurring more upstream in the process. Such a candidate is the gene *sgk*, which was first identified from a rat mammary gland tumour cell line (Webster et al. 1993). Its product is a serine / threonine kinase which can be induced by aldosterone in *Xenopus* kidney A6 cells, the renal collecting duct of rabbits and the colon of rats (Chen et al. 1999; Naray-Féjes-Toth et al. 1999; Shigaev et al. 2000). When the sgk protein from the A6 cells is coexpressed with ENaC channels in *Xenopus* oocytes, Na^+ transport is increased. The activity of the sgk protein is dependent on the presence of a lipid kinase, phosphatidylinositol3-kinase (P13K) (Kobayashi and Cohen 1999; Wang et al. 2001). The latter's activity can be increased by insulin. This effect may account for the syngergism that has been observed between the actions of aldosterone and insulin (André and Crabbé 1966; Wang et al. 2001).

The regulatory effects of aldosterone in promoting potassium excretion have been observed in the mammalian renal tubule and colon. It is not seen in the amphibian urinary bladder, possibly reflecting a deficiency of potassium channels in the epithelial apical plasma membrane. Such potassium channels are present in this membrane in the mammalian epithelia, and aldosterone appears to promote a secretion of K^+ through them (Giebisch 1998). Two indirect mechanisms exist for a secretion of K^+ and may contribute to aldosterone-stimulated K^+ transport. An increase in the activity of Na-K-ATPase could produce a higher intracellular concentration of K^+, thus facilitating its diffusion out of the cell. Concomitantly, the increase in uptake of Na^+ across the apical plasma membrane appears to result in its partial depolarization. This change in electrical potential would also be expected to favour the diffusion of K^+ out through the potassium channels. However, aldosterone may have a more direct effect, as it has been observed to increase the transcription of mRNA coding for ROMK potassium channels in the plasma membrane of rat distal renal tubules (Beesley et al. 1998).

Chapter 3

The Mammals

1 General Introduction

About 1050 contemporary genera of mammals are named by Walker (1964) in his monograph *Mammals of the World*. Three of these belong to the order Monotremata (egg-laying mammals), 90 to the order Marsupialia (typically undeveloped young kept in pouches), and the remainder are eutherians (placentals). Monotremes are known only from the Australian region and are represented by the echidna and platypus. This order probably represents a separate line of mammalian evolution, beginning in the Triassic period. The marsupials and placentals probably diverged from a common stock early in the Cretaceous period, about 85 million years ago. The Australian (75 genera) and American (15 genera) marsupials subsequently diverged late in the Cretaceous. The placental mammals today contain about 950 genera, of which 360 are rodents (Rodentia) and 180 are bats (Chiroptera). The other generically numerous groups are the Insectivora (hedgehogs, shrews), 66 genera; Primates (man and monkeys) 62; Cetacea (whales) 35; Carnivora 103; Pinnipedia (seals) 20 and Artiodactlya (pigs, camels cattle, sheep) 82. The mammals have a wide geographic and osmotic distribution; they are predominantly terrestrial and are even found in the driest desert regions, though some are aquatic and live in rivers and lakes or in the sea. Only one genus of the monotremes (*Ornithorhynchus*, the platypus), and one marsupial (the South American water opossum, *Chironectes*) live in fresh water, and these groups have no marine representatives. Most mammalian orders have representatives that live in desert regions.

The ability to osmoregulate in such contrasting environments depends on physiological and behavioural adaptations, many of which are uniquely suited to the particular needs of mammals. In the absence of such specializations most temperate terrestrial species find it difficult to survive in areas where water is more restricted, though there are some exceptions such as the dog and European rabbit, which thrive in the desert conditions of Australia. Dispersal of mammals is, nevertheless, relatively restricted by deserts, and especially limited by salt-water barriers (Darlington 1957). Bats, due largely to their aerial capabilities, have the widest distribution among the mammals, reaching isolated oceanic islands, such as the Azores, New Zealand and Hawaii, that are otherwise inaccessable to all but marine species. The rodents also have often managed to cross saltwater barriers (probably on rafts), but if the aid of man can be excluded, this probably occurs only across the short distances to islands lying adjacent to the continents. Such journeys, whether as aviators or mariners, nevertheless constitute an osmoregulatory feat of some substantial magnitude considering the rather unfavourable conditions experienced by terrestrial mammals in oceanic regions.

2 Osmoregulation

Two evolutionary events have had profound effects on the osmoregulatory problems of tetrapods. The first was the adoption of a terrestrial, instead of an aquatic, habitat. Evaporation of water then made its initial appearance as a vertebrate problem. In addition, there arose the novel necessity to replenish the body water periodically from the sporadic sources available in a terrestrial environment. The second major event to influence tetrapod water metabolism was the adoption of homoiothermy by the birds and mammals (or possibly an ancestor of these). Potential evaporation from the body surface is increased at the relatively high, and sustained, body temperatures of such animals, while the added metabolic need for respiratory gas exchanges also increases evaporative water loss from the respiratory tract. Heat loss accompanying evaporation of water is utilized for cooling in hot environmental conditions, when additional evaporation is facilitated by the secretion of sweat or by panting. Viviparity also increases the needs for water and salts, especially in placental mammals, which retain the young in utero until they are relatively well developed. The secretion of milk to nourish the young places yet another periodic osmoregulatory burden on the mother, but has obvious advantages for the offspring. The mammals thus have the usual tetrapod problems with respect to their water and salt metabolism, and a few additional ones as part of the price that they must pay for their unique physiology. However, the high metabolic rate of mammals allows them to travel considerable distances for food and water, while their viviparity and ability to suckle their young foster reproduction in situations in which it could otherwise be difficult, both nutritionally and osmotically. It should be remembered that while the high rate of metabolism in mammals increases their requirements for osmoregulation, such adjustments may be facilitated by the increased energy produced by the cells of such animals.

The relative abundance or scarcity of salts may limit the distribution of mammals. An excess of water or salts in the food can be tolerated by some species, but not others. Marine mammals like whales and seals live in a hyperosmotic salt solution, equivalent in concentration to 3.3% sodium chloride. Many marine mammals eat invertebrates that are isosmotic to sea-water. Such mammals do not appear to drink sea-water normally but, when feeding on invertebrates, take in large quantities of salts, far more than most other mammals could tolerate with the limited quantities of osmotically-free water that are available. Most terrestrial mammals cannot survive if provided with drinking water that contains 2 to 3% sodium chloride, but there are exceptions. The North African sand rat, *Psammomys obesus*, normally eats halophytic plants with a salt concentration higher than that of sea-water (Schmidt-Nielsen 1964a), while the western harvest mouse, *Reithrodontomys r. halicoetes*, of the United States eats similar plants and can drink sea-water (Haines 1964). Other species may drink sea-water sporadically, and this has been shown among the Australian marsupials, including the Tammar wallaby, *Macropus eugenii* (Kinnear et al. 1968). In many areas of the world there is a deficiency of sodium in the soil, a condition that appears most often in the interior of continental areas and high mountainous regions. The vegetation reflects the low sodium content of the soil, so that herbivorous animals may have restricted amounts of salt available. Thus, the herbage in grazing areas of central Australia may contain only 1 to 3 mEq. kg^{-1} dry weight of sodium, compared to 100 to 350 mEq. kg^{-1} dry weight in coastal areas (Denton 1965, 1982). The popularity of "salt licks" among mammals in such regions is well known.

2.1 Water Loss

2.1.1 Cutaneous

Cutaneous water loss takes place through the skin by diffusion, and as a result of secretion by the sweat glands. This water is evaporated from the body surface, resulting in a loss of heat, a process that can be used physiologically for cooling in hot environments. In some species, including humans, cattle, camels and some antelope, sweating is the principal mechanism for thermal cooling. However, in other mammals, such as horses, dogs and small herbivores, it is generally utilized only during exercise. Evaporation from the respiratory tract by panting is otherwise the predominant cooling mechanism (see below). Sweating is considered to be a relatively inefficient way of cooling, as in some mammals, such as horses, it mainly occurs from the surface of the fur. When it does take place directly from the skin the surface temperature declines and facilitates further gain of heat by conduction from the environment. In man, the eccrine sweat glands are usually considered to be associated principally with thermal cooling while the apocrine glands, which are associated with hair follicles, respond to other stimuli. Sweating is initiated by exercise in the horse, a response that is stimulated by elevated levels of adrenaline in the blood and one which contributes to heat and water losses (Lovatt Evans et al. 1956). The sweat glands can secrete large volumes of fluid; a man working in the desert loses up to 15 l of water a day in this manner. Sweat glands are typically mammalian structures though their presence, and relative importance for thermal cooling is not the same in all species. MacFarlane (1964) found that the maximal rate of thermal sweating in merino sheep was 32 g m^{-2} h^{-1}, while the camel can form 260 g m^{-2} h^{-1} (Schmidt-Nielsen et al. 1957). Many mammals, such as the rabbit and rat, do not possess sweat glands, though water loss can still take place across the skin. Dehydration may result in a decreased rate of sweating. The camel after 24 h without water perspires at 60 to 70% of the rate seen when it is normally hydrated (Schmidt-Nielsen et al. 1957). Thermal cooling responses of sweat glands are under the control of the sympathetic nervous system. In man, the final transmitter is acetylcholine. However, in bovidae, like domestic ox and various wild species like the oryx, eland, wildebeest and buffalo which live in East Africa, these nerves are adrenergic (Findlay and Robertshaw 1965; Robertshaw and Taylor 1969). The injection of adrenaline in these species as well as in the horse, sheep and camel results in the secretion of sweat. This type of response may arise physiologically during exercise rather than in response to heat stress. Noradrenaline is relatively ineffective. Thermal sweating is abolished in the dehydrated oryx, though sweat glands can still respond to injected adrenaline (Taylor 1969). Kangaroos utilize sweating as a cooling mechanism during exercise (Dawson 1973; Dawson et al. 1974). As soon as the activity stops, the sweating ceases and the animals continue to pant in order to cool. The sweat glands, like those in other mammals, have an adrenergic innervation. The injection of adrenaline also results in sweating, but such a physiological role for this hormone has not been established. It seems likely that reductions in the volume of sweat secretion during dehydration are mediated by nervous, rather than endocrine, mechanisms, but changes in the level of circulating adrenal medullary hormones possibly could have an effect in some species.

2.1.2 Respiratory

The respiratory tract is a major route of water loss in tetrapods and more especially in the mammals and birds. This loss takes place by evaporation, as an unavoidable consequence of the uptake of oxygen and excretion of carbon dioxide, but this avenue can also be utilized physiologically for thermal cooling by panting. The normal overall obligatory losses of water from the respiratory tract increase or decrease with parallel changes in metabolic rate and oxygen consumption. Normally, mammals extract about 30% of the oxygen present in the inspired air, but in certain circumstances this may be altered, with reflected changes in the accompanying water loss. Dick (C. R.) Taylor (1969) showed that an African antelope, the oryx, which lives in desert areas, may reduce its oxygen consumption by 30% when it is dehydrated, and, by breathing more deeply, can extract additional oxygen from the inspired air. The net result is a considerable reduction in water loss. Similarly, Schmidt-Nielsen and his collaborators (1967) have shown that the camel reduces its oxygen consumption when dehydrated. Such a decrease in metabolism is not invariably seen in dehydrated mammals. Thus, the East African waterbuck, *Kobus defassa ugandae*, fails to change its rate of oxygen consumption when deprived of water (Taylor et al. 1969). Largely as a result of the high rate of accompanying evaporative water loss, these antelope withstand water deprivation poorly, even when compared with domestic Hereford cattle. It seems possible that the ability to reduce the metabolic rate under such conditions constitutes a physiological reaction that could adapt the animals more readily to the changed circumstances. The thyroid gland influences the metabolic rate in mammals, but it is not known whether it is concerned with the changes that are observed in dehydrated animals. Such a possibility should be simple to test.

2.1.3 Renal Water Loss

Obligatory losses of water occur in the urine in response to the homeostatic need to excrete solutes that are either obtained from the diet or are formed as a result of metabolism. In mammals, the principal product of the latter process is urea.

In many herbivores, including cattle, sheep and kangaroos, the fermentative microorganisms in their guts can reutilize this urea to make proteins (Stevens and Hume 1995). Such urea recycling may increase under conditions of water restriction and a low protein diet. The urea enters the gut in the saliva and by diffusion across its lining mucosa. The necessity to excrete urea in the urine can thus be reduced in such mammals. The magnitude of water loss in the urine of mammals depends, apart from excesses of dietary salt and urea, on their ability to concentrate the urine. The proportion of the total water loss each day varies in different mammals and reflects the environmental conditions, which usually determine the extrarenal losses. In normal circumstances renal water excretion is less than 50% of the total water loss, but it may be less than 10% under hot dry conditions when evaporative losses are increased.

The mammalian kidney, like that of birds and reptiles, is a metanephric one (Braun and Dantzler 1997). However, unlike in the other phyletic groups, it lacks a renal portal blood system. The nephrons of mammals are generally larger than those in most other species. The mammalian kidney can secrete urine that has a concentration that is hyperosmotic to the plasma. Among vertebrates, the birds are the only other species that have such an ability, but they generally cannot concentrate their urine as highly as mammals do (Table 1.6). Nevertheless, the capacity to concentrate the urine is quite variable in different species of mammals. Thus, mountain beaver can only secrete urine with a maximum concentration of about 500 mosmol l^{-1} while in Australian hopping mice it can exceed 9000 mosmol l^{-1}. The ability to form hyperosmotic urine is related to the structural arrangement of the nephron which includes the renal tubular loop of Henle. The latter is present only in mammals and birds. The renal capillaries also form long loops, which descend into the renal medulla (vasae rectae). The renal tubules and the vasae rectae form, respectively, a countercurrent multiplier system and a countercurrent exchanger for solutes. These morphological arrangements contribute the mechanisms necessary for the formation of hyperosmotic urine. In 1944, Sperber observed that the length of the loop of Henle in different mammals, as reflected by the thickness of the inner renal medulla, could often be correlated with their ability to concentrate the urine. This ability also appeared to be related to the degree of aridity of their normal habitats. Such correlations have often been observed, but there are a number of exceptions (Greenwald 1989; Beuchat 1996). The "metabolic intensity", or "capacity", of the renal tissue also appears to be an important non-morphological factor that can contribute to an ability to concentrate the urine. The weight of the kidney in mammals increases with their body weight and generally reflects an increase in the numbers of nephrons. This increase in kidney size usually follows a predictable formula. However, it is especially interesting to observe that marine mammals, including whales, dolphins and seals, have exceptionally large kidneys (Beuchat 1996). Nevertheless, their urine-concentrating ability is unremarkable compared to many other mammals.

In mammals, an absence of vasopressin or an inability of the distal renal tubule-collecting duct system to respond to this hormone results in the formation of large volumes of dilute urine (the disease called diabetes insipidus). This condition is mainly due to a decreased reabsorption of water from the distal renal tubular system. As described in Chapter 1, vasopressin, by promoting the reabsorption of urea and sodium from the renal tubules, also helps to maintain the renal medullary osmotic concentration gradient that is necessary for the formation of a hyperosmotic urine. It could also be contributing in this way to the urea-recycling mechanism in some herbivorous mammals.

2.1.4 Faecal Water Loss

Water loss in the faeces depends on their water content and the total mass of faeces that are formed (Degen 1997). Rodents absorb about 90% of the content of a diet of seeds, and carnivores appear to exhibit a similar efficiency in the utilization of their food. However, most herbivores utilize only about 65 to 70% of their fibrous diet of grasses and shrubs. Faeces usually contain 60 to 70% water. However, depending on the species, this concentration can often be reduced when water is not freely available. The faeces of camels normally contain about 55% water, but after a day without drinking the water content decreases to about 45% (Schmidt-Nielsen 1964a). The faecal water content of jackrabbits living in the Mohave Desert can be as low as 38% while that of wild rabbits is about 48%. Several species of rodents, especially those living in arid areas, can form faeces containing about 40% water. In sheep, the water content of the faeces varies from 75 to 45%. However, cattle cannot reduce the water content of their faeces as efficiently (McKie et al. 1991). Water is mainly removed from the faeces in the descending (distal) colon. In sheep, the mucosal epithelium lining the colon is considered to be "tight", with narrow paracellular pathways. In cattle these intercellular pathways are described as being "leaky", which may contribute to their inability to achieve high levels of faecal dehydration. It appears that the reabsorption of salt from the colon is more effective in cattle than in sheep, but the higher osmotic concentration of gradients necessary for the formation of drier faeces cannot be maintained. The mechanism that is utilized to dehydrate the faeces in rats has been described in Chapter 1. In rats, the establishment of the osmotic gradients involved in maintaining the necessary suction tension and fluid absorption across the distal colonic crypt barriers is increased by aldosterone (Naftalin and Pedley 1999). Mammals that form dry faeces have been observed to have a longer distal colon than those that excrete wetter faeces. This relationship has been observed in East African antelopes, rodents and kangaroos (Osawa and Woodall 1992; Woodall and Skinner 1993; Murray et al. 1995).

2.1.5 Water Loss Associated with Reproduction

Reproduction results in increases in the need for water (and salt) in mammals. This additional requirement reflects pregnancy and the growth of the foetus and, subsequently, lactation. Mammals that live in regions of seasonal aridity may synchronize their breeding seasons to coincide with the availability of adequate drinking water and food. However, an unexpected subsequent drought can result in the loss of the young. Kangaroo rats living in the Mohave Desert can normally survive without drinking on a diet of dry seeds. However, during periods of reproduction they may seek more succulent vegetation (Nagy and Gruchacz 1994). The milk of desert species often contains high concentrations of fat and less water than in species from more temperate regions. A recycling of water from the young to the mother, and sometimes even the father, can result from anogenital licking (Degen 1997). This retrieval of water from the urine and faeces has been studied most often in rodents but also occurs in other species, including the young in the pouch of marsupials (Dove et al. 1989). It has been calculated that during early lactation in rats up to 70% of water in the milk is recovered in this way. The urine of young mammals is much more dilute than in the adults and they are relatively unresponsive to vasopressin. The adult kidney with its greater concentrating ability thus provides a more efficient organ for conserving water under such conditions. In mammals living in areas where there is deficiency of salt in the diet, such a mechanism could also be useful for the conservation of sodium.

2.2 Salt Loss

2.2.1 The Sweat Glands

The secretion of the sweat glands contains dissolved solutes, principally sodium, potassium, chloride and bicarbonate. In mammals that utilize sweat secretion for thermal cooling the losses of these solutes, especially sodium, may be considerable. The sodium concentration of sweat in humans varies from 5 to $60\,\mathrm{mEq\,l^{-1}}$, being greatest when the rate of sweat secretion is high (Thaysen 1960). During sodium depletion the amount of this solute in the sweat decreases more than five fold, while the potassium level rises. Such changes require several days to occur. Schmidt-Nielsen (1964a) suggested that in herbivorous animals, like the donkey, little sodium is lost in sweat, and measurements by MacFarlane (1964) on sheep and camels support this view. The sweat of the camel contains $9.5\,\mathrm{mEq\,l^{-1}}$ sodium and $40\,\mathrm{mEq\,l^{-1}}$ potassium. In humans, a decrease in the Na/K ratio in sweat has been related to the action of the adrenocortical hormones (Conn et al. 1947), and it seems likely that these steroids may contribute to the acclimatization in the salt levels of the sweat that is seen in other species.

2.2.2 The Gut

The gut is an avenue of entrance and exit of salts from the external environment. Salts are taken into the gut with the food and water, while large volumes of fluids, rich in electrolytes, are continually secreted into the lumen of the alimentary tract in order to facilitate digestion. These fluids include saliva, various gastric and intestinal secretions, the bile and the pancreatic juices. In order to maintain adequate salts in the body, all of the electrolytes must ultimately be reabsorbed. Inadequate intestinal absorption, which may occur, for instance, as a result of infection, can result in death from water and electrolyte deficiency.

The composition of the secretions derived from the different *salivary glands* varies, and depends on the species and the physiological condition of the animal. Salivary secretion in the sheep has been extensively studied by Denton and his

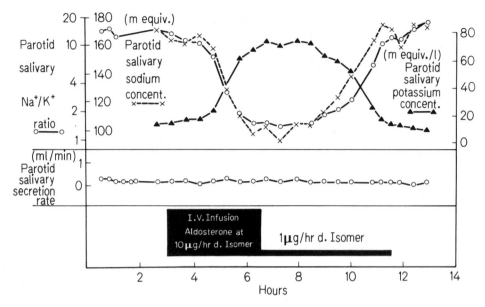

Fig. 3.1. Sheep Zachary (adrenally insufficient, Na$^+$ deficit 250 mEq.). Effect of intravenous infusion of aldosterone on parotid salivary Na$^+$ concentration ×---×; K$^+$ concentration ▲—▲; and parotid salivary Na$^+$: K$^+$ ○—○. Parotid secretion rate is also shown. (Blair-West et al. 1967)

collaborators (Blair West et al. 1967). The total volume secreted by this species in a day is 6 to 16 l, principally by the parotid glands. Normally, in a sheep, with adequate dietary sodium, the parotid secretion contains 180 mEq sodium l^{-1} and 5 mEq. potassium l^{-1} (Na/K is 36). If sheep are deficient in sodium, the levels can change to give a Na/K ratio of 10/170 mEq l^{-1}, or 0.06. Adrenalectomized animals exhibit little adjustment in their response to sodium depletion. The infusion of aldosterone can, however, restore this deficiency (Fig. 3.1).

The rate of secretion of aldosterone from the adrenal cortex can increase 13-fold in sheep depleted of 300 to 500 mEq of sodium, and the levels of this steroid are consistent with the amounts which must be infused in order to change similarly the composition of the saliva. Cortisol and corticosterone can also alter the electrolyte level in the saliva, but the necessary concentrations are large, and probably only of pharmacological significance. Vasopressin has no effect on salivary secretion.

The site(s) of action of the corticosteroids on the salivary glands is uncertain; they conceivably could act on the primary secretory mechanism in the acinar cells, or alter reabsorption from the ducts. Blair West et al. (1967) suggest that aldosterone probably acts at both sites.

The biological importance of the action of aldosterone on the electrolyte composition of saliva is more readily apparent in animals that secrete large volumes of saliva. These include many of the Artiodactyla, which digest their herbivorous diet with the aid of microbial fermentation. The volume of the fluid in the digestive tract of these animals is largely derived from the salivary glands, making up

10 to 15% of the body weight, and containing sodium equivalent to half that present in the extracellular fluid. A physiological mechanism to limit such sequestration of sodium could well be vital for survival when the amounts of this ion in the body are depleted, and its dietary availability is limited. In addition, some mammals drool saliva from the mouth when the rectal temperature is elevated in hot situations. Such salivation has been observed in several marsupials and placentals, including some rodents, rabbits, cats and cattle. If this saliva is spread over the fur it may aid in cooling. In rodents a substantial loss of fluid may occur in this way (Degen 1997). The forearms of kangaroos have a plentiful blood supply and they spread saliva over this area when environmental temperatures are high (McCarron and Dawson 1989). This response does not occur when the kangaroos are dehydrated. The use of saliva for evaporative cooling may constitute an emergency reaction that could be life-saving, but its efficiency is uncertain (Schmidt-Nielsen 1964a). The accompanying loss of salts in the saliva possibly could be reduced by aldosterone, as observed in the saliva of sheep (Fig 3.1).

The absorption of salts from the diet and the secretion that occurs into the gut takes place continually in the small intestine. A role for hormones in the processes of such conservation is not readily apparent. However, an increased secretion of Cl^- into the intestine can be promoted by guanylins that are secreted by enterochromaffin cells in the lining mucosa (Forte et al. 2000). A hormonal role for these peptides is considered likely but is not established. However, in conjunction with their diuretic and natriuretic actions on the kidney they could be contributing to a postprandial excretion of excess salt (Fig. 3.2). The colon is the site of a hormonal mechanism for the fine-tuning of sodium and potassium excretion in the faeces. Such a role for adrenocorticosteroids had long been suspected as a result of changes in the salt concentrations of the faeces that accompanied human Addison's disease. This condition results from a destruction of the adrenal cortex. In 1965, Cofré and Crabbé demonstrated that aldosterone had a direct effect in vitro, in promoting Na^+ absorption from the colon of the toad *Bufo marinus*. Such an effect was also demonstrated in vivo in humans and rats, where it is usually accompanied by an increased excretion of K^+ in the faeces (Levitan and Ingelfinger 1965: Edmonds and Marriott 1970; Binder and Sandle 1994).

2.2.3 The Kidney

Large volumes of plasma are filtered across the glomerulus; this ultrafiltrate passes into the renal tubule, where most of its constituents are reabsorbed. The quantities of solutes filtered are enormous, and usually more than 99.9% of the sodium is reabsorbed back into the plasma. These processes are relatively greater in mammals than in cold-blooded species. If the diet contains an excess of sodium, more sodium will be excreted. Most vertebrates can form a hyposmotic urine which may contains only small amounts of sodium. In animals on a normal diet such urinary losses are easily replenished, but in some circumstances even these may be physiologically significant. The formation of large volumes of urine, due to excessive drinking, or to feeding on a very succulent diet, exaggerates such

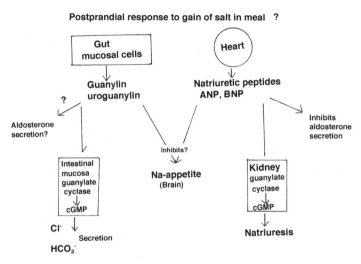

Fig. 3.2. A suggested model for the endocrine control of salt excretion following feeding in mammals. (After Forte et al. 2000, with some additions)

losses. The role of the adrenal cortex in renal sodium conservation and potassium excretion has already been described (Chap. 2); aldosterone promotes reabsorption of sodium and secretion of potassium across the wall of the distal renal tubule and collecting ducts. The absence of a functional adrenal cortex results in an excess loss of sodium in the urine and failure to excrete adequate potassium. Changes in the sodium content of the diet are reflected in the sodium content of the urine, and the circulating levels of the adrenocortical hormones, especially aldosterone. Thus, rabbits living in the sodium-poor areas of the plateau of the Australian Snowy Mountains have urinary sodium levels as little as one-thirtieth of those rabbits living in adjacent sodium-replete grasslands (Blair West et al. 1968). The adrenal glands of the sodium-deficient rabbits are enlarged and the levels of aldosterone and renin in the blood are three to six fold greater than in the sodium-replete animals.

2.3 Accumulation of Water and Salts

Water and salts are obtained:

1. **In the food.** The water content of plants is often equivalent to more than 95% of their weight, and a carnivorous diet contains about 70% water. Even dry seeds contain water that is taken up hygroscopically from humid air. In addition to such directly available fluid, the metabolism of fats, carbohydrates and proteins results in formation of additional water; the yield from oxidation of 1 g of fat is 1.1 ml, while the same amount of carbohydrate gives 0.6 ml and protein 0.4 ml. There has been much conjecture about the importance of fat reserves in assist-

ing animals to maintain a positive water balance. The camel's hump was probably the earliest recipient of such speculation, but this has been discounted by Knut Schmidt-Nielsen. It must be remembered that the oxidation of an energy substrate requires an exchange of gases in the respiratory tract, and this may result in evaporative water loss. If the animals are breathing dry air, oxidation of such substrates will result in a net loss of water, due to saturation of the expired air with water vapour. Inspiration of more humid air will decrease this loss, so that in marine mammals like the whale, which breathe air already saturated with water vapour, a net gain of water from metabolism can be expected.

As already described, the sodium content of plants may vary considerably and in some inland continental and alpine regions the sodium levels may be low. This is the prime determinant of the salt status of all herbivorous animals living in the area, while carnivorous species obtain adequate sodium from the prey on which they feed. The diet of some mammals may consist predominantly of halophytic plants with a high concentration of salts such as eaten by African sand rats (Schmidt-Nielsen 1964a), possibly Australian hopping mice (MacMillen and Lee 1969) and kangaroos and sheep living in the Australian outback (Dawson et al. 1975).

Many mammals, especially herbivores, that live in inland areas of the world where the vegetation has a low concentration of sodium, actively seek, and regularly visit, mineral deposits containing sodium chloride (Denton 1982). They replenish their depleted body sodium at such sites, which are called salt licks. This behaviour has been observed more often during pregnancy and lactation, when it may involve a response to reproductive hormones. Humans living in such regions may also be affected and often place a high commercial and monetary value on salt.

2. In the drinking water. Drinking supplies the water requirement beyond that obtained from the food. O'Connor and Potts (1969) have emphasized the predominant importance of this process in regulation of a positive water balance. The urinary volume and concentration of dogs on a standard diet, in contrast to what they drink, does not vary over a wide range of environmental circumstances. This suggests that it is drinking, rather than kidney function, that is responsible for regulating the positive water balance. Fresh water, containing little dissolved salt, is usually the beverage of choice. However, in some situations the only drinking water that is available may contain significant quantities of salt that are apparent on tasting (brackish water). Such water supplies are regularly utilized and can help sustain animals providing that the sodium chloride concentration does not exceed the capacity of the kidney to concentrate the salt in the urine. In some unusual circumstances animals may even drink sea-water (= 3.3% NaCl). However, in most instances when this occurs, diuresis and diarrhoea follow, resulting in further dehydration.

The sensation of thirst, which results in drinking, and an appetite for sodium are initiated in the brain, especially in regions near the base of the third ventricle and the hypothalamus (Fitzsimons 1998). Thirst is a complex process that can be initiated by increases in the concentration and decreases in the volume of the extracellular fluid. The administration of angiotensin II, into either the brain or the peripheral circulation, can evoke drinking behaviour in many vertebrates, including mammals. Extracellular hypovolemia initiates activation of the renin-angiotensin system. The angiotensin II that is formed in the plasma contributes to the thirst-drinking mechanism. It activates AT_1 receptors in regions of the brain, including the subfornical organ and the organum vasculosum of the lamina terminalis. These sites are accessible to the peripheral circulation. Natriuretic peptides also have receptors in such tissues and can antagonize this effect of angiotensin II.

Angiotensin II can also induce sodium appetite (Weisinger et al. 1987). However, this effect is slower and it is delayed in onset compared to the drinking response. Sodium appetite is mediated through a "sodium-appetite centre" that lies near, but is separated from, the thirst centre (Weisinger et al. 1993). This effect on sodium appetite has been observed in rats, mice, sheep, wild rabbits, cattle and pigeons. Sodium appetite can also be promoted by the administration of ACTH (Weisinger et al. 1980). Its principal effect appears to be mediated by adrenocortical steroids, including aldosterone. Angiotensin II and the adrenocorticosteroids may have synergistic effects on sodium appetite. A release of ACTH in response to stress in mice has been shown to evoke sodium appetite (Denton et al. 1999).

2.4 Effects of Dehydration on the Body Fluids

The water content of terrestrial species of mammals is usually about 70% of their body weight. However, there is a range between 60 and 70% (Degen 1997). The differences reflect such factors as the fat content, the season and the volume of fluids that are sequestered in the gut. The latter may be substantial in herbivorous animals when the rumen, caecum and large intestine can provide capacious chambers for microbial fermentation (Stevens and Hume 1995). This gastrointestinal solution is considered to be part of the interstitial fluid compartment. The blood plasma in mammals is usually about 5% of the body weight and its volume is maintained within the vascular space by the interactions of the capillaries, the osmotic pressure of the plasma proteins and the hydrostatic pressure provided by the beating heart (see Chap 1). An excessive loss of water from the plasma results in an increased viscosity of the blood, which hampers its circulation (Horowitz and Samueloff 1988). This effect can retard the loss of body heat, especially in hot external environments, and can result in fatal increases in body temperature (heat stroke). Some species, especially those adapted to life in desert conditions, can maintain their plasma volume during periods of dehydration (Schmidt-Nielsen 1964a; Horowitz and Borut 1970; Degen 1997). Such plasma volume conservers include camels, the burro, kangaroos and some antelope and rodents. Species that do not conserve their plasma volume under such conditions include humans, dogs, merino sheep and laboratory rats. Species that conserve their plasma volume during dehydration may initially lose more water from the interstitial and intracellular fluids than from the blood plasma, and this can help maintain the blood circulation. This response can be influenced by the degree and speed of dehydration. It appears to be related to the ability of the animals to maintain the concentrations of plasma proteins (Horowitz and Adler 1983; Horowitz and Samueloff 1988). The latter may be influenced by such factors as a decreased leakage of proteins across capillaries, an increased protein synthesis and the shunting of blood to vascular beds that are less permeable. Cardiovascular adjustments undoubtedly contribute to the maintenance of the circulation during dehydration.

3 The Monotremes (Egg-Laying Mammals)

The monotremes are confined to Australia and New Guinea, where they occupy a variety of contrasting habitats. The platypus is aquatic, living in lakes and rivers, while echidnas live in areas ranging from deserts to tropical rainforests.

Most of our knowledge about the physiology of monotremes is based on observation of echidnas, especially *Tachyglossus aculeatus*. This mammal weighs about 3 kg and in its native habitat is insectivorous, eating ants and termites. The echidna, in common with the platypus, has a body temperature of only 30 °C (Martin 1902) and a metabolic rate about half that expected in a placental mammal of similar size (Schmidt-Nielsen et al. 1966b). The extraction of oxygen from the inspired air is similar to that in other mammals (Bentley et al. 1967b) so that its respiratory losses of water are expected to be less. Although the echidna can maintain a constant body temperature at low environmental temperatures, its thermal evaporative cooling at high temperatures is poor, since it neither sweats nor pants under such conditions (Schmidt-Nielsen et al. 1966b). The overall water balance of echidnas with a natural diet of termites has been calculated (Bentley and Schmidt-Nielsen 1967); it appears that, in their natural habitat, they probably require little or no water to drink. An investigation on echidnas maintained under natural conditions has shown this prediction to be remarkably close (Griffiths 1978). However, when they were maintained in dry air drinking occurred. The ability of the kidney to form a hyperosmotic urine is critical to the maintenance of a positive water balance under such conditions. Dehydration results in a decreased volume of urine, that may have a concentration of 2300 mosmol l^{-1}. This is achieved by tubular reabsorption of water and a decreased rate of glomerular filtration. The manner of controlling changes in urine volume and concentration is unknown, but the pituitary contains the placental antidiuretic hormone, arginine-vasopressin, as well as oxytocin (Sawyer et al. 1960). The estimated (in a 3-kg echidna) concentration of vasopressin in the pituitary is about 10^{-9} M kg^{-1} body weight, which is less than that usually found in mammals (see Follett 1963; Table 3.2).

The adrenal glands in monotremes do not show the distinct delineation into a cortex of interrenal tissue and a medulla of chromaffin cells. Indeed, the echidna adrenal is somewhat akin in its morphological arrangement to that in reptiles, while the platypus has the form seen in marsupials and placentals (Wright et al. 1957). The adrenal is relatively small in the echidna in relation to its body weight (Sernia 1980). Corticosteroid hormones are present in the peripheral circulation, but at a very low concentration compared to other mammals. Corticosterone is the major such steroid, while cortisol and aldosterone are present, but at even lower concentrations (Weiss and McDonald 1965; Sernia 1980). The predominance of corticosterone is not a characteristic of the monotremes as cortisol is the major such steroid in the platypus (McDonald et al. 1988). A physiological role for corticosteroids in regulating salt metabolism in the echidna is doubtful. When they are not stressed they can survive for at least 20 weeks following adrenalectomy (McDonald and Augee 1968). No detectable electrolyte or metabolic disturbances were observed in such animals. However, when exposed to cold environmental conditions, they cannot adapt and they then die (Augee and McDonald 1969). Nevertheless, they can be protected under such conditions by the administration of glucocorticoids, which facilitate the mobilization of fatty acids from fat stores. The synthesis of corticosteroids can be increased by the administration of ACTH in the echidna. Angiotensin II increases the synthesis of aldosterone. However, renin has not been detected in the plasma of these animals (Reid 1971). Nevertheless, a juxtaglomerular apparatus and renin are present in the kidney. The adrenal cortex in the echidna appears to be principally involved in metabolic responses and the adaptation to stressful conditions. The absence of mineralocorticoid effects may be a unique situation in mammals.

4 The Marsupials (Pouched Mammals)

Two families of marsupials live in the Western Hemisphere: the Didelphidae (opossums) and the Caenolestidae (shrew-like animals living in South America). Most marsupials, however, live in the Australian region, where the contemporary families occupy the variety of ecological niches on the continent, in a fashion that parallels the manner in which placentals occupy similar places elsewhere. Various species have adopted insectivorous, herbivorous and carnivorous diets. They may be arboreal, like the possums, fossorial, like marsupial moles, or they may live in a more conventional way on the surface. Large areas of Australia are desert with little and irregular rainfall, combined with a high environmental temperature and a low humidity. Other regions on the continent are covered with tropical rainforest, while there are also many temperate areas with more equitable climates. The marsupial fauna has occupied all these regions with success, often to the chagrin of the sheep farmers, whose flocks may fare poorly compared to the indigenous species. In many of the more barren regions, sheep and cattle, in contrast to marsupials, require supplements of trace elements, protein and water in order to survive and breed.

Physiological knowledge about most of the 75 genera of Australian marsupials is lacking, but a modest amount of information is available for some species, especially members of the family Macropodidae. The macropods consist of 17 genera of kangaroos and wallabies, nearly all of which are herbivorous (the diet of musky rat kangaroos includes worms and insects) and have a grazing manner of life. They range in size from 500 g in the musky rat kangaroo to 70 kg in the larger members of the genus *Macropus*.

Martin in 1902 reported that the metabolic rate of several marsupials was only about one-third as great as would be expected in placentals of comparable size.

That marsupials do indeed generally have a lower basal rate of oxygen consumption than placentals has been shown in a variety of species (Dawson and Hulbert 1969; Arnold and Shield 1970). The basal rate of oxygen consumption is about one-third less than predicted in a placental of the same size. This is accompanied by a lower body temperature. It thus seems likely that evaporative water losses from marsupials may be less than in placentals. The daily water consumption of kangaroos is only about two-thirds that of placental mammals of a similar size, which is consistent with such a water saving (Denny and Dawson 1975).

Some marsupials have labile body temperatures that change in a characteristic cycle over the course of each day. This is well shown in the chuditch or western native cat, *Dasyurus geoffroii*. The chuditch is a carnivorous marsupial weighing 1 to 2 kg, which has a wide geographic range throughout Australia, where it inhabits even the hot desert regions. It is nocturnal in its behaviour, coming out to hunt for food in the evenings. The body temperature of these marsupials changes in a parallel manner, declining to a minimum in late afternoon and rising as much as 4 °C in the evening (Arnold and Shield 1970). During the period of inactivity the rate of oxygen consumption was found to be only two-thirds as great as predicted in a placental of the same size. Red kangaroos commence their feeding activity in the early morning, when their body temperature may be 3 °C below normal. This adjustment appears to delay the need for evaporative cooling (Brown and Dawson 1977). Such changes in the metabolic rate must result in a conservation of calories and also a reduction in the rate of evaporative water loss. These physiological adjustments could well assist the animal's survival, especially in its desert habitats (Arnold and Shield 1970).

The kidneys of marsupials, like those of placental mammals, possess a loop of Henle and can form hyperosmotic urine. The functioning of their kidneys appears to be regulated also by hormones from the adrenal cortex and the neurohypophysis. The morphology of these endocrine glands, including the zonation of the adrenal cortex, is similar to that of placental mammals. As in the latter mammals, the principal secreted adrenocorticosteroids are cortisol and aldosterone, which are usually accompanied by smaller amounts of corticosterone (Oddie et al. 1976; McDonald and Martin 1989). However, the compound F / compound B ratio (cortisol / corticosterone) varies from 40 in the pademelon wallaby to 0.7 in the common wombat. It also varies considerably in placental mammals; it is 20 in a monkey and 0.1 in rabbits and the laboratory rat (Bush 1953). Such variations can reflect genetic differences in the enzymes involved in their synthesis. However, the ratio of their secretion is also influenced by the prevailing level of adrenocortical activity and the presence of ACTH. Some marsupials also secrete significant quantities of 11-deoxycortisol, which in placental mammals is a precursor of cortisol. A secretion of 11-deoxycorticosterone, which is a precursor of aldosterone, may also occur. Marsupials possess ACTH and the renin-angiotensin system and they contribute, as in placental mammals, to the regulation of the activity of the adrenal cortex (Reid and McDonald 1969; Simpson and Blair-West 1971; Blair-West and Gibson 1980).

The neurohypophysial hormones in marsupials are, rather surprisingly, more diverse in their structures than those in placental mammals or the echidna (Table 3.1). Anomalies in the ratio of vasopressin / oxytocin had earlier suggested the

Table 3.1. Phyletic distribution of the neurohypophysial hormones in marsupials

Family and species	AVP	LVP	PP	OT	MT
Didelphidae					
Opossums (American)					
Didelphys virginiana	+	+	–	+	+
D. marsupialis	+	+	–	+	–
Macropodidae					
Kangaroos and wallabies					
Macropus rufus	–	+	+	–	+
M. eugenii	–	+	+	–	+
Setonyx brachyurus	–	+	+	–	+
Phalangeridae					
Brushtail possum					
Trichosurus vulpecula	+	–	–	–	+
Peramilidae					
Bandicoot					
Isoodon macrourus	+	+	–	+	+
Dasyuridae					
Native cat					
Dasyurus spp.	+	–	–	–	+
Phascolarctidae					
Koala					
Phascolarctos cinereus	+	–	–	–	+

AVP = 8-arginine vasopressin; LVP = 8-lysine-vasopressin; PP = 2-phenylalanine-8-arginine-vasopressin (phenypressin); OT = oxytocin; MT = 8-isoleucine-oxytocin (mesotocin). (After Chauvet et al. 1983; Rouillé et al. 1988; Bathgate et al. 1992; Acher 1996).

possibility that such hormones may be different in marsupials (Bentley 1971a). Most placentals and echidnas typically secrete 8-arginine-vasopressin (AVP, ADH) and oxytocin. Pigs and the Peru strain of mice secrete 8-lysine-vasopressin (LVP) instead of AVP. The marsupial neurohypophysis has been shown to contain up to four different such peptide hormones in a single individual. In the northern bandicoot (family Peramilidae) the neurohypophysis contains AVP and oxytocin, but there is in addition also LVP and a homologue of oxytocin, 8-isoleucine-oxytocin (mesotocin) (Rouillé et al. 1988). The latter observation was of considerable interest, as this peptide, which has a biological activity similar to oxytocin, was formerly only known to occur in non-mammalian tetrapods and lungfish. The duplication of the antidiuretic hormones and oxytocin appears to have arisen as a result of gene duplication. A similar situation has been observed in the North American Virginia opossum (Table 3.1). Lysine-vasopressin, but not AVP, is present in kangaroos (family Macropodidae). In this family of marsupials it is accompanied by a novel second antidiuretic hormone, 2-phenylalanine-AVP (phenypressin). When tested for their antidiuretic effects in various species, AVP, LVP and phenypressin all display significant activity, which has probably contributed to their evolutionary survival. Thus, marsupials appear to have a plethora of antidiuretic hormones. Receptors for vasopressin and oxytocin have been identified in the brushtail possum (Bathgate and Sernia 1995). Their pharmacologi-

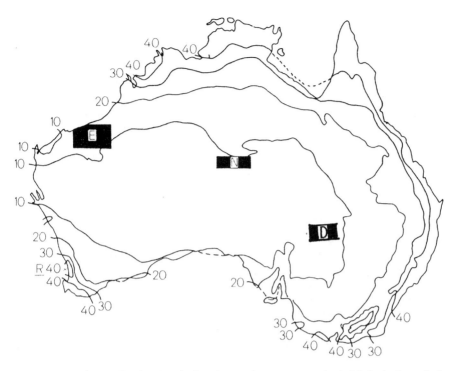

Fig. 3.3. Map of Australia showing the locations and average annual rainfall (in inches; 1 inch = 2.54 cm) in the areas where the quokkas (*Setonix brachyurus*), euros (*Macropus robustus*) and red kangaroos (*Megaleia (Macropus) rufus*) have been studied under natural field conditions. *E* Ealey (euros); *N* Newsome (reds); *D* Dawson (reds and euros) areas. *R* Rottnest Island (quokkas)

cal binding to various neurohypophysial peptides differs from those of the receptors in a placental mammal (the laboratory rat).

Information relevant to marsupial osmoregulation is available from four population groups, one living in a temperate and three in desert areas.

Professor Harry Waring initiated such studies on a small (2- to 5-kg) wallaby, the quokka, *Setonix brachyurus*, that lives in a temperate, but seasonally dry, habitat on Rottnest island, 18 km from the coast of southwest Australia (see Fig. 3.3). Today, *Setonix* is principally confined to this island, though a few isolated pockets of the population remain on the mainland. The annual rainfall on Rottnest is 75 cm, of which 68 cm falls between March and October, so that in the warm summer months, when the temperature may rise to 38 °C, little rain falls. The vegetation dries out during this period. There are two distinct populations of quokkas on the island (Jones et al. 1990). One occupies the extreme west end where there are no perennial supplies of drinking water. The other, larger, population lives in the region of several salt lakes on the eastern side of the island, where fresh and brackish water soaks are present. By 1990, only one freshwater pool still existed, the remainder being brackish. The quokkas living under seasonally arid conditions on Rottnest Island appear to be the descendants of mainland animals that originally lived near swamps. However, the Rottnest population has adapted to life in more arid circumstances. Their physiology has been studied over a period of more than 50 years both in the laboratory and in their native island refuge.

Martin in 1902 reported that marsupials do not regulate their body temperature adequately in hot environments, an observation that is relevant to their water loss in such situations. However, it was found that *Setonix brachyurus* regulated its body temperature as efficiently as placental mammals at temperatures up to 40 °C (Bentley 1955; Bartholomew 1956). This regulation is accompanied by sweating, salivation and panting, and results in a substantial loss of water. When *Setonix* is dehydrated, such evaporation of water is reduced by 40% (Bentley 1960) an adaptation reminiscent of that in placentals like the rat, camel and oryx. It seems likely that this decrease in water loss is mediated by a reduction in the metabolic rate, but neither this, nor its possible relationship to the secretions of the thyroid gland has been investigated. *Setonix* is not exceptional among marsupials in its ability to regulate its body temperature; members of this group are as efficient in this respect as placental mammals.

During the dry summer period, the concentrations of urine of the quokkas increase to levels two to three times as great as those observed in the wet winter period. Dehydrated quokkas can form a hyperosmotic urine with a concentration of 2000 mosmol l^{-1} (Bentley 1955), which is comparable to the ability of placental mammals. Exposure of *Setonix* to heat, with accompanying dehydration, was found to result in the appearance of an antidiuretic substance, presumably lysine-vasopressin, in the plasma (Robinson and MacFarlane 1957). The injection of vasopressin into hydrated quokkas reduces the urine flow (Patricia Woolley, quoted by Waring et al. 1966), and increases its concentration (Bentley and Shield 1962). Severing the supraopticohypophysial tract in the tammar wallaby results in the secretion of copious, dilute, urine (Bakker and Waring 1976). There is thus strong presumptive evidence to suggest that urine flow in this marsupial is regulated, as in placentals, by the release of antidiuretic hormone that acts on the kidney.

The water metabolism, kidney function and concentrations of lysine-vasopressin in the plasma of Rottnest Island quokkas has been measured in late summer prior to seasonal rains (Jones et al. 1990; Bradshaw 1999). The two population groups on the island were studied; one from the west end, where there is no drinking water, and the other from the area around the salt lakes, where brackish water was available. The turnover of water was nearly 2.5 times greater in the animals from the salt lakes area. Their production of urine was also greater. The quokkas from the salt lakes area secreted urine with a concentration of about 1000 mosmol l^{-1}. In the west end animals this concentration was 1250 mosmol l^{-1}. The plasma of the latter was also more concentrated. Plasma concentrations of lysine-vasopressin were measured. The mean concentration was 89 pg ml^{-1} in the west-end quokkas and 36 pg ml^{-1} in those from the salt lakes area. In the late summer period the west-end quokkas were in poor condition but, nevertheless, they can survive without drinking water. It was suggested that the availability of brackish water, combined with its efficient use, is important for the maintenance of the better condition of the salt-lakes quokkas.

It has been suggested that the quokkas on Rottnest Island may drink sea-water during the summer drought period and so possibly acquire a considerable excess of salt. These animals can concentrate urinary electrolytes to about $800\,mEq.l^{-1}$, which is higher than the level of $560\,mEq.l^{-1}$ in sea-water. Some quokkas, in extreme circumstances when no other fluid is available, drink small amounts of sea-water and make a net gain of osmotically free water (Bentley 1955). However, it seems unlikely that this species normally drinks sea-water, though a similar small wallaby, the tammar, *Macropus eugenii*, apparently can do so (Kinnear et al. 1968).

The concentrations of lysine-vasopressin in the plasma of two species of desert-dwelling wallabies have been measured during the dry and wet seasons (Bradshaw et al. 2001). The spectacled hare-wallaby, *Lagochestes conspicillatus*, is now confined to Barrow island off the NW coast of Australia while Rothschild's rock-wallaby, *Petrogale rothschildi*, is endemic to the adjacent Pilbara region of the mainland. Summers are very hot in this region and seasonal drought conditions often prevail. During the daytime the rock-wallabies shelter in cool caves among rocky outcrops while the hare-wallabies seek refuge in clumps of *Spinifex* grass. The environmental temperatures often reach 40 °C in the latter. Both species can form a highly concentrated urine: up to $3600\,mosmol\,kg^{-1}$ water. In the dry season plasma lysine-vasopressin levels increase considerably in the hare-wallabies but they remain low in the rock-wallabies. The urine volumes in both species decline at this time. The change in urine volume and increases in its concentration are related to the plasma vasopressin levels in the hare-wallabies but not in the rock-wallabies. Instead, the latter experience a decrease in renal plasma flow and the GFR. This change in renal haemodynamics and the selection of a cooler microhabitat may then obviate a role for vasopressin in regulating the composition of the urine.

The first studies to be carried out on the endocrine regulation of sodium and potassium metabolism in Australian marsupials were those of Buttle et al. (1952). Earlier work on the American opossum, *Didelphis virginiana*, suggested that the adrenal cortex may not be as important to marsupials as to placental mammals, since this species could, in certain circumstances, survive adrenalectomy for prolonged periods, with little or no change in electrolyte level. Jenny Buttle and her collaborators found that adrenalectomy in the quokka was usually fatal within 2 days. Death was accompanied by elevated plasma potassium and decreased plasma-sodium concentrations. The mean survival time could be increased two or three times by either giving the quokkas 1% sodium chloride solutions to drink, or by injecting them with a corticosteroid, deoxycorticosterone acetate. The subsequent administration of newly available preparations of cortisol and aldosterone to adrenalectomized quokkas showed that they could be maintained indefinitely by these corticosteroids (McDonald and Bradshaw 1993). When given together, they maintained plasma Na^+ concentration, renal plasma flow and the GFR, which otherwise fall in their absence. Cortisol also maintained plasma glucose and increased food intake. Similar observations have been made on several other marsupials (McDonald and Martin 1989). They include the brush-tail possum (Reid and McDonald 1968) and the red kangaroo (McDonald 1974). The latter develops a sodium appetite following adrenalectomy and if a saline solution is provided for them to drink, they survive indefinitely without hormone supplements.

When quokkas are maintained on a sodium-replete diet, the concentration of aldosterone in the plasma is low, about 2 ng $(100 \, \text{ml})^{-1}$, but after 7 days on a sodium-deficient diet the concentrations increase to about 17 ng $(100 \, \text{ml})^{-1}$ (Miller and Bradshaw 1979). Urinary Na^+ excretion is very low in the latter quokkas, but the concentrations in the plasma were maintained at normal levels. The concentrations of aldosterone in the plasma were also measured over a 2-year period in quokkas living in their native habitat on Rottnest Island. The values did not vary in the different seasons and were generally maintained at a level of 7 to 8 ng $(100 \, \text{ml})^{-1}$, which is somewhat higher than in sodium-replete animals. Urinary Na^+ excretion in the quokkas living on Rottnest Island averaged 8 mEq. $kg^{-1} \, day^{-1}$ compared to 23 mEq. $kg^{-1} \, day^{-1}$ in sodium-replete laboratory animals. Under natural conditions the quokkas appear to be utilizing aldosterone to limit urinary losses of sodium.

The water metabolism of two large species of kangaroo has been studied in far hotter and drier areas than those occupied by *Setonix*. Tim Ealey led a series of ecological and physiological investigations on a wild population of the hill kangaroo, or euro, *Macropus robustus*, living in a hot dry desert region of northwest Australia (Fig. 3.3). Alan Newsome carried out a similar investigation on the red kangaroo, *Megaleia* (or *Macropus*) *rufus*, living in an arid central region of the continent near Alice Springs. The average annual rainfall in both the areas is about 25 cm, most of it falling during the summer months. The rain, however, is sporadic and unreliable, and both areas may undergo periods of drought when little rain falls for periods of 3 or 4 years. The summer temperature is high, the average daily maximum often being greater than 40 °C for many weeks, and may rise to 49 °C. The relative humidity usually is low, so that the potential evaporation is high. Despite these osmotically adverse conditions the kangaroos in these areas continue to survive and multiply. A man lost in these regions in the height of summer can expect to survive for less than 1 day (Adolph 1947), so the adaptability of the kangaroos is impressive. Both the euro and the red kangaroo may form a highly concentrated urine, which can be 2700 mosmol l^{-1} in the red kangaroo (A. Newsome, quoted by Schmidt-Nielsen 1964a) and 2200 in the euro (Ealey et al. 1965). Such an ability must aid water conservation by these animals but is not a predominant factor in their survival. Ealey and his collaborators observed that euros can withstand dehydration equivalent to 30% of their body weight, and that they reduce evaporation by sheltering in rocky caves during the heat of the day. Euros drink periodically, and even during hot dry periods only come in to obtain water on an average of once every 3 days. During the intervening period they incur a water deficit equivalent to about 12% of their body weight. Newsome (1965a,b; 1971) found that the red kangaroos also drank only sporadically, apparently gaining most of their water from their diet, which consists preferentially of succulent green shoots of the herbage. The animals limit evaporation during the day by seeking the shelter of bushy woodlands and avoiding the open plains. During the dry season kangaroos often utilize drinking water at man-made sites provided for livestock. The degradation of the surrounding natural rangelands has been partly attributed to the presence of kangaroos. They can, with such water available, continue to breed during hot dry seasons, when under more natural conditions such activity would cease. Limiting their access to artificial drinking points has been utilized as a strategy to control their proliferation (Norbury and Norbury 1992).

Terry Dawson (Dawson et al. 1975) and his collaborators carried out a similar survey of the water metabolism and diet in natural populations of red kangaroos and euros living in a seasonally arid region in eastern Australia (Fig.3.3). They compared these marsupials to domestic sheep and wild goats. The water turnover of the placental mammals was three to four times greater than that of the kangaroos. The latter also needed to drink much less often. The principal reasons for the differences appear to be metabolic, physiological and behavioural adaptations of the native marsupial fauna to the local conditions.

The grasses of the central Australian regions are deficient in sodium compared to those in coastal areas (Denton 1965), so that conservation of this solute could be of particular importance to many kangaroos. However, there is no definitive information about this for the red kangaroos living in the Newsome study area. The urine (Newsome, quoted by Schmidt-Nielsen 1964a) of these animals has a Na/K ratio of 0.13, while the euros in the Ealey study area exhibit a ratio of 0.33. This suggests that relatively more sodium may be available to the euro than the red kangaroo. Two populations of the grey kangaroo, *Macropus giganteus*, living on grasses with contrasting contents of sodium, have been studied by Coghlan and Scoggins (1967). One group of these kangaroos lives in the coastal areas of southeast Australia, where the grasses contain most than 150 mEq sodium kg^{-1} dry weight, and the other occupies the Snowy Mountain plains, where the herbage contains less than 10 mEq sodium kg^{-1} dry weight. Aldosterone, cortisol and corticosterone have been identified in the blood of the red kangaroo (Weiss and McDonald 1967) as well as of the grey kangaroo. The red kangaroos were studied in the laboratory and, while aldosterone was secreted at rates comparable with those seen in placentals, the cortisol levels were somewhat lower. Rates of corticosteroid secretion were measured in grey kangaroos caught in the field and it was found that animals living in the sodium-deficient Snowy Mountains area have blood aldosterone levels eight times as great as those from the sodium-replete coastal regions. There was little difference in the blood levels of cortisol or corticosterone. The zona glomerulosa (which secretes aldosterone) in the adrenal cortices of the animals from the Snowy Mountains plains was larger than that in the other group.

5 The Placental Mammals

Generically and numerically, the placentals are the predominant group of mammals. All of the marine and most aquatic mammals belong to this group. They are present in all of the world's deserts. The placentals have thus adapted to a wide range of differing osmotic circumstances.

The Rodentia make up about one-third of all placental genera. This order has a very cosmopolitan distribution, occupying diverse habitats including the deserts of Africa, Asia, America and Australia, where, with the possible exception of the latter area, they are the principal mammals present. Osmoregulatory problems in such dry hot areas result in the exclusion of most mammals, except certain rodents that are equipped with behavioural and physiological adaptability. Knut and Bodil Schmidt-Nielsen investigated the water metabolism of the banner-tailed kangaroo rat, *Dipodomys spectabilis*, and revealed the manner in which this little rodent lives in a desert without drinking water. The kangaroo rat belongs to the family Heteromyidae, which is confined to the New World; it lives in the deserts of the western United States and parts of Mexico. Its ability to survive and thrive on a diet of air-dried seeds and vegetation, without requiring water to drink, is the result of several adaptations that have subsequently been shown to exist in other rodents occupying comparable ecological situations elsewhere. The principal adaptation allowing the kangaroo rat to live under such conditions is its small evaporative water loss. Rodents do not sweat or pant, and thus have a limited ability to undertake thermal cooling; they may allow their body temperature to rise somewhat in hot conditions (Schmidt-Nielsen 1964b). Obligatory losses of water that take place by evaporation from the respiratory tract are reduced in the kangaroo rat and white rat by utilizing a countercurrent heat-exchange mechanism in the nasal passages to cool the expired air (Jackson and Schmidt-Nielsen 1964). Such cooling reduces its content of water vapour. Rodents living in hot desert conditions have adopted patterns of behaviour that can reduce their need for evaporative cooling. Aided by their relatively small size, they usually seek refuge in underground burrows, crevices in rocks and nests in dense vegetation. The environmental temperature is usually lower and the relative humidity higher under such conditions than they are in the outside habitat. Many such rodents are nocturnal and so avoid the heat of the day. Species that are active in the daytime may make only brief forays out of such refuges. After their body temperature rises, they return to them to cool off. Such activity is described as shuttling or crepuscular behaviour. These rodents thus reduce the need for evaporative cooling. Some rodents reduce their metabolic rate and body temperature, and aestivate in their burrows in late summer. This period of dormancy may be prolonged to become one of hibernation over the winter period. It occurs in some underground squirrels (genus *Citellus*) such as those that live in the deserts of the southwest United States. A daily torpor, when the temperature decreases by only a few degrees, has been observed in a number of Australian marsupials (Geiser 1994). It also occurs in bats that live in arid regions (Carpenter 1968). Under such conditions the energy reserves of the animals are extended and evaporative water loss is reduced (Schmidt-Nielsen 1964a).

Rodents, especially those living in arid areas, conserve additional water by forming a highly concentrated urine and relatively dry faeces. Not all desert rodents can, however, achieve a positive water balance while eating a dry diet, and some must obtain food containing a substantial amount of free water. The pack rat, *Neotoma*, like the kangaroo rat, lives in the deserts of the western United States but eats cactus plants, while the North African sand rat, *Psammomys obesus*, subsists on succulent halophytic vegetation (Schmidt-Nielsen 1965a,b).

Neurohypophysial peptides, arginine-vasopressin and oxytocin, have been identified in the four species of rodents that have been examined, including the white rat, *Rattus norvegicus*, the kangaroo rat, *Dipodomys merriami*, and the guinea pig, *Cavia porcellus* (see Sawyer 1968). Arginine-vasopressin has also been identified in five strains of mice (*Mus musculus*), while in the Peru strain, lysine-vasopressin has been found (Stewart 1968).

The neural lobe of rats can be destroyed by cutting the supraoptico-hypophysial tract and allowing the peripheral tissue to degenerate. Such animals form large volumes of dilute urine which they are unable to concentrate (diabetes insipidus). Valtin and his colleagues (1962) have identified a strain (Brattleboro) of rats with hereditary diabetes insipidus due to the specific absence of vasopressin, although oxytocin is still present (Sawyer et al. 1964). The injection of vasopressin into such laboratory rats with diabetes insipidus results in a reduction of their urine volume and increases in its concentration. Laboratory rats are a mutant strain of the brown rat, *Rattus norvegicus*, which in its indigenous habitat in East Asia lived in wetlands along the banks of streams and rivers. Vasopressin has been identified in the plasma of a variety of other rodents. Apart from laboratory rats and mice, they include several species that live in the deserts of the United States and the Middle East, such as kangaroo rats (Ames and van Dyke 1952), jerboas, gerbils (El-Husseini and Haggag 1974; Baddouri et al. 1984) and spiny mice (Castel et al. 1974). The concentrations of vasopressin in the plasma of the desert species are usually maintained, under natural conditions, at far higher levels than those that occur in dehydrated laboratory rats (Dicker and Nunn 1957). Such concentrations of this hormone can, however, be reduced by hydrating the animals (Cole et al. 1963). Dehydration of laboratory rats results in a ten fold rise in the concentration of vasopressin in the plasma, an increase in its excretion in the urine and a depletion of its stores in the neurohypophysis (Dicker and Nunn 1957). Kangaroo rats, despite normally maintaining plasma levels of vasopressin that are far greater than in laboratory rats, maintain stores in their neurohypophyses that are five times greater (Table 3.2). Vasopressin is also released more readily in response to increases in plasma concentration in the kangaroo rats than in the laboratory rats (Stallone and Braun 1988). Mammals that normally live under xeric conditions appear generally to maintain larger stores of vasopressin in their neurohypophyses. There are also associated differences in their neural lobes, which may be larger (Fig. 3.4). Chronic water deprivation has also been observed to produce a hypertrophy of the neural lobe in rodents (Castel and Abraham 1969). The long-term regulation of vasopressin synthesis appears to have become adjusted to the everyday needs of xeric species.

The ability to limit urinary water loss may vary among species in relation to the dryness of the normal habitat (Table 1.6). Extrarenal water loss, however, predominates, and in white rats kept in the laboratory without water it is five times greater than the urinary loss, while in the kangaroo rat it is 2.5 times as great (see Dicker and Nunn 1957). Under such conditions, these rodents form hyperosmotic urine, which in the white rat may be nine times more concentrated than the plasma, and in the kangaroo rat 14 times that level (Schmidt-Nielsen 1964a). In the absence of neurohypophysial antidiuretic hormones, urine that is isosmotic to plasma is formed, so that the total water loss under such conditions would be expected to increase considerably. Urinary loss would then be more than twice as great as the extrarenal loss.

Table 3.2. Quantities of vasopressin in the neurohypophysis of rodents

	wt (g)	nmol kg^{-1} body wt
Mus musculus[a] (mouse)	20	7
Rattus norvegicus[a] (rat)	250	3.7
Aplodontia rufa[b] (Mountain beaver)	1000	0.8
Cavia porcellus[c] (Guinea pig)	400	3.2
Dipodomys merriami[d] (kangaroo rat)	40	18

[a] Follett (1963).
[b] Dicker and Eggleton (1964).
[c] Dicker and Tyler (1953).
[d] Ames and Van Dyke (1950).

Removal of the adrenal gland from laboratory rats results in death, usually within a few days. This results largely from a depletion of the sodium and an excessive accumulation of potassium, due to an inadequate integration of renal and colonic electrolyte excretion in the absence of the adrenal corticosteroids. Adrenalectomy has also been carried out in the kangaroo rat, *Dipodomys spectabilis*, with the same results, that are characteristic of the mammals (Cole et al. 1963). The desert gerbil, *Gerbillus gerbillus*, dies within 7 days of adrenalectomy, but this period can be extended by injecting the animals with cortisone and

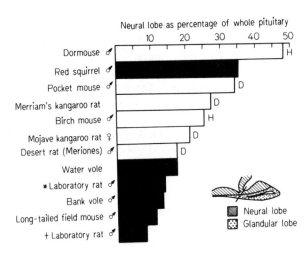

Fig. 3.4. Comparison of the relative size of the neural lobe in several species of rodents. *H* Hibernating; *D* desert-living species; ■ non-hibernating rodents of temperate zone habitat. (Howe and Jewell 1959. Redrawn from Enemar and Hanström 1956)

feeding them a diet with a high content of sodium chloride (Burns 1956). The adrenal cortex of laboratory rats secretes aldosterone and corticosterone, but little, if any, cortisol. Bush (1953) found that the ratio in the secretion of cortisol / corticosterone (compounds F / B) was less than 0.05. The banner-tailed kangaroo rat, on the other hand, also forms cortisol and has an F / B ratio of about 1.4. Bush, after comparing F / B ratios in a number of species of placentals which normally have a differing diet and way of life, concluded that such differences are probably genetically determined variations of the biosynthetic pathways in the adrenal cortex, and cannot be related to differences in the role of the corticosteroids.

Plasma 17-hydroxyxcorticosteroid concentrations (principally consisting of cortisol and corticosterone) have been measured in two rodents, a jerboa, *Jaculus jaculus*, and a gerbil, *Gerbillus gerbillus*, that live in arid areas in the Middle East (Haggag and El-Husseini 1974). The concentration of these corticosteroids was greater in the hot summer months than in the winter, and increased when the animals were maintained on dry diets. Their release in response to ACTH or dehydration was greater in the jerboa, which is a more desert-adapted species than the gerbil. Such secretion of corticosteroids may be related to stress resulting from the hot conditions and dehydration. The renin-angiotensin system in mammals can be activated by dehydration (Blair-West et al. 1983). This response appears to occur following a decrease in blood volume and results in a rise in the concentration of angiotensin II in the plasma. This hormone may initiate compensatory responses including a secretion of aldosterone, peripheral vasoconstriction, a stimulation of thirst, drinking and sodium appetite. Such effects are well known in laboratory rats. However, species of rodents that live under xeric conditions, including kangaroo rats, Mongolian gerbils and Australian hopping mice, apparently do not appear to utilize the RAS to help maintain their body fluids during dehydration (Wright and Harding 1980; Weaver et al. 1994). The drinking response to injected angiotensin II in such desert rodents is also often poor (Kobayashi et al. 1979). Species that were found to be insensitive to administered angiotensin II include the house mouse, Syrian hamsters, chipmunks and Mongolian gerbils. Wild rabbits, captured in Australia, also failed to exhibit a drinking response to injected angiotensin II (Denton et al. 1985). (However, sodium appetite was stimulated.) Marsupial mice, *Antechinus stuartii*, normally do not drink in nature, and also fail to exhibit a drinking response to angiotensin II (Blair-West et al. 1983). Nevertheless, despite being unable to detect a role for the RAS in response to dehydration, the normal, basal, levels of angiotensin II in the plasma were found to be much higher in kangaroo rats and Mongolian gerbils than in laboratory rats (Wright and Harding 1980). Dehydration results in a five fold increase in the concentration of angiotensin II in the plasma of laboratory rats, but it is still then only $50\,pg\,ml^{-1}$ compared to $110\,pg\,ml^{-1}$ in water-deprived kangaroo rats. The possible role of such high concentrations of angiotensin II in the plasma of kangaroo rats invites further investigation.

The sea provides an ecological niche for many mammals, including the Cetacea (whales and dolphins), Sirenia (manatees and dugong) and several families of the Carnivora (seal, sea lions and sea otters). There is even a fishing-bat (*Pizonyx vivesi*) which, although not a regular swimmer, manages to subsist, without fresh water, on the fringes of a marine habitat (Carpenter 1968). Seals and sea lions spend prolonged sojourns on land during their breeding seasons, though such visits by sea otters are quite brief. The other orders of marine mammals are entirely aquatic, though some, such as the West Indian manatee (*Trichechus manatus*), live mainly in fresh water. (Ortiz et al. 1999). Indeed, some species live entirely in fresh water, such as dolphins that live in the Amazon, Ganges and Yangtze rivers.

The mammals that live in the sea clearly do so without access to fresh drinking water. Many species eat fish, which have a relatively low salt content. However, many whales and the sea otters eat invertebrates that are isosmotic to sea-water, while the sirenids are herbivores. Whales and seals live for prolonged periods, during migrations and breeding, by utilizing large reserves of body fat. The milk of marine mammals has a high fat content. Such dietary observations and measurements of the salt and urea content of their urine indicate that feeding animals can gain substantial amounts of preformed water, as well as metabolic water, from the consumption of fish. Invertebrates and sea grasses provide a more frugal source of such water due to their relatively high salt content. Fasting marine mammals and their suckling young appear to gain substantial amounts of metabolic water from the metabolism of body fat. (Fat has been called a water storage depot in such animals.) Seals and manatees are known to drink fresh water when it is available. However, whether marine mammals regularly drink sea-water is uncertain. Measurements of the salt contents of their urine and their water turnover rates suggest that such behaviour (mariposia) is rare. However, the drinking of sea-water has been observed in seals (Gentry 1981; Skalstad and Norday 2000), sea otters (Costa 1982), fish-eating bats (Carpenter 1968) and dolphins (Hui 1981). Seals and dolphins may drink some sea-water "accidentally", such as during feeding and in response to stress, but it is generally considered unlikely that they normally indulge in this behaviour in order to maintain a positive water balance. It has been suggested that the consumption of small amounts of sea-water may "expand the urinary osmotic space" and so facilitate the excretion of urea (Carpenter 1968; Hui 1981; Costa 1982). It may also help them to maintain mineral balance (Skalstad and Norday 2000). Urinary salt concentrations in marine mammals usually do not exceed that of sea-water (Smith 1936; Bentley 1963), but such concentrations have sometimes been observed (Carpenter 1968; Costa 1982). Gains of osmotically free water could occur on such occasions. Marine mammals appear to have a relatively impermeable skin. Small percutaneous movements of water, though not sodium, have been observed in dolphins (Hui 1981). Respiratory water losses are expected to be low in an aquatic environment. It is generally considered that marine mammals can usually maintain their water balance without drinking (Krogh 1939; Smith 1951).

The kidney appears to be the principal organ for regulating water and salt excretion in marine mammals. No extrarenal salt excretory mechanisms, such as exist in fish, marine reptiles and sea birds, have been identified. Hyperosmotic urine appears to be habitual in marine mammals except in manatees when they enter fresh water (Ortiz et al. 1998). It has been observed that the kidneys of marine mammals are very large as compared to those of terrestrial species

(Beuchat 1996). The physiological significance of this observation is not known but possibly reflects their habitual excretion of large amounts of salt. Seals (Smith 1951) and dolphins (Malvin and Rayner 1968) have been observed to have an especially labile GFR and renal plasma flow. This capacity may be principally related to the ability to shunt blood from the viscera to the brain during diving. However, large increases in GFR and renal plasma flow have been observed in seals and dolphins following feeding. Such changes could be facilitating salt excretion.

Sporadic observations have been made on the possible roles of vasopressin, renin and adrenocorticosteroids in regulating renal water and salt excretion in marine mammals. Arginine-vasopressin and oxytocin have been identified in the neurohypophyses of fin-back whales (Chauvet et al. 1963). Vasopressin has also been measured in the pituitaries of dolphins (*Tursiops truncatus*) (Malvin et al. 1971). Cortisol, corticosterone and aldosterone have been identified in the adrenal cortex of whales (Race and Wu 1961; Carballeira et al. 1987) and the California sea lion (DeRoos and Bern 1961).

Vasopressin has been measured in the plasma of grey seals (Skog and Folkow 1994), elephant seal pups (Ortiz et al. 1996), West Indian manatees (Ortiz et al. 1998) and bottlenose dolphins (Ortiz and Worthy 2000). Early attempts, using bioassays, to measure it in the plasma of dolphins and a killer whale found only concentrations that were barely detectable (Malvin et al. 1971). Injected vaso-pressin has been observed to decrease the urine volume of harbour seals (Bradley et al. 1954). The infusion of mannitol into grey seals was found to increase the plasma concentrations of vasopressin (Skog and Folkow 1994). The plasma con-centration of vasopressin has also been shown to increase in the grey seals when they are deprived of food and water for 5 days. The levels of this hormone in plasma rise in manatees after they enter sea-water (Ortiz et al. 1998). Thus, vaso-pressin could be contributing to the regulation of urine flow in at least some marine mammals. Its site of action in the kidney is unknown. However, the administration of this hormone had no effect on the GFR in seals, suggesting that a renal tubular response is occurring (Bradley et al. 1954).

Plasma renin activity increases during fasting in dolphins and decreases after feeding them a sodium-enriched diet (Malvin et al. 1978). In West Indian mana-tees the concentration of renin in the plasma is lower when they are in sea-water than fresh water (Ortiz et al. 1998). The sodium concentration in their urine declined when they were in fresh water. Plasma renin activity has also been observed to increase in Northern elephant seal pups during their prolonged post-weaned fast (Ortiz et al. 2000). The higher renin activity was correlated with increases in the levels of aldosterone in the plasma in all these species. The plasma concentration of aldosterone in harp seals and hooded seals decreases when they are kept in sea-water and rises in fresh water (Skalstad and Norday 2000). These observations suggest that the renin-angiotensin system and the release of aldos-terone are responsive to the sodium levels in marine mammals.

Marine mammals appear to have retained the hormonal mechanisms for regulating renal water and salt excretion that were developed in their terrestrial forebears.

Chapter 4

The Birds

Birds are the most cosmopolitan of the vertebrates. The ability to fly confers on them a mobility that is matched in other vertebrates only by the bats. Some avian species have, however, lost the ability to fly, or do so to little effect. Flightless birds include the ostriches, kiwis, cassowaries, emus and rheas, while many other species like the domestic fowl are aviatorially inept. Some birds, like the penguins and auks, use their wings for propulsion under water. Birds thus may live a predominantly terrestrial or marine existence or they may also utilize the skies above such areas. Geographically, birds are found on the most remote oceanic islands, in dry continental desert regions, and in tropical and temperate areas where water is abundant. They can indeed live, and thrive, in a variety of osmotic environments.

The osmotic problems of birds are basically like those of mammals; their osmotic anatomy is similar, they occupy the same geographical regions and habitats and they are also homoiothermic. Their needs for water and salts are thus potentially comparable, but differences in their physiological and morphological characteristics may modify their respective requirements. Most birds are smaller than mammals, so that they more often experience the difficulties inherent in a large surface area to body weight ratio. The body temperature in birds is often in the region of 40 to 42 °C, which is 3 or 4 °C higher than that of most mammals. This reflects a metabolic rate that is higher than that of mammals, and so modifies heat exchanges with the environment. Flight results in metabolic burdens which are completely unfamiliar to mammals (except possibly the bats), and this is reflected in increased respiratory gas exchange, with an accompanying additional water loss. The mobility conferred by flying may, however, be a considerable osmotic advantage, as it allows birds to travel rapidly for long distances to suitable feeding and watering places. This may be a relatively local commuting, or it may take place between major geographical regions, in which seasonal changes in environmental temperature and available food and water may occur.

The birds exhibit a number of other physiological features relevant to their osmoregulation, that they share with either the mammals or their phyletic progenitors, the reptiles. Both the birds and mammals, in contrast to the reptiles, can form a hyperosmotic urine. However, in birds, the maximal concentrations are not as great as usually seen in mammals, but they conserve additional renal water by converting most of their catabolic nitrogen to uric acid. Most reptiles are similarly uricotelic. While in the mammals the kidney is the principal route for excretion of salts, some birds and reptiles can also secrete such solutes as hyperosmotic solutions from cephalic "salt" glands. Birds, like reptiles, but in contrast to mammals (except the monotremes), are oviparous, and it can be conjectured that this could influence the pattern of their osmoregulation. The rapid production of a clutch of eggs containing all of the water and salt necessary for an extended period of embryonic growth would seem to result in a more acute need for water and salt than embryonic development in utero. Tending and incubating such eggs temporarily restricts the movements of birds so that breeding can only occur in a place and time of adequate proximal supplies of food and water.

Erik Skadhauge provided a landmark monograph on the osmoregulation of birds (Skadhauge 1981). An excellent collection of works in progress was presented in 1989 by Hughes and Chadwick.

Birds have the same complement of osmoregulatory hormones as mammals, but their chemical structures in the two phyletic groups often differ. Their physiological roles in regulating water and salts, which mainly involve the kidneys and gut, are similar. However, there are also some notable differences in their functioning.

Like most species of vertebrates, the birds secrete two peptide hormones from their *neurohypophysis* (Table 4.1). Both of these differ in their structures from

Table 4.1. Neurohypophysial hormones in birds

	Vasotocin	Mesotocin*	V/M ratio	Vasotocin $m\mu\, mol\, kg^{-1}$ BW
Galliformes				
Domestic fowl	+[a]	+[a]	6.7	1.2[b]
Domestic turkey[c]	+	+	1.2	0.2
Coturnix japonica[d] (Japanese quail)	+	+	4.2	Approx. 2
Columbiformes				
Columba livia[b] (pigeon)	+	+	7	0.4
Anseriformes				
Domestic duck[e]	+	+		
Psittaciformes				
Melopsittacus undulatus[f] (budgerygah)	+			
Charadriiformes				
Larus canus[g] (gull)	Pressor and oxytocic activity		8.6	9.5
Passeriformes				
Zonotrichia leucophrys gambelii[h] (white-crowned sparrow)	+	+	3.5	

* Originally identified as oxytocin, but now known to be mesotocin (Acher et al. 1970; Acher 1996).
[a] Munsick et al. (1960); Chauvet et al. (1960).
[b] Heller and Pickering (1961).
[c] Munsick (1964).
[d] Follett and Farner (1966).
[e] Hirano (1966).
[f] Hirano (1964).
[g] C. Tyler quoted by Follett (1963).
[h] D. S. Farner and W. H. Sawyer; quoted by Sawyer (1968).

those present in most mammals, but they conform to the pattern of those secreted by the other tetrapods. They are 8-arginine-vasotocin (vasotocin, AVT) and 8-isoleucine-oxytocin (mesotocin, MT). Vasotocin functions as the antidiuretic hormone in birds, where it replaces vasopressin. It differs from the latter by a single amino acid substitution, isoleucine replacing phenylalanine at position 3 in the peptide. Mesotocin differs from mammalian oxytocin by the replacement of leucine by isoleucine at position 8. (It will be recalled that mesotocin is also present in some marsupials.) Mesotocin has negligible antidiuretic activity. Both hormones can contract the bird oviduct, but vasotocin is more potent. A physiological role for mesotocin has not been established in birds. Vasotocin is released into the circulation of birds in response to hyperosmotic concentrations in the plasma and also following stress, haemorrhage and during oviposition.

The principal secretions of *adrenocortical tissue* in birds are corticosterone and aldosterone (Fig 2.1). Both of these hormones can influence salt metabolism, but aldosterone is much more active. However, corticosterone is present at considerably higher concentrations in the plasma and may also contribute to the regulation of sodium and potassium in the body. The principal sites of such actions in birds are the kidneys and the large intestine. Corticosterone is considered to principally act as a glucocorticoid hormone and replaces cortisol, which usually regulates such functions in mammals. However, in some mammals, such as rats, mice and rabbits, corticosterone is also the main glucocorticoid hormone. Aldosterone and corticosterone are also the principal corticosteroid hormones in reptiles and amphibians. Secretion of aldosterone into the blood usually occurs when birds are provided with a diet that has a low content of sodium (Rice et al. 1985; Rosenberg and Hurwitz 1987). The release of corticosterone can occur in conditions that may result from stress. Some such circumstances may be related to osmoregulation. Such a release occurs in ducks provided with an excess of salt in their diet or in drinking water (Harvey et al. 1984; Klingbeil 1885). Dehydration also increases the release of corticosterone, as well as aldosterone, in ducks (Klingbeil 1985; Árnason et al. 1986). Hormones that influence the synthesis of adrenocorticosteroids appear, as in mammals, to include ACTH, angiotensin II and natriuretic peptides. However, precise information of the relative importance and respective roles of these hormones in regulating adrenocortical function in birds is not available.

Birds possess a *renin-angiotensin system*. The juxtaglomerular apparatus appears to be similar to that in mammals. It includes granulated renin-secreting cells and a macula densa. Renin promotes the formation of angiotensin, which can contribute to the regulation of corticosteroid synthesis, blood pressure and thirst. However, such functions have not been established in all the birds that have been studied. For instance: angiotensin II can promote corticosteroid synthesis in ducks, quail and turkeys, though apparently not in domestic fowl (Holmes and Cronshaw 1993). The structures of renin and angiotensin in birds differ from those in mammals. Renin from domestic fowl is specific to birds and interacts only with avian plasma angiotensinogen to produce angiotensin I (Nolly and Fasciola 1973). Angiotensin I in birds differs from that in mammals (cattle) and reptiles (turtles and alligator) by a single amino acid substitution at position 9 (Fig. 2.6). However, avian angiotensin II has a structure identical to that in cattle and reptiles. *Natriuretic peptides* have been identified in birds. They include the type-B (BNP-29) (Miyata et al. 1988) and type-C (CNP-22) peptides (Takei 1994). As in mammals, they can inhibit the synthesis of aldosterone in birds (Gray et al. 1991).

Prolactin is released into the blood of birds in response to a hyperosmotic plasma. However, a role for this hormone in avian osmoregulation is controversial (Harvey et al. 1989). Avian prolactin differs in its structure from that of other vertebrates, but when administered to other species it can still elicit characteristic actions, such as the secretion of milk by mammary glands. In pigeons it stimulates the secretion of "pigeon's milk" from the crop sac epithelium. In some fish, prolactin can influence the movement of salt and water across various epithelia, including the gills (see Chap. 7). Such observations appear to have encouraged a search for analogous actions in the kidneys and gut of birds.

1 Water Loss

Water loss in birds takes place by evaporation, and in the urine, faeces and nasal salt gland secretions (Maclean 1996). Increased amounts of water are also used periodically for egg laying, and in some species can be disgorged from the crop sac to feed the young. The contribution of each of these avenues to the total water loss can vary, but principally depends on the prevailing climatic conditions. Evaporative water loss is usually predominant. It occurs from the respiratory tract and the skin. Water loss in the urine and faeces may vary from the equivalent of 2 to 20% of the body weight each day. However, it usually represents about 5% of the body weight in a day. Nasal salt gland secretion can be substantial, especially in the marine birds. As these secretions have high osmotic concentrations, they usually provide a net gain of water from imbibed salt solutions.

Few birds can survive indefinitely on a regimen of dry food, such as plant seeds, with no water to drink. The water gained from the metabolism of such diets and the small amounts of free water present usually do not compensate for other losses. Under controlled laboratory conditions with moderate temperatures and relative humidity, and freed from the need to actively search for food, budgerygahs and zebra finches have been observed to survive prolonged periods without drinking water. However, under natural conditions, birds on such a regimen progressively lose weight and eventually would die from dehydration. California quail without water to drink die after losing about 50% of their body weight, and the mourning dove after a loss of 37%. The house finch cannot survive a loss exceeding 15% of its weight (Bartholomew and Cade 1963). The reasons for such differences are not clear, but they may be related to the rates of dehydration. The house finch loses water five-times as rapidly as the California quail and the mourning dove at an intermediate rate. During dehydration, the haematocrit is initially maintained at relatively constant levels in some birds, such as galahs and feral chickens (Roberts 1991a,b). This behaviour is reminiscent of that which occurs in some desert mammals in which the blood volume is maintained at the expense of intercellular and intracellular fluids.

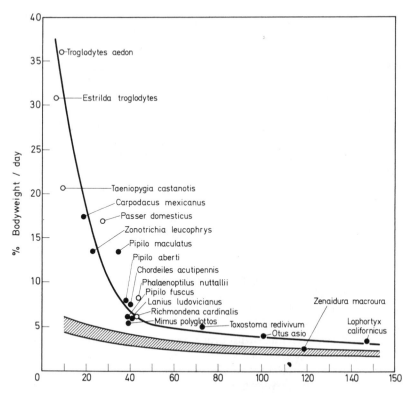

Fig. 4.1. The relation of weight-relative evaporative water loss to body weight of birds at ambient temperatures near 25 °C. The *cross-hatched curve* indicates the theoretical relation between body weight and the production of metabolic water by birds in basal conditions utilizing exclusively carbohydrates (*upper boundary*) or exclusively proteins (*lower boundary*). The values for fats are intermediate. (Bartholomew and Cade 1963)

1.1 Evaporative

Evaporative water loss in birds is inversely related to their body weight (Fig. 4.1). This observation reflects their relative surface areas and rates of metabolism. This water loss is increased at high body and environmental temperatures and when the water vapour content of the air is low.

Water loss occurs from the respiratory tract as an unavoidable consequence of the need to exchange oxygen and carbon dioxide. The body temperature of birds is normally about 2 °C higher than that of most mammals (Prinzinger et al. 1991). Their rates of metabolism are also usually higher. This avenue of water loss can be substantial. Small birds, like small mammals, can utilize a countercurrent heat exchanger in their nasal passages to cool expired air to a temperature that is less than deep body temperature (Schmidt-Nielsen et al. 1970). The water losses in the expired air are thus reduced, though the magnitude of the saving is not as great in small birds as in mammals. Evaporation of water from the avian respiratory tract, as in mammals, increases in response to homoiothermic need at elevated ambient temperatures. This process can be seen to occur when the birds pant. Birds can also accelerate movements of the hyoid apparatus in the throat to promote evaporation. This activity is called gular flutter.

Birds lack sweat glands and so it was once thought that they lose little water by evaporation from their skin. However, direct measurements of such water loss from the skin of a variety of birds have shown that over 60% of their total water loss can occur by this route (Bernstein 1971; Lasiewski et al. 1971). Such cutaneous water loss can be regulated and increase in response to the need to regulate the body temperature. Inhibition of the homoiothermic response of the respiratory tract of mourning doves results in a rapid increase in cutaneous evaporation (Hoffman and Walsberg 1999). Dehydration has been shown to decrease evaporation from a number of birds, including the skin of zebra finches (Lee and Schmidt-Nielsen 1971). The mechanisms involved in regulating such changes in cutaneous evaporation are uncertain. In long-term hot conditions it may involve changes in skin structure, such as the laying down of lipids that retard water loss (Peltonen et al. 1998). Vascular changes could also be contributing. In mammals, the circulation is under the control of the sympathetic nervous system and α-adrenergic receptors, which mediate vasoconstriction. However, in heat-acclimated pigeons the situation appears to be more complex and may involve β-adrenergic receptors both peripherally and in the brain (Ophir et al. 2000). Blockade of such receptors with drugs such as propranolol increases cutaneous evaporation. The details of such a control system remain to be elucidated. However, it may be relevant to observe that propranolol, via its β_1-adrenergic blocking effect, can prevent the activation of the renin-angiotensin system and the formation of angiotensin II. This hormone has a powerful vasoconstrictor action.

Evaporative water loss and metabolic rate in birds generally appear to be less in species that live in deserts than those from more temperate regions (Tieleman and Williams 2000). They reduce their evaporative water losses by utilizing a number of strategies. Behavioural adjustments are very important (Davies 1982; Dawson 1984). Most birds are diurnal in their activity so that under desert conditions, in contrast to many mammals, they cannot avoid the heat of the day. In such circumstances, birds may seek refuge in the shade of trees, bushes and rock crevices. Very few birds are fossorial. Some birds avoid heat by soaring to heights where the temperatures are lower. Some birds can enter states of torpor, but this behaviour usually occurs under cold conditions when a conservation of energy results. Corticosterone appears to be involved in the restoration of normal energy metabolism following such torpor in humming birds (Hiebert et al. 2000). Under hot conditions, birds may limit their activity and search for food to the cooler hours in the early morning and late afternoon. To facilitate evaporative cooling from the skin they can expose lightly feathered areas and spread their wings.

Non-evaporative heat exchanges involve conduction, radiation and convection. The net flow of heat is from the hotter to the cooler region. As described earlier, the body temperature of birds is generally somewhat higher than that of mammals. Their lethal temperature is generally about 5 to 6 °C higher than the normal body temperature. They can utilize this latitude for a controlled hyperthermia, which provides a strategy to delay the need for evaporative cooling (Tieleman and Williams 1999). This process is most effective in small birds, when savings of as much as 50% of their total evaporative water loss can occur. Such birds, like mammals, may undergo alternate periods of nocturnal hypothermia and hyperthermia in the daytime (King and Farner 1964). Flying in birds is associated with considerable increases in metabolic rate and, potentially, evaporative water loss. Budgerygahs flying in a wind tunnel at 20 °C have an evaporative water loss which is five-times as great as at rest (Tucker 1968). At 36 °C, water loss during flying was three-times as great as at 20 °C. Water loss at the latter temperature was equivalent to about 7% of their body weight each hour. Under natural conditions, birds can make long, often uninterrupted flights, especially during periodic breeding migrations. They apparently do not arrive at their destinations in a dehydrated sate. Measurements of evaporative water loss in starlings flying in a wind tunnel at different environmental temperatures indicate that radiation and convection are the major mechanisms for heat loss during flying (Torre-Bueno 1978). By ascending to higher altitudes, where the air is colder, birds can apparently maintain their hydration with water produced by the oxidation of body fat. It appears that long-distance migrants fly at sufficiently high altitudes to dissipate heat principally by convection.

1.2 Urinary and Faecal

Birds lack a urinary bladder, so that the urine and faeces are usually accumulated together in the cloaca and rectum. Except in ostriches, in which the urine is stored separately, the faeces and urine are voided at the same time. The composition of the urine can be modified during such storage in the posterior large intestine as a result of transmucosal movements of water and solutes. Due to their intermixing separate measurements of urinary and faecal water and salt losses are usually not feasible under natural conditions.

The excreta, urine plus faeces, of normally hydrated budgerygahs and zebra finches contain about 80 g water $(100 g)^{-1}$ wet weight. When drinking water is restricted, this value decreases to nearly 50 g $(100 g)^{-1}$ (Cade and Dybas 1962; Calder 1964). In these birds, the total water loss each day in the urine and faeces is about 2% of their body weight in a day. The water content of the excreta of other birds has generally been found to be similar, though even lower concentrations of water have sometimes been reported (Maclean 1996). Whether birds utilize osmotic and hydraulic forces in the crypts of intestinal villi to dehydrate the faeces, as occurs in mammals (see Chap. 1), is unknown.

Urinary water loss is influenced by the water and salt content of the diet, the availability and composition of drinking water and the prevailing environmental temperature and relative humidity. It may also be related to the concurrent activities of the gut and, if present, nasal salt glands. The volume of the urine will depend on the amount of solutes to be excreted, including salts and uric acid, and the state of the birds' hydration. Like mammals, birds can form a hyperosmotic urine, but the concentrations that are achievable are usually not as great (Table

1.6). The urine/plasma concentration ratio usually does not exceed 2.5 (Poulsen and Bartholomew 1962). However, there are exceptions, and in the savannah sparrow this ratio can be 4 to 5 (about 2000 mosmol l^{-1}). This small bird lives in salt marshes and can drink brackish water. The ability of birds to form hyperosmotic urine reflects the presence of mammalian-type nephrons that form a countercurrent multiplier system (Poulsen 1965). However, these nephrons, which are present in the renal medulla and have a loop of Henle, make up only about 20 to 30% of the total of those present in the bird kidney (Dantzler 1997). The remaining cortical nephrons are smaller and lack a loop of Henle. They are described as reptilian-type nephrons and do not contribute to the formation of the hyperosmotic urine. Although they cannot usually concentrate their urine as much as mammals, birds reduce their urinary water losses by excreting waste nitrogen as uric acid. It has been estimated that 1 g of nitrogen then only requires 10 ml of water for its excretion (Smith 1951). In contrast, a mammal, which utilizes urea, would require as much as 50 ml of water for this excretion. The uric acid in the urine forms smooth spheres that contain Ca^{2+} and K^+. With the aid of proteins, they form a colloidal suspension that exerts a negligible osmotic pressure while maintaining the patency of the renal tubules (Braun 1999a). The protein present in the urine, and much of the uric acid, is not excreted but reprocessed and absorbed from the large intestine (see later).

1.3 Renal Water Conservation and the Neurohypophysis

The kidneys of birds adjust the volume and concentration of the urine in response to their need to maintain normal bodily hydration. The administration of water results in the formation of an appropriate volume of dilute urine. When water is restricted, a smaller volume of more concentrated urine is formed. In mammals, changes in the volume of the urine principally depend on regulating the process of its reabsorption from the renal distal tubular-collecting duct system. However, in birds, such renal conservation of water involves changes in both the glomerular filtration rate (GFR) and tubular water reabsorption. This involvement of glomerular filtration is similar to that which occurs in reptiles and most other non-mammalian vertebrates. In birds, changes of the GFR involve variations in the numbers of functioning glomeruli rather than the filtration across individual, single, glomeruli. Such a process is called glomerular intermittency or recruitment (Dantzler 1997). The reptilian-type nephrons are principally involved in this process. When hyperosmotic saline is administered to desert quail, the reptilian-type nephrons are completely closed down while the functioning of the mammalian-type ones is little changed (Braun and Dantzler 1972). Dehydration has been found to decrease the GFR in all birds that have been studied. The inhibition varies from 20% in domestic fowl to 58% in starlings (Roberts and Dantzler 1989). It is the predominant mechanism for antidiuresis in desert-dwelling Gambel's quail (Williams et al. 1991a).

Neurohypophysectomy in the domestic fowl results in the formation of large volumes of dilute urine (Shirley and Nalbandov 1956). This condition of experimental diabetes insipidus is due to the lack of the avian antidiuretic hormone vasotocin. In 1933 E. K. Marshall and his colleagues (Burgess et al. 1933) injected commercial preparations of mammalian vasopressin (ADH, pitressin) into domestic fowl and observed an antidiuretic response that was due to *both* an increased reabsorption from the renal tubule and a decrease of the GFR. In 1960, it was established that the antidiuretic hormone in birds was, in fact, vasotocin and it elicited this same dual effect on the kidney (Munsick et al. 1960). Erik Skadhauge in 1964 carefully analyzed the antidiuretic response to vasotocin in the domestic fowl. He injected small doses of this hormone into the renal portal veins so that only the tubules were exposed to effective concentrations of the peptide. An antidiuretic effect was observed. Much larger doses of vasotocin were required to promote the glomerular response. The administration of different doses of vasotocin, over the physiological range, confirmed that in domestic fowl the renal tubular response was the primary one (Stallone and Braun 1985). However, in kelp gulls it has been observed that the glomerular response shows a sensitivity to vasotocin similar to the tubular one (Gray and Erasmus 1988a). These gulls live in coastal areas of South Africa where osmotic stresses are expected to be greater than in domestic fowl, which originated in tropical jungles. The greater emphasis on changes in the GFR in the gulls in response to vasotocin may reflect an adaptation to more arid conditions. The glomerular response to vasotocin is due to its vasoconstrictor effect on the afferent glomerular arteriole (Braun 1976).

The role of vasotocin in regulating water excretion in birds was initially inferred from changes in its storage in the neurohypophysis and the observed antidiuretic effects of its administration in low doses. Giving pigeons 0.5 M sodium chloride solutions to drink, or depriving Japanese quail of drinking water, resulted in an 80 to 90% decline in the amounts of vasotocin stored in the neurohypophysis (Ishii et al. 1962; Follett and Farner 1966). The presence of vasotocin in the blood, at appropriate concentrations, was demonstrated only much later, following the introduction of radioimmunoassays for the hormone. The release of vasotocin into the blood of domestic fowl was shown to occur incrementally over a range of changes in plasma osmotic pressure of physiological magnitude (Stallone and Braun 1986a,b). In contrast to mammals, changes in plasma volume, including haemorrhage, were much less effective stimuli for such hormone release.

The changes in osmotic concentration in the plasma that promote release of vasotocin have been compared in a number of birds. There are large differences that appear to be related to the normal osmotic needs of the birds in their natural habitats. The results have been summarized by Roberts (1991b) and are shown in Table 4.2. The house sparrow displays the greatest osmotic sensitivity, releasing vasotocin in response to the smallest change in the concentration of the plasma, while the galah is the least sensitive. Feral "domestic" fowl living on an island off the Australian coast were found to be more sensitive than their more domesticated relatives. The galah is a parrot that is well adapted to life in the arid regions of Australia. Possibly its needs for vasotocin have been reduced and the kidney has evolved a greater sensitivity to this hormone. The concentration of vasotocin in the plasma of pigeons has been measured following 5 h of continuous flight.

Table 4.2. Osmotic sensitivity of release of vasotocin (AVT) into the plasma of birds

	pg AVT ml^{-1} per mosmol kg^{-1} water increase in concentration
Domestic fowl[a]	0.5, 0.28
Feral domestic fowl[a]	1.75
Kelp gull[b]	1.13
Cape garnet[b]	0.36
Jackass penguin[b]	0.44
Pekin duck[c]	0.39
Galah[d]	0.16
House sparrow[e]	4.4
Ostrich[f]	0.54

[a] Roberts (1991b). [b] Gray and Erasmus (1988b). [c] Gray and Simon (1983). [d] Roberts (1991a). [e] Goldstein and Braun (1988) calc by[d]. [f] Gray et al. (1988).

They were found to increase to three to eight-times greater than those levels that were present before the flight (Giladi et al. 1997). This change was accompanied by a three-fold increase in urine concentration. The release of the hormone apparently reflected the effects of exercise and dehydration.

The effect of vasotocin on the GFR in birds involves a constriction of the afferent glomerular arteriole, but the nature of its action on the renal tubule is not clear. Studies on the effects of synthetic analogues of vasotocin on the kidneys of house sparrows indicate that the vascular effect is mediated by a V_1-type receptor (Goecke and Goldstein 1997). This receptor appears to be similar to the vascular receptor utilized by vasopressin in mammals. However, the renal tubular effect of vasotocin in birds does not appear to involve the mammalian-type V_2 receptor that mediates a change in permeability to water. Indeed, a direct renal tubular effect of vasotocin has been difficult to demonstrate in birds (Osono and Nishimura 1994; Goldstein 1995; Nishimura et al. 1996). Based on experiments using isolated renal tubules in Japanese quail, Nishimura and her collaborators have suggested that the effect of vasotocin may be indirect. It could, for instance, be due to a reduced delivery of fluid to the medullary collecting duct system.

2 Conservation and Excretion of Salts

The principal avenues for the excretion of salt in birds are the kidneys, gut and, in some species, nasal salt glands. The amounts of Na^+ and K^+ present in the urine and faeces are mainly regulated by the adrenocortical steroids. Their synthesis is controlled by ACTH, the renin-angiotensin system and, probably, natriuretic peptides. The composition of the urine is further modified, again by adrenocorticosteroids, after it has passed from the ureters into the cloaca. However, the secretory activity by the nasal salt glands in birds, despite some early protestations, does not appear to be directly influenced by hormones. The parasympathetic nervous system is the primary regulator of this process. The salt glands in some birds may indirectly interact with the physiological activities of the kidneys as salt-excreting organs. The diets and natural habitats of birds have a bearing on the relative importance of these organs on the regulation of salt in the body. The diet of seed-eating birds usually has a low content of Na^+ and a high one of K^+. Carnivorous birds obtain adequate Na^+ in their food. The diet of marine and estuarine birds also contains adequate Na^+ and may even provide an excess. Many such species feed on invertebrates and even drink sea-water. These birds usually utilize salt glands for the excretion of sodium chloride. Salt glands are also present in a number of non-marine birds, especially carnivorous species. However, their state of development varies and may not always be adequate to make a major contribution to salt excretion.

2.1 Kidneys

As the ability of the kidneys in birds to form a hyperosmotic urine is not as great as in mammals, their capacity to excrete excess salt by this route is somewhat limited. However, as described earlier, some birds, such as the savannah sparrow, can excrete sufficient amounts of salt in the urine to maintain osmotic balance on diets containing relatively large quantities of salt. The sodium chloride content of the urine of birds on low-salt diets indicates that concentrations as low as 5 mmol l^{-1} are attained. Indirect estimates of tubular reabsorption, involving measurements of the rates of glomerular filtration of solutes into the nephron and the amounts excreted in the urine, indicate the presence of a highly efficient reabsorptive process for salts and water. These processes include tubular absorption and secretion of K^+. Direct measurements of such tubular processes involving the collection of fluid in micropipettes and the perfusion of selected segments of the nephron are not widely available in birds. However, the available information indicates that the mammalian pattern of salt and water reabsorption in the proximal tubule, followed by a more selective modification of the filtrate in the distal tubule, also appears to be present in birds (Laverty 1989; Osono and Nishimura 1994; Nishimura et al. 1996). Nevertheless, there are some differences, such as involve the mechanism of action of vasotocin on water reabsorption (see above).

Adrenalectomy in ducks results in an excessive loss of Na^+ in the urine that reflects a deficiency of adrenocorticosteroids (Thomas and Phillips 1975). Normal function can be restored by the administration of these hormones. Similar effects of adrenalectomy have been observed in domestic fowl and pigeons (Parkins 1931; Miller and Riddle 1942). Aldosterone is the most active mineralocorticoid hormone in birds, but corticosterone also can exhibit substantial such activity. However, in domestic fowl, the former steroid is released more predictably in response to a depletion of body sodium (Rice et al. 1985; Rosenberg and Hurwitz 1987). Aldosterone, as well as corticosterone, has been observed directly, in vitro, to stimulate Na$^+$ transport across the large intestine of domestic fowl (see below). It seems likely that it is exerting a similar effect on the renal tubule. Aldosterone also promotes K$^+$ secretion in the urine of mammals, but such an effect has not been clearly demonstrated in birds (Simon and Gray 1991).

The infusion of isosmotic sodium chloride solutions into birds evokes a diuresis. In domestic fowl and ducks, this effect appears to be mediated by atrial natriuretic peptide. This hormone promotes an increased GFR and a decreased reabsorption of Na$^+$ from the distal renal tubule (Gray 1993, 1994). The latter effect appears to reflect a reduced synthesis of aldosterone, as observed in mammals, but a direct inhibition of Na$^+$ transport could also be occurring (Rosenberg et al. 1988).

2.2 Nasal Salt Glands

Excessive accumulations of sodium chloride are excreted solely by the kidneys of many birds. However, some species also utilize nasal salt glands for such excretion. Their secretions usually contain sodium chloride at far higher concentrations than are present in the urine (Table 1.5). (Avian salt glands, in contrast to those of some reptiles, do not appear to contribute significantly to K$^+$ excretion.) Concurrent water loss is accordingly much less than that in urine. Many marine and estuarine birds can subsist on a high salt diet and even fruitfully drink seawater by utilizing nasal salt glands for such excretion. These glands have been mainly studied in marine birds, but they have also been identified in many carnivorous species, especially Falconiformes, including hawks, eagles and vultures (Cade and Greenwald 1966). They are also present in the roadrunner, *Geococcys californianus*, which is carnivorous and lives in desert areas of the southwest USA. However, the nasal salt glands in such birds do not always appear to be in a highly functional condition. It has been suggested that their activity may be more important in their relatively immobile exposed nestlings than in the adults (Ohmart 1972). Nasal salt glands do not appear to be present in the Passeriformes (perching birds). Laboratory investigations have mainly utilized gulls, Pekin ducks and geese. The functioning of these glands was first described by Knut Schmidt-Nielsen and his colleagues in 1958. They studied cormorants, brown pelicans and herring gulls (Schmidt-Nielsen and Fänge 1958; Schmidt-Nielsen et al. 1958; Fänge et al. 1958). The administration of hyperosmotic solutions of sodium chloride rapidly evoked a voluminous secretion from the nasal glands of these birds. Hyperosmotic sucrose, but not urea or glucose, was also effective, suggesting that the response involved osmoreceptors. Such receptors have been identified in the hypothalamus (Gerstberger et al. 1984). Whether or not volume receptors that respond to an expansion of the extracellular fluid space are also involved has been controversial. A response to the infusion of isosmotic solutions has been described in Pekin ducks and Elder ducks (Hammel et al. 1980; Bøkeness and Mercer 1995; Bennett et al. 1997). Volume receptors appear to be present in the heart or are associated with the large blood vessels. However, the response to extracellular volume expansion is not as great as that to osmotic stimuli. It may even be an indirect effect due to a lowering of the threshold for the stimulation of the osmoreceptors.

Secretion by nasal salt glands in birds can be abolished by severing their parasympathetic nerve supply (Schmidt-Nielsen 1960). Pharmacological blockade can also be achieved by administering atropine, which blocks acetylcholine (muscarinic) receptors at the nerve junctions. General anaesthesia also inhibits secretion. The secretory process appears to be initiated in the brain in the region of the hypothalamus. It is accompanied by a large increase in the blood flow to the nasal gland. The administration of adrenaline inhibits secretion, apparently by its vasoconstrictor action.

Vasoactive intestinal peptide (VIP) is a neuropeptide that was first identified in the gut of mammals. It has since been found in association with other neurotransmitters at many nerve endings. VIP has been identified in the terminals of the parasympathetic nerves supplying the salt glands of Pekin ducks (Lowy et al. 1987). The infusion of VIP into these ducks can induce secretion from the nasal glands. Normally, VIP appears to be released, concurrently with acetylcholine, from the nerves supplying the salt glands. It may act by enhancing the local blood supply and even potentiate the response to acetylcholine (Shuttleworth and Hildebrandt 1999).

Hormones do not appear to have a direct role in the functioning of the nasal salt glands in birds. However, experimental manipulation of the endocrine system and the administration of some hormones can influence their secretion. Such effects have generally been described as indirect, secondary or modulating (Butler et al. 1989; Shuttleworth and Hildebrandt 1999). Adrenalectomy in ducks was found to inhibit the responsiveness of the salt glands, and their function could be restored by the administration of corticosterone. However, this hormone deficiency appears to result in a general debilitation of the birds and the blood supply to the salt glands. The administration of vasotocin is without effect on secretion, and a possible action of prolactin is considered to be of doubtful significance. The injection of angiotensin II into Pekin ducks and kelp gulls inhibits salt gland secretion (Hammel and Maggert 1983; Wilson et al. 1985; Gray and Erasmus 1989a). This effect appears to be mediated in the hypothalamus as it is not seen following direct stimulation of the gland with methacholine (an analogue of acetylcholine) (Butler 1999). Whether angiotensin II has a physiological role in modulating the initiation of the secretion of the salt glands remains to be established. Binding sites, possibly receptors, for melatonin (normally secreted by the pineal gland) have been identified in the salt glands of Pekin ducks (Ching et al. 1999). The infusion of melatonin delays the secretory response and decreases its rate. The administration of atrial natriuretic peptide into ducks has been shown to increase secretion from the salt glands (Schütz and Gerstberger 1990). Specific binding sites (receptors?) for ANP have also been identified in this gland. It was suggested the ANP acts as an "emergency hormone" and may facilitate secretion following a rapid expansion of the extracellular fluid space.

The size of nasal salt glands appears to be related to the feeding habits of some birds. If young gulls, *Larus glaucescens*, and ducklings are provided with salt solutions to drink, the salt glands enlarge compared to birds provided with fresh water (Holmes et al. 1961; Schmidt-Nielsen and Kim 1964). Birds that inhabit marine environments also have larger salt glands than those species that habitually drink fresh water (Technau 1936). The increase in the size of the salt glands of eider ducks adapted to drinking sea-water has been shown to be due to an increase in the size, but not the number, of the cells (Bøkenes and Mercer 1998). This hypertrophic response, like secretion, is initiated by stimulation of the parasympathetic nerve supply to the gland (Shuttleworth and Hildebrandt 1999). Local growth factors could ultimately be involved in this response.

3 Conservation of Water and Salt by the Cloaca and Large Intestine

Birds possess a cloaca consisting of three chambers, the proctodaeum, urodaeum and coprodaeum. The ureters usually open into the urodaeum. (In ostriches they open into the coprodaeum.) The large intestine, or colon, opens into the coprodaeum. One or two caecae open into the large intestine of many birds. This occurs at its junction with the small intestine. These "rectal caecae" are blind sacs, which are lined by intestinal mucosa cells. They are often quite large in herbivorous and omnivorous birds, but may be small, vestigial or absent in carnivorous and some granivorous species, such as the budgerygah. Morphologically, these caecae are considered to be part of the large intestine. The organ complex of the large intestine and cloaca is often referred to as the lower intestine. It provides a fermentative chamber for nutrients and the recycling of nitrogen derived from urinary uric acid and protein (Braun 1999b). The contents of the lower intestine can be mixed longitudinally so that urine may pass in a retrograde direction, by reverse peristalsis, into the caecae. This process appears to be influenced by the concentration of the urine and could involve neural, and even hormonally controlled, mechanisms. This passage of urine into the large intestine was first observed in the domestic fowl using radiographic contrast media and X-ray examination (Akester et al. 1967; Nechay et al. 1968). Skadhauge (1968) observed traces of uric acid, from the urine, far up into the large intestine and caecae of domestic fowl and ducks.

The possibility that changes in the water and salt content of the urine may take place after it passes into the cloaca was first investigated in domestic fowl in 1942 by Hart and Essex. They found that surgical transplantation of the ureters to the exterior, thus bypassing the cloaca, resulted in an additional dietary need for salt. Dicker and Haslam in 1966 also demonstrated that such birds drank twice the quantity of water as normal ones. These observations suggested that additional losses of salt and water were occurring following the surgical by-passing of the cloaca. This loss could be due to a failure in their reabsorption from the cloaca and large intestine. A reabsorption of sodium and water has since been demonstrated both in vivo and in vitro from the coprodaeum, large intestine and caecae. Such observations have been made on domestic fowl and a variety of more exotic birds (Skadhauge 1967; Bindslev and Skadhauge 1971; Chosniak et al. 1977; Thomas et al. 1980; Rice and Skadhauge 1982; Laverty and Skadhauge 1999). The lower intestine of birds has been described as an integrator of renal and intestinal excretion (Thomas 1982; Braun 1999b). In domestic fowl, the caecae make the greatest contribution to the reabsorption process, followed by the large intestine and then the coprodaeum.

Reabsorption of Na^+ from the avian lower intestine occurs as a result of its active transepithelial transport. An electrical potential difference (PD), lumen usually electronegative, and transmural solute concentration gradients are generated by this process. The latter can influence osmotic movements of water while the PD can affect the movements of other ions, including Cl^- and K^+. At least two active Na^+ transport mechanisms have been identified in the avian lower intestine. There is a linked cotransport of either amino acids, glucose or acetate to Na^+. There is also an amiloride-inhibitable transport involving specialized epithelial sodium channels (ENaC). The latter process can be increased, as in many other tetrapod vertebrates, by aldosterone (Thomas and Skadhauge 1979; Thomas et al. 1980; Grubb and Bentley 1987, 1992). A diet with a low content of sodium also increases such Na^+ transport in the domestic fowl. As described earlier, such a regimen is known to increase the secretion of aldosterone in these birds. Aldosterone has also been shown to stimulate the secretion of K^+ by the large intestine of the domestic fowl (Thomas and Skadhauge 1979, 1988, 1989).

The possible actions of other hormones on ion and water movements across the avian large intestine have also been investigated (Laverty and Skadhauge 1999). Vasotocin has no effect on osmotic water transfer (Skadhauge 1967). Ovine prolactin has been shown, in vitro, to decrease Na^+ and water absorption from the large intestine of the domestic fowl (Morley et al. 1981). This hormone is released in response to dehydration in these birds (Árnason et al. 1986). A possible physiological role for such an effect of this hormone has not been established. Corticosterone is normally present in the plasma of domestic fowl in concentrations which, in vitro, are sufficient to stimulate Na^+ transport across the caecum (Grubb and Bentley 1992). However, it is possible that its local concentration, in vivo, in the region of its effectors is reduced by 11β-hydroxysteroid dehydrogenase. Aldosterone is not inactivated by this enzyme so that its effect would then be expected to predominate. Thus, in mammals its contribution to a specific mineralocorticoid-mediated effects is unaffected. It is possible that corticosterone contributes to the maintenance of basal levels of Na^+ transport in such avian tissues (Grubb and Bentley 1989; Árnason and Skadhauge 1991).

The functioning of the cloaca and large intestine of birds may be integrated with that of the nasal salt glands (Schmidt-Nielsen et al. 1963). Birds can secrete sodium chloride from the salt glands in much higher concentrations than those in the urine. Salt in the urine may be reabsorbed from the lower intestine and be subsequently secreted as a more concentrated solution by the salt glands. However, attempts to confirm this ingenious hypothesis by direct measurements has yielded conflicting results (Goldstein et al. 1986; Thomas 1997; Hughes et al. 1999). Such a recycling of salt and water could be occurring in some birds. It is interesting to observe that ligation of the caecae in ducks undergoing acclimation to drinking sea-water results in a decrease in the size of their salt glands (Hughes et al. 1992). It appears that they do not then detect or respond to the additional salt in their diets. A more direct interaction of salt gland and kidney function could also arise from a controlled increase in the renal tubular absorption of excess salt. This salt may then be preferentially excreted by the salt glands.

4 Sources of Water and Salts

4.1 Food

Salts and water are obtained from the food of birds. The adequacy or excess of such accumulation will depend on the diet. A succulent herbivorous diet may supply a bird with all the water and salts it needs. However, in some geographic areas, the sodium content of such vegetation may be low. Halophytic plants are common in some areas, but their contribution to the diets of birds does not appear to have been recorded. Nectarivorous birds usually have an excess of water, which can result in a persistent diuresis. Avian carnivores and insectivores generally obtain adequate salts and water from their food. However, many birds, including species that live in dry desert regions, subsist on air-dried seeds. This food contains little sodium and only about 10% preformed water. Water formed by the oxidation of fats, carbohydrates and proteins (metabolic water) makes an important contribution to the needs of granivorous birds. Birds gain a little more water than mammals from the oxidation of proteins. This difference reflects the ultimate formation of uric acid instead of urea. Birds produce 0.499 g water g^{-1} protein oxidized compared to 0.396 g water g^{-1} protein in mammals. Cade and Dybas (1962) have calculated that commercial birdseed yields water of oxidation that is equivalent to 45% of its weight. Estimates are similar for the seeds that birds eat in the wild. However, even under temperate conditions, this water is inadequate for the bird's normal needs. Some birds exhibit signs of an enhanced salt appetite and actively seek sources of salt. The red crossbill, *Loxia curvirostra*, has even been observed to eat rock salt (Dawson and Shoemaker 1965).

4.2 Drinking

Many birds, especially granivorous species, depend on the availability of drinking water to maintain adequate bodily hydration. This need is inversely related to their body weight and is greatest in hot dry conditions and following flight, especially at low altitudes. In some instances, the amount of water required may be equivalent to 100% of their body weight each day (Bartholomew and Cade 1963). Drinking may be sporadic, but birds often exhibit regular patterns of drinking behaviour. In deserts they may frequent water holes only during the cooler early morning and evening hours (Fisher et al. 1972; Maclean 1996). Some birds drink large volumes of water rapidly, even, as observed in budgerygahs, "on the wing". Such behaviour appears to reduce the risk of predation. Birds often utilize their natural aerial ability to travel long distances to drink (Dawson 1984). Sandgrouse have been observed to travel 50 km to water. Young ostriches can store substantial volumes of water in their glandular stomach (Degen et al. 1994). This allows them to drink only once a day. Ostriches usually range within 20 to 30 km of water, but they have been observed 80 km away.

Sea birds, such as albatross, cormorants, penguins and gulls, drink sea-water and utilize their nasal salt glands for the excretion of the excess salt. Some terrestrial birds have brackish water available, and this can occur in desert environments and salt marshes (Maclean 1996). As such birds usually lack functional salt glands and have a limited ability to concentrate the urine, the acceptable concentrations of salt in such drinking water is usually less than 300 mmol l^{-1} (=50% sea-water). Given a choice of fresh or salt water, few birds will drink solutions with a greater concentration than 100 mmol l^{-1} (Robinson et al. 1980). Sometimes, under stringent laboratory conditions, zebra finches and savannah sparrows can be persuaded to drink solutions approaching the concentration of sea-water.

A decrease in the extracellular fluid volume, such as occurs during dehydration, results in hypovolemic thirst. As described in Chapter 1, this behaviour results from an activation of the renin-angiotensin system and the formation of angiotensin II. This hormone acts in the hypothalamus to induce drinking behaviour in many vertebrates, including birds (Kobayashi and Takei 1982, 1997). Dehydration has been shown to increase the concentration of this hormone in the plasma of various avian species, including Japanese quail (Okawara et al. 1985), ducks (Gray and Simon 1987), ostriches (Gray et al. 1988) and several marine birds (Gray and Erasmus 1988b). Some birds, especially omnivorous and granivorous species, readily exhibit drinking behaviour in response to the administration of low doses of angiotensin II. Carnivorous birds are generally less sensitive. Birds that have evolved in dry desert conditions are also generally less responsive to the administration of angiotensin II (Kobayashi 1981; Kobayashi and Takei 1997). Three species of Australian parrots belonging to the genus *Barnadius* display such differences in sensitivity to this hormone. The twenty-eight parrot lives in temperate areas, while the Port Lincoln parrot occupies arid regions. However, it is considered to have only recently evolved from a temperate-living ancestor. Both of these parrots readily drink in response to administered angiotensin II. In contrast, the mallee ringneck parrot, which evolved long ago in desert areas, does not exhibit drinking behaviour in response to angiotensin II. Possibly the threshold for the activation of this thirst mechanism has increased in the desert species. Such a change could result in a less incessant thirst sensation and greater intervals between periodic drinks.

The antidiuretic hormones, vasopressin in mammals and vasotocin in birds, are released in response to dehydration. The possibility that they may exhibit dipsogenic effects is controversial (Fitzsimons 1998). In ducks, angiotensin-like responsive neurons in the hypothalamus can be stimulated, in vitro, by vasotocin (Schmid and Simon 1996). However, such an effect is apparent only with high, non-physiological, concentrations of this hormone. The stimulation of salt appetite in mammals can involve interactions between angiotensin II, ACTH and adrenocortical steroids (see Chap. 2). A synergistic effect between angiotensin II and mineralocorticoids in promoting salt appetite has been described in pigeons (Massi and Epstein 1990).

5 Reproduction, Migration and Osmoregulation

Reproduction results in an increased need for water and salts in birds as a result of their increased activity, the production of eggs and, in many instances, the nurture of the young. The timing of reproduction in birds is well coordinated and usually involves responses to changes in the length of the daylight hours. Such photoperiodic cues are received and interpreted by the hypothalamus. A secretion of gonadotropin-releasing hormone (GnRH) takes place, resulting in the release of gonadotropins from the pituitary gland. The latter hormones promote the development of the gonads that secrete androgens and estrogens. Birds usually breed in spring in response to increases in day length, but some respond to shortening of the days and reproduce later in the year. Such times are predicted to be those that are the most favourable for the survival of the young. Breeding may be preceded by the migration of birds to areas where favourable conditions are likely to occur. Such migration may be transcontinental and even transhemispheric.

Bird migration involves the actions of various hormones, but their possible roles as initiators of these events are not clear (Biebach 1990; Wingfield et al. 1990). Preparations for migration include increased feeding activity and the storage of body fat, which may increase from normal values of 2 to 10% of the body weight to 50%. Apart from supplying energy for such flights, this fat also indirectly provides water storage. Yapp (1956) considered that water loss is the greatest single limiting factor in bird migration. However, long-distance avian migrants do not usually arrive at their destinations in seriously dehydrated condition. Providing that they fly at altitudes where the temperature does not exceed 7 to 10 °C, they appear to be able to exist on the water provided by the oxidation of the fat that they metabolize (Torre-Bueno 1978; Biebach 1990). This strategy is not feasible for all birds, such as those that fly at lower altitudes. However, it has been calculated (Tucker 1968) that a budgerygah can fly for at least 500 km without becoming seriously dehydrated. Even with fat storage equivalent to 4% of their body weight, they possibly could fly non-stop, under natural conditions, for about 100 km (Wyndham 1980). Vasotocin levels in the plasma have been shown to increase markedly in pigeons following flight (Giladi et al. 1997). This hormone could be contributing to the water conservation of birds on such occasions. Corticosterone and growth hormone levels in the plasma were found to be elevated during a stopover in the autumnal migration of sonderlings and semipalmated sandpipers in Delaware Bay in the USA (Tsipoura et al. 1999). The corticosterone concentrations could be reflecting stress. The elevated growth hormone levels may be the result of rapid lipolysis during the flight. Corticosterone has also been measured in the plasma of garden warblers during a stopover in the Sahara Desert in Algeria (Schwabl et al. 1991). It was found to be low and inversely related to the remaining fat stores.

Deserts provide relatively unpredictable habitats for successful breeding due to the hot dry conditions and the sporadic nature of the rainfall. In the Southern Hemisphere, especially in the central and western Australian deserts, photoperiodic cues for breeding appear to have been replaced by signals related to rainfall (Marshall 1961; Immelman 1971; Serventy 1971; Maclean 1996). Courting behaviour can commence within minutes of rain falling in such areas and copulation may occur 2 h later. Such "opportunistic" breeding may also occur in response to the associated increases in food supply, humidity and even the green colour of the vegetation. Such birds appear to maintain a tonic pituitary gonadotropic activity which, when further activated, results in a rapid recrudescence of the gonads. Rainfall in such areas can result in the immigration of nomadic species and even water birds. Birds living in the deserts of the Northern Hemisphere, such as the Sonoran Desert in the USA, appear to still utilize photoperiodic signals to control the time of their breeding. However, their behaviour at this time may be modified in response to a limited water supply (Vleck 1993). Thus, drought and dehydration can suppress reproductive activity (Cain and Lien 1985; Vleck and Priedkalns 1985; Wingfield et al. 1992). It is possible that this response is mediated by increased secretion of corticosterone.

The clutch size in desert birds is usually smaller than that in species that live in temperate regions. This may reflect a more tentative investment in reproductory success. The eggs often hatch asynchronously, which facilitates a reduction in the brood size in times of nutritional hardship. The older, larger nestlings can then more readily engage in siblicide. If conditions are favourable, the breeding season may be extended so that as many as four successive broods are produced. Sexual maturity is often achieved rapidly so that the earliest chicks may breed towards the end of the same season. The sexes of desert birds frequently exhibit cooperative behaviour to assure the survival of the eggs and young. They are also often monogamous.

Desert birds exhibit some interesting adaptations related to the survival of the eggs and young in hot dry conditions. Some birds, such as sandgrouse and emus, incubate the eggs in exposed situations. They protect them from overheating by raising the feathers, thus increasing the insulation and providing an enveloping heat sink. The male emu incubates the eggs. He has a reduced metabolism and loses less than 50% as much water by evaporation as non-incubating dehydrated emus (Buttemer and Dawson 1989). The physiological basis for this effect is unknown, but could possibly involve a depression of thyroid gland function. Many birds feed their young with food that has been moistened in the crop. As described earlier, columbiforme birds can, under the influence of prolactin, secrete "milk" from the crop sac with which they feed their young. Male sandgrouse (genus *Pterocles*) that live in deserts in Africa and the Middle East have a remarkable way of collecting water for the young (Cade and Maclean 1967; Thomas and Robin 1977). In 1896, the British aviculturist E. G. B. Meade-Waldo (quoted by Cade and Maclean 1967) observed these birds saturating their breast feathers with water while they were drinking. On returning to the nest the young imbibe this water by "stripping" it from the feathers. The feathers have a special design enabling their rapid hydration and the ability to hold large amounts of water. Possibly angiotensin II promotes such stripping?

6 Osmoregulation and Hormones in Free-Living Birds

The observations that have been described in this chapter were usually made on domestic or captive wild birds maintained under laboratory conditions. The experiments were often performed in vitro, or in vivo using anaesthetized and restrained birds. Normal physiological processes can only be inferred from such experiments. Observations on free-living wild birds may be illuminating, but are few in number. Exchanges, or turnover, of body water and salts can sometimes be measured in free-living birds with the aid of radioisotopes. Implanted devices, such as miniosmotic pumps, can be used to measure the GFR. Following "non-stressful" apprehension, samples of tissue can be promptly collected and analyzed. The results of a variety of such observations and their ecological significance have been reviewed by Goldstein (1997) and Thomas (1997).

Plasma concentrations of aldosterone from wild house sparrows collected in Ohio were found to be similar to those of captive birds maintained on a diet which had a low sodium content (Goldstein 1993). The plasma concentrations of vasotocin, aldosterone and the turnover of body water and sodium have been measured in five species of Australian honeyeaters (family Meliphagidae) (Goldstein and Bradshaw 1998). These birds live in a variety of geographic areas. Quantitative differences between the values of such physiological measurements could be related to their natural diets, the season, body size and habitat. Measurements of the body water turnover, haematocrit, and plasma osmolality and corticosterone concentration have been made in a small passerine, the silver eye, *Zosterops lateralis* (Rooke et al. 1983, 1986). This bird lives in a temperate but seasonally variable region in south Western Australia. The measurements indicated that the birds experienced dehydration in summer and potentially fatal stressful conditions in late summer and early winter. Similar measurements were made to define the seasonal physiological condition of a free-ranging subspecies of scrub wrens, *Sericornis frontalis* (family Acanthizidae) (Ambrose and Bradshaw 1988). These birds occupy arid, semiarid and temperate environments in Western Australia and were found to experience different levels of osmotic stress related to their natural environments. Extrarenal water loss has been compared in several species of wild Australian parrots (Williams et al. 1991b). Water loss was lowest in the desert-adapted species. Recently captured wild emus have provided in vitro preparations of the lower intestine for the study of its ability to actively transport Na^+ (Skadhauge et al. 1991). Measurement of transmucosal electrical parameters, such as the potential difference, in this preparation indicated that it had a high capacity for active Na^+ transport. As this process could be inhibited by the drug amiloride, it probably is under the control of aldosterone. The lower intestine could be playing an important role in the adaptation of these birds to their life in arid conditions. Other observations have been made in wild birds, especially in species living in the deserts of the USA, Africa and the Middle East. However, there are few direct measurements of the hormones that contribute to their osmoregulation.

Chapter 5

The Reptiles

Reptiles are considered phyletically to represent the first truly terrestrial verte-brates. They originated in the early Mesozoic period from an amphibian-like ancestor. In those times they became the predominant tetrapod vertebrates, living not only on the dry land, but also in the fresh water of lakes and rivers, and in the sea. Four principal groups of reptiles have persisted to the present day. The Chelonia (turtles and tortoises) have changed little since their origin in early Triassic times and today are represented by about 50 genera. These reptiles have a worldwide distribution; they are usually aquatic (or more strictly amphibious) in their habits. Five species live in the sea, although they must return to dry land in order to lay their eggs. A number of chelonians have adopted a life in arid desert regions, and these include the North American desert tortoise, *Gopherus agassizii*, and the Mediterranean tortoise, *Testudo graeca*. The Crocodilia (nine genera) have existed in a relatively unchanged form since they first appeared in the late Triassic period. They are mostly aquatic, living in the vicinity of fresh water, but at least one species, *Crocodilus porosus*, ventures into the sea for periods of uncertain duration. The Squamata, numerically the principal contemporary reptiles, consist of two main groups; the Lacertilia (lizards) and Ophidia (snakes), which originated in the Jurassic and Cretaceous periods, respectively. Today, they are each represented by about 300 genera. The lizards have the widest geographic distribution of the reptiles and are even found on many oceanic islands. The marine iguana of the Galapagos islands, *Amblyrhynchus cristatus*, spends much of its time feeding on the algae in the sea. The snakes also have a wide distribution and one family, the Hydrophiidae (15 genera), lives principally in the sea. Terrestrial snakes and lizards live in habitats ranging from tropical rainforests to dry desert regions. The remaining major group is the Rhynchocephalia, a relict branch of the reptiles with one surviving species, *Sphenodon* (the tuatara), which is now confined to a wet temperate environment on a few small islands situated off the coast of New Zealand.

Compared with their amphibian ancestors, reptiles are considered to be relatively independent of the proximity of fresh water. Many species, however, continue to live in the vicinity of lakes and rivers, though, unlike almost all amphibians, they produce an egg that, by virtue of its shell and amnionic membrane, can be laid on land. Some of the sea snakes, *Pelamis*, however, do not even need to return to land for this purpose, as they give birth to live young in the sea. As compared to the amphibians, the reptiles have a relatively impermeable skin which limits exchanges of their water and salts with the environment. Many reptiles, especially those living in arid areas, can form uric acid (instead of urea or ammonia) as the principal end product of nitrogen metabolism, so that they require little water for the renal excretion of catabolic nitrogen. The three afore-mentioned factors, amniotic egg, relatively impermeable integument and uri-cotelism, distinguish the reptiles osmotically from their ancestral amphibian forms, and must contribute substantially to the cosmopolitan radiation of this group into regions of potential osmotic hostility.

The geographic dispersion of many terrestrial species across large expanses of the oceans is often difficult to envisage, but this has undoubtedly occurred among the reptiles (Darlington 1957), a feat that must largely reflect their relative osmotic independence from their environment. Reptiles are thought to have originated in the Old World and migrated to the New World, with a subsequent pilgrimage from South America to Australia. Darlington states that these movements can be explained "without resorting to special land bridges or continental drift", pre-sumably by making transoceanic voyages. Fossil representatives of an extinct group of land tortoises (Meiolaniidae) from South America have been found on Lord Howe Island and Walpole Island in the eastern Pacific Ocean. It seems likely that these reptiles floated or possibly "rafted", across the many hundreds of kilo-metres of ocean to occupy these remote islands. Turtles have a remarkable propensity to survive such conditions. Among the snakes, the genus *Natrix* has a remarkably wide geographic distribution and is found in Europe, Asia, Africa, Australia, America and islands such as Cuba and Madagascar. These snakes are primarily aquatic, but have probably attained their cosmoplitan status by cross-ing the seas. Pettus (1958) compared the fluid metabolism of two races of *Natrix sipedon* living in the southern United States. One of these groups normally lives in fresh water and the other in brackish water; but the former cannot usually survive transfer to a saline medium. The success of such adaptation seems to reside solely in a predisposition to refrain from drinking salt solutions. Provid-ing that reptiles maintain the relative impermeability of their integument, dis-persal of terrestrial species through the seas is physiologically conceivable.

The reptiles utilize the same complement of hormones for their osmoregula-tion as their amphibian progenitors and the birds that evolved from them. Some differences in the precise structures of these hormones have emerged and a few of their physiological actions may differ. The *neurohypophysis* of reptiles, like that of other nonmammalian tetrapods, stores and secretes 8-arginine vasotocin (AVT, vasotocin) and mesotocin (Table 5.1). The former peptide functions as an antid-iuretic hormone, as occurs in birds and the amphibians. The principal hormones secreted by *adrenocortical tissue* in reptiles are aldosterone and corticosterone (Table 5.2), which is also similar to the pattern in the latter two groups of tetrapods. The actions of these steroidal hormones are not as well defined as in birds, but they also appear to contribute to the regulation of sodium and, pos-sibly, potassium metabolism. The activity of adrenocortical tissue in reptiles, like that in most other vertebrates, is regulated by *ACTH* and the *renin-angiotensin system*. However, the evidence for such effects is somewhat fragmentary in the reptiles. Removal of segments of the adenohypophysis of the chameleon lizard, *Anolis carolinensis*, resulted in the disappearance of corticosterone from the plasma (Licht and Bradshaw 1969). Conversely, the administration of mammalian ACTH and extracts of the adenohypophyses of various reptiles increased plasma corticosterone concentrations. Renin-like activity has been identified in the

Table 5.1. Distribution of neurohypophysial hormones in reptiles

	Vasotocin	Mesotocin
Crocodilia		
Caiman (*Caiman sclerops*)[a]	+	−
Chelonia		
Mediterranean tortoise (*Testudo graeca*)[b]	+	+*
Green sea turtle (*Chelonia mydas*)[a,c]	+	+*
Slider turtle (*Pseudemys scripta*)[d]	+	
Ridley turtle (*Lepidochelys kempi*)[c]	+	+*
Loggerhead turtle (*Caretta* sp.)[c]	+	+*
Lacertilia		
Green iguana (*Iguana iguana*)[e]	+	+
Ophidia		
Grass snake (*Tropidonotus natrix*)[b,c]	+	+*
Rattler (*Crotalus atrox*)[f]	+	+
Cobra (*Naja naja*)[g,i]	+	+
Viper (*Vipera aspis*)[h,i]	+	+
Elaph (*Elaph quadrivergata*)[i]	+	+

* Not a definitive distinction from oxytocin.
[a] Sawyer et al. (1961). [b] Heller and Pickering (1961). [c] Follett (1967). [d] Sawyer (1968). [e] Acher et al. (1972). [f] Munsick (1966). [g] Pickering (1967). [h] Acher et al. (1969b). [i] Acher et al. (1969a).

Table 5.2. Distribution of adrenocortical steroid hormones in reptiles

	Corticosterone	Aldosterone
Crocodilia		
Alligator mississippiennsis[a]	+	+
Crocodylus nilotocis[b]	+	+
Chelonia		
Pseudemys spp.[c]	+	+
Emys orbicularis[c]	+	+
Chrysemys picta[d]	+	+
Testudo hermanni[e]		+
Lacertilia		
Lacerta viridis[e]	+	+
Anolis carolinensis[f]		+
Uromastix acanthinurus[g]	+	+
Tiliqua rugosa[g]	+	+
Varanus gouldii[h]	?	+
Ophidia		
Natrix natrix[c,i]	+	+
Hydrophysis cyanocinctus[j]	+	+
Rhyncocephalia		
Sphenodon punctatus[k]	+	

[a] Gist and DeRoos (1966). [b] Balment and Loveridge (1989). [c] Macchi and Phillips (1966). [d] Sandor et al. (1964). [e] Phillips et al. (1962a) . [f] Licht and Bradshaw (1969). [g] Bradshaw and Grenot (1976). [h] Bradshaw and Rice (1981). [i] Phillips and Chester Jones (1957). [j] Duggan and Lofts (1979). [k] Tyrrell and Cree (1998).

kidneys of several reptiles (Capelli et al. 1970). While renin from mammals and birds interacts only with plasma angiotensinogen from their own phyletic groups, that of the reptiles is less fastidious. It can also react, to produce angiotensin, with the plasma substrate from birds and amphibians (Nolly and Fasciola 1973). Angiotensin II from turtles and an alligator is identical to that from cattle, the domestic fowl and the bullfrog (Fig. 2.6). However, its decapeptide precursor, angiotensin I is different and has an amino acid substitution at position 9. Reptilian *prolactin*, when administered to other vertebrates, displays a similar spectrum of activities, such as milk let-down in mammals and crop sac secretion in pigeons. Nevertheless, its structure differs. Prolactin from a sea turtle has a 75% homology in its amino acid sequence to that of the human hormones (see Bentley 1998). It is uncertain if prolactin has a role in the osmoregulation of reptiles, but the possibility has been considered.

Reptiles have a water content equivalent to about 70% of their body weight, an amount similar to that of birds and mammals, though less than that of the Amphibia. The concentration of the body fluids conforms to the usual tetrapod pattern, being about 250 to 300 mosmol l^{-1}. Variations in the concentration and electrolyte content of the body fluids do, however, occur; reptiles in a marine environment have a slightly higher plasma concentration than those living on land (Table 1.2) though they remain hyposmotic to sea-water. Reptiles can tolerate quite large changes in the concentrations of their body fluids; the sodium concentration in the plasma of the lizard, *Trachysaurus rugosus*, may rise from 150 to nearly 200 mEq l^{-1} during the dry Australian summer (Table 1.2) and comparable increases have also been observed in the lizard, *Amphibolurus ornatus*, and the desert tortoise (Dantzler and Schmidt-Nielsen 1966; Bradshaw and Shoemaker 1967; Bradshaw 1970). Such osmotic behaviour has been called "ahomeostasis" (Peterson 1996). Decreased plasma electrolyte levels can also be readily tolerated; the softshell turtle, *Trionyx spinifer*, normally has a plasma sodium concentration of about 150 mEq l^{-1}, but in healthy hibernating individuals this may decrease to 80 mEq l^{-1} (Bentley, unpubl observ). Birds and mammals do not seem to be able to withstand such variations in the concentrations of their body fluids, but in some reptiles such tolerance may be an important factor in their ability to survive adverse conditions when the possibility of osmotic regulation is limited.

Reptiles are ectotherms and so, unlike birds and mammals, do not usually utilize evaporative water loss for thermal cooling. Nevertheless, they do attempt to maintain their body temperature at levels that are optimal for their activity and discreetly utilize external heat sources for this purpose. This is performed principally by conforming to a certain pattern of behaviour: warming by periodic basking in the sun and opposing the ventral body surface to warm surfaces, and cooling by seeking shade and refuges, such as rock crannies and burrows, where the temperature is more moderate. Modification of the body temperature may be assisted physiologically by changing the conductivity of the body by means of circulatory adjustments. Some species can change the colour of their skin so as to alter the rate of absorption of solar radiation (see Schmidt-Nielsen 1964a). Such adjustments are probably mediated principally by the nervous system, though in some reptiles melanocyte-stimulating hormone mediates colour change (Waring

1963). Evaporation of water from the respiratory tract and skin increases when the temperature of the body and the environment rises, but this is physically obligatory and cannot be considered as part of a regulatory response to facilitate cooling.

The principal avenues for the exchange of water and solutes between the reptile and its environment are basically the same as in mammals and birds. The physiological significance and magnitude of the exchanges through the different channels differ however. The main osmoregulatory organs are the kidneys, as in all vertebrates, but in certain reptiles the action of these organs may be augmented by cephalic "salt" glands, and the cloaca-colon and urinary bladder.

1 Water Exchanges

1.1 Skin and Respiratory Tract

In a terrestrial environment reptiles lose water by evaporation from the external integument, and the lungs and pulmonary passages. This loss increases when body temperature and environmental temperature increase; thus, the bobtail goanna, *Trachysaurus rugosus*, when kept in dry air at 25 °C, loses water by evaporation at the rate of 0.7 g $(100\,g)^{-1}$ day^{-1} while at 37.5 °C it is 2.9 $(100\,g)^{-1}day^{-1}$ (Warburg 1965a). This water loss is small relative to that in birds, mammals and the Amphibia, and makes little contribution to thermal cooling; the temperature of the goannas when equilibrated to such conditions being only 1 or 2 °C less than that of the environment.

The increased evaporation of water at high environmental temperatures is due to the decrease in the saturation deficit for water in the surrounding air, and an increase in the body temperature of the animal. The metabolic rate of reptiles increases two to three times for every 10°C rise in body temperature (Benedict 1932), so that there is an increased rate of gas exchange in the lungs that further facilitates water loss.

Different species of reptiles have been observed to lose water by evaporation at various rates (Bogert and Cowles 1947). The reasons for such differences were not initially clear, as they could involve several factors such as: the rate of oxygen consumption, the ratio of surface area to body weight or the particular properties of the skin of the different species. This was not further investigated until Schmidt-Nielsen (1964a) suggested, after examining the fragmentary information that was then available, that cutaneous evaporation in reptiles may be greater and more variable than was hitherto considered likely. When cutaneous and pulmonary water losses were measured separately in a variety of species of reptiles (at 23 °C), the movement through the integument was indeed found to contribute from 66 to 87% of the total evaporative loss in a desert lizard, the chuckwalla, to 88% in the brown water snake (Schmidt-Nielsen and Bentley 1966; Schmidt-Nielsen et al. 1966a; Prange and Schmidt-Nielsen 1969). At higher temperatures, 35 to 40 °C, pulmonary water loss increases relative to that from the skin. However, the latter still amounts to about 50% of the total. Similar observations have been made on a variety of lizards (Claussen 1967). Some legless burrowing worm-lizards (family Amphisbaenidae) have even greater rates of evaporative water loss. They have a cutaneous permeability to water that approaches that of amphibians (Krakauer et al. 1968). The rates of such evaporative water loss appear to be related to the ecological habitat that is normally occupied by each species of reptile. Thus, the aquatic brown water snake loses water by evaporation 13 times more rapidly than the desert-dwelling chuckwalla.

Water loss from the *respiratory tract* depends on the rate of oxygen consumption and the ability to extract this gas from the inspired air. The metabolic rate varies considerably in different reptiles, reflecting their activity, size and whether they happen to be in states of aestivation or hibernation. Such differences are partly reflected in their rates of water loss. For instance, at 23 °C the desert gopher tortoise, which has an oxygen consumption of $0.26\,ml\,g^{-1}\,day^{-1}$, loses $0.4\,mg$ water $g^{-1}\,day^{-1}$ from the respiratory tract. On the other hand, the forest-dwelling *Iguana iguana* has an oxygen consumption of $2.6\,ml\,g^{-1}\,day^{-1}$ and loses $3.6\,mg$ water $g^{-1}\,day^{-1}$ in this manner (Bentley and Schmidt-Nielsen 1966; Schmidt-Nielsen and Bentley 1966). However, the relationship between oxygen consumption and water loss varies in different reptiles. This difference may be due to the proportion of oxygen extracted from the air in the lungs. There is usually a reduction of oxygen from the atmospheric levels of 20.9% to 17 to 20%. However, some reptiles only breathe sporadically (breath holding) and then can extract more oxygen from the inspired air. The chuckwalla sometimes breathes only two to three times an hour and then reduces the oxygen level in its lungs to as little as 5% of that present in the atmosphere (Schmidt-Nielsen et al. 1966a). This behaviour has also been observed in hibernating reptiles and three species of Australian lizards (genus *Varanus*) (Thompson and Withers 1997). In the latter, at 25 °C, respiratory water loss can then decline to about 6% of the total water lost by evaporation. Respiratory water loss in reptiles can also be reduced in some species by utilizing a countercurrent heat exchange process in the nasal passages. This mechanism reduces the temperature, and hence the water vapour content of the expired air (Murrish and Schmidt-Nielsen 1970a). This strategy, as described earlier, also occurs in small mammals and birds.

Evaporative water losses in reptiles are probably temporally regulated, in conjunction with the body temperature, by changes in behaviour. Reptiles may be diurnal or nocturnal in their habits and avoid extreme temperatures by seeking the shelter of cool refuges. Warburg (1965a) found goannas, *Trachysaurus rugosus*, in burrows 2 to 3 m deep. The temperature in these holes never exceeded 28°C, even though the external air temperature was as high as 40 °C and that of the surface soil 50 °C.

Many species of reptiles live an aquatic life submerged in fresh water or sea-water. The air that they breathe is nearly saturated with water vapour, so that little respiratory loss is expected to occur. However, the osmotic gradients between their body fluids and the external solution will tend to favour a net water uptake in fresh water and a loss in sea-water. These exchanges of water are usually quite small in reptiles. Such gains of water in freshwater turtles, the softshell turtle, *Trionyx spinifer,* and the slider turtle, *Pseudemys scripta,* were found, respectively, to be 26 and $6\,mg\,cm^{-2}\,day^{-1}$ (Bentley and Schmidt-Nielsen 1970). The difference appears to be due to the nature of their skin. When these turtles were placed in sea-water they lost water at the rate of, respectively, 38 and 10 mg cm^{-2} day^{-1} (sea-water is an unnatural environment for these turtles and it is possible that it changes the permeability of their skin). Juvenile salt-water crocodiles in sea-water were found to lose water across their integument at the rate of 2.4 mg cm^{-2} day^{-1} (Taplin 1984). However, young caiman, which normally live in fresh water, lost water at a much greater rate, 14 mg cm^{-2} day^{-1}, when in an equivalent salt solution (Bentley and Schmidt-Nielsen 1965). Measurements of water exchanges, in sea-water, have been made in freshwater and marine snakes using tritiated water to measure the influx, from the sea-water and the efflux from the snakes' body fluids (Dunson 1978). Such fluxes of water were both very low. However, they were greater in the freshwater species.

There are folkloric tales that some terrestrial reptiles may be able to absorb water across their integument from dew or damp surfaces. However, closer scrutiny of such reports in the Australian desert lizard *Moloch horridus* indicated that, while water uptake could occur from such sources, it takes place through the mouth (Bentley and Blumer 1962). The water is absorbed into narrow open channels on the skin surface and it travels along these to the lips, where it is collected and swallowed. Small accumulations of water may also occur across the buccal membranes of aquatic snakes (Dunson 1978) and crocodilians (Taplin 1988).

1.2 Urinary and Faecal Water Loss

Reptiles, like birds and amphibians, possess a cloaca into which the colon, ureters and reproductive ducts open. It is thus not always easy to separately measure urinary and faecal losses of water and solutes. However, tortoises and turtles, some lizards and the tuatara (*Sphenodon*) have a urinary bladder. This receptacle is absent in crocodilians, snakes and most lizards. However, in some lizards, the colon may store substantial amounts of urine and have a bladder-like appearance. The cloaca, colon and urinary bladder are the sites of water and salt exchanges that modify the composition of the urine and faeces. As reptiles cannot form hyperosmotic urine, such processes may make important contributions to their salt and water economy. The water content of the faeces of reptiles appears to be usually about 50 to 60 g (100 g)$^{-1}$ wet weight (Minnich 1970; Nagy 1972). The loss of water in the urine and faeces in reptiles varies considerably. It has been found to be as little as 5% of the total water loss each day in the carnivorous side-blotched lizard *Uta hesperis* (Claussen 1967) to 60% in the desert iguana *Dipsosaurus dorsalis* (Minnich 1970). Carnivorous reptiles generally lose less water in the faeces than herbivorous ones, who may excrete large masses of incompletely utilized plant material. Separate urinary losses have been measured in some reptiles and vary from 15% of the total water loss in the desert snake, *Spalerosophis cliffordii* (Dmi'el and Zilber 1971), to 6% in the gecko, *Hemodactylus* (Roberts and Schmidt-Nielsen 1966). The relative water loss in the urine can be much higher in aquatic species, such as the caiman. In small juvenile animals in fresh water it was equivalent to nearly 30% of their body weight each day (Bentley and Schmidt-Nielsen 1965).

1.2.1 Role of the Kidney

Reptiles, like birds and mammals, have a metanephric kidney. The nephrons usually begin with a glomerulus and Bowmans capsule, which is followed by a ciliated neck segment, proximal tubule, intermediate segment and distal tubule that opens into a collecting duct system. They lack a loop of Henle and countercurrent concentrating mechanism and hence cannot form hyperosmotic urine. A venous renal portal system is present that supplies the tubules, but not the glomeruli. Some reptiles, including some snakes and the desert lizard *Ctenophorus ornatus* have some aglomerular nephrons (O'Shea et al. 1993). This morphological loss appears to be an adaptation to life in arid environments. The formation of urine follows the usual vertebrate pattern of plasma filtration across the glomeruli and subsequent reabsorption of water and solutes during passage of the filtrate along the renal tubule (Braun and Dantzler 1997). Tubular secretion of solutes, and in aglomerular nephrons presumably water, also takes place. Such processes can result in the formation of a dilute, hyposmotic, urine and the conservation of essential solutes such as Na^+, K^+, Cl^- and HCO_3^-. Excesses of such ions, uric acid and water appear in the urine. When water is not abundant, urine that approaches isosmoticity with the plasma may be formed. The excretion of uric acid involves both its glomerular filtration and secretion, which occurs mainly across the proximal tubule. Uric acid is the principal excretory product of nitrogen metabolism in most reptiles, especially snakes and lizards. The crocodiles can also utilize ammonia for such excretion and tortoises and turtles, urea. As described in Chapter I, uricotelism can result in a substantial reduction of urinary water loss that is especially important in predominantly terrestrial and desert-living species. Uric acid can also form insoluble sodium and potassium salts which are deposited in the cloaca of reptiles, thus facilitating a more osmotically economic way for excreting these ions (Minnich 1972).

The glomerular filtration rate in reptiles is usually quite labile, and changes can contribute to the regulation of water and salt excretion (Braun and Dantzler 1997). Decreases in the GFR take place in response to dehydration and increases following the ingestion of excess water (Dantzler 1992). The experimental administration of sodium chloride usually, though not always, decreases the GFR. Changes in the GFR in reptiles are mainly due to increases or decreases in the number of functioning glomeruli (glomerular intermittency). However, small changes in the filtration rate across individual, single, glomeruli may also occur. Such variations in glomerular activity are principally due to vascular responses, which appear to involve the afferent glomerular arteriole.

The *neurohypophysis* contributes to the regulation of the urine flow in reptiles. Vasotocin is their antidiuretic hormone. Information about the precise mechanism of its antidiuretic effect is more fragmentary than in birds and amphibians. In 1933, Burgess et al. showed that the administration of the mammalian neurohypophysial hormones vasopressin and oxytocin had an antidiuretic effect in an alligator (*Alligator mississippiensis*). However, in contrast to mammals, this decrease in urine flow was solely due to a reduction in the GFR with no detectable change in renal tubular water reabsorption. Sawyer and Sawyer in 1952 confirmed this observation. Extracts of the neurohypophysis of the lizard *Tiliqua (Trachysaurus) rugosa* were observed to have an antidiuretic effect when injected into this reptile (Bentley 1959). Subsequently, it became apparent that this action principally reflected the presence of vasotocin in the neurohypophysis of reptiles. The administration of purified vasotocin was also shown to have an antidiuretic effect in water snakes, *Nerodia (Natrix) sipedon* (Dantzler 1967). Mesotocin, the other reptilian neurohypophysial peptide, also has an antidiuretic action, but only when administered in supraphysiological doses. An antidiuretic effect of vasotocin has also been demonstrated in the painted turtle *Chrysemys scripta* (Butler 1972; Dantzler 1982). In the water snakes, low doses of vasotocin were observed, by measuring the clearance of water relative to the GFR, to increase the water reabsorption across the renal tubule. A decrease in the GFR was also observed, but only when using higher doses of vasotocin. A similar difference in the sensitivity of the kidney has been observed in the slider turtle following the administration of extracts of its neurohypophysis (Dantzler and Schmidt-Nielsen 1966). The differential sensitivity of the glomerular and renal tubular responses is similar to that described in the domestic fowl (Chap. 4). Whether or not it applies to most reptiles, including the alligator, is unknown. Attempts to demonstrate a direct effect of vasotocin on isolated renal tubules of reptiles have, so far, been unsuccessful (Stolte et al. 1977; Beyenbach 1984). However, a vasotocin-sensitive adenylate cyclase, which mediates water permeability responses in mammals and amphibians, has been identified in the collecting ducts of the lizard *Ctenophorus ornatus* (Bradshaw and Bradshaw 1996). Receptors for vasotocin were also identified, but in the intermediate segment of the renal tubule. A mechanism that could be mediating a permeability response to vasotocin thus appears to be present. The role of vasotocin receptors in the intermediate segment is unknown.

Neurohypophysial function in the lizard *Ctenophorus ornatus* can be blocked by placing electrolytic lesions in the hypothalamo-hypophysial tract (Bradshaw 1976). The GFR was increased. Urine volume, in response to administered saline, was also increased. The administration of vasotocin decreased the urine volume and increased renal tubular water reabsorption. The results suggested a physiological role for vasotocin in controlling urine volume in this reptile.

Direct measurements of vasotocin in the blood of the lizard *Varanus gouldii* were made, using a radioimmunoassay, by Rice in 1982. Release of the hormone from the neurohypophysis was related to the osmotic concentration of the plasma. The concentrations of vasotocin were found to be 3.9 pg ml^{-1} in dehydrated lizards as compared to 1.6 pg ml^{-1} in the hydrated ones. The osmotic concentrations of the plasma that induced release of vasotocin (osmotic sensitivity) were similar to those in other vertebrates. The antidiuresis that occurred in response to dehydration of these lizards was not accompanied by a decrease in the GFR. It appears to involve only a tubular response (Bradshaw and Rice 1981). However, following saline loading, the plasma vasotocin levels were increased to 7.1 pg ml^{-1} and a decrease in GFR was then observed. Vasotocin has also been identified in the plasma of the snake *Bothrops jaraca* (Silveira et al. 1992), the green turtle *Chelonia mydas* (Figler et al. 1989) and the lizard *Tiliqua rugosa* (Fergusson and Bradshaw 1991).

It is noteworthy that neurohypophysial peptides exert other actions in reptiles. They may reduce the blood pressure, as in *Trachysaurus rugosus* (Woolley 1959) and *Natrix sipedon* (Dantzler 1967), but high doses are required, so that these effects are probably only of pharmacological significance. Vasotocin (in vitro) contracts the oviduct of the turtle, *Pseudemys scripta*, when it is in an ovulating condition (Munsick et al. 1960). LaPointe (1969) has shown that vasotocin also contracts the oviduct of the viviparous island night lizard, *Klauberina riversiana*, in vitro. Oxytocin and mesotocin were very much less effective than vasotocin. It is possible that this action of vasotocin reflects a physiological role for this peptide in reptilian oviposition and parturition, just as oxytocin assists the latter process in mammals.

1.2.2 Role of the Cloaca-Colon Complex

The deposition of solid pellets of uric acid in the cloaca of reptiles suggests that changes in the composition of the urine are occurring at this site. Differences in the composition of urine collected directly from the ureters and that expelled from the cloaca confirm that such changes are occurring. They involve not only the cloaca, but also the adjoining colon. In some lizards, large accumulations of urine can be seen in the colon, which then takes on a bladder-like appearance. Although intestinal caecae are present in some reptiles, they lack the recta caecae, which contribute to the modification of the urine in birds. The ureteral urine of reptiles is usually hyposmotic, while that excreted from the cloaca is isosmotic. This observation suggests that absorption of water is occurring at a postrenal site. An absorption of salts also takes place. These processes have been studied in several crocodilians (Bentley and Schmidt-Nielsen 1965; Schmidt-Nielsen and Skadhauge1967; Kuchel and Franklin 1998), snakes (Junqueira et al. 1966) and lizards (Braysher and Green1970; Murrish and Schmidt-Nielsen 1970b; Bentley and Bradshaw 1972; Bradshaw 1975; Skadhauge and Duvdevani 1977; Bradshaw and Rice 1981). The processes of such absorption have been studied in vivo, in vitro and using in vivo perfusion techniques. A comprehensive summary of the available information has been compiled by Don Bradshaw (1997).

The calculated reabsorption of fluid from the cloaca-colon complex of reptiles varies from 20% in a crocodile (*Crocodylus acutus*) to 96% in a lizard (*Scleloporus cyanogenys*). This absorption appears to be due to an initial osmotic equilibration with the plasma and a reabsorption of Na^+, which, in turn, establishes further osmotic gradients. Potassium may be either reabsorbed or secreted. In the desert iguana, the colloidal osmotic pressure of the plasma proteins may also promote water reabsorption (Murrish and Schmidt-Nielsen 1970b). In the Israeli desert lizard, *Agama stellio*, fluid reabsorption from the cloaca-colon complex increases from $2.4\,ml\,kg^{-1}\,h^{-1}$ when they are normally hydrated to $3.3\,ml\,kg^{-1}\,h^{-1}$ during dehydration (Skadhauge and Duvdevani 1977). In the Australian desert lizard, *Varanus gouldii*, reabsorption of fluid from the urine in the cloaca-colon increases from 22% in hydrated lizards to 40% in dehydrated ones (Bradshaw and Rice

1981). However, a hormonal basis for such absorption has not been established (Bradshaw 1997). A possible role for vasotocin in promoting osmotic water movement has been investigated on several occasions. However, with a single exception (Braysher and Green 1970), such an action has not been demonstrated. Local changes, such as involve osmotic gradients, may be determining the reabsorption.

1.2.3 Role of the Urinary Bladder

The urinary bladder of reptiles can be quite voluminous, especially in turtles and tortoises. It is osmotically permeable to water and so may act as a reservoir for the storage of water. However, snakes and many lizards that live in deserts do not have a urinary bladder, so that such an opportunity is not available to all reptiles. Nevertheless, tortoises utilize the urinary bladder as a water-storage organ, especially during times of drought and when there are extended periods of time between access to drinking water (Nagy 1988; Jørgenson 1998). Naturalists and explorers, including Charles Darwin (1839), have commented on the large volumes of fluid held in the urinary bladders of species such as the Indian giant tortoise and the Galapagos tortoise. These quantities of fluid can be equivalent to as much as 20% of their body weight. Initially, the fluid appears to be quite limpid and, as commented on by early travellers, can even provide a beverage for thirsty humans. However, as observed in the tortoise *Gopherus agassizii* in the Mohave Desert, the solute concentration rises considerably in times of drought. The contained fluid then approaches isosmoticity with the plasma. It mainly contains potassium salts obtained from their herbivorous diet and is retained until fresh drinking water becomes available. In these tortoises, the fluid in the bladder appears to act as a large "sink" (or cesspool) for the storage of unwanted solutes. The osmotic permeability of the chelonian urinary bladder has been measured, in vitro, in the Mediterranean tortoise *Tesudo graeca* (Bentley 1962b) and the freshwater turtle *Pseudemys scripta* (Brodsky and Schilb 1965). Water absorption was not found to change in the presence of vasotocin. These observations are in contrast to the effects of this hormone on the urinary bladder in the amphibians (see Chap. 6). Water absorption has also been measured from the urinary bladder of the gopher tortoise in vivo. Water introduced into the bladder was found to be absorbed at the rate of $20 \, \mathrm{ml \, h^{-1}}$ in either hydrated or dehydrated tortoises (Dantzler and Schmidt-Nielsen 1966). Salt can be reabsorbed from the reptile urinary bladder (see the next section), which may enhance the osmotic gradient for water reabsorption. Evaporative water loss is relatively slow in reptiles, so that the normal rate of its absorption from the urinary bladder may be adequate for their needs. A hormonally induced response may not be advantageous.

Reptiles that live in hot desert environments where adequate food and water is available only sporadically may resort to *aestivation* to reduce water loss and the consumption of energy. Such a state of dormancy occurs in hot conditions. (Hibernation, which some reptiles also undergo, occurs in response to cold in winter.) Aestivation appears to be initiated by a reduction in the food supply and the drying out of the vegetation, soil and pools of water. It occurs in various reptiles including turtles, lizards, snakes and crocodiles (Gregory 1982). Oxygen consumption may decline by as much as 70% (Halley and Loveridge 1997) below levels that normally occur at a particular body temperature. There is an accompanying decrease in evaporative water loss, especially from the respiratory tract (Seidel 1978). Physiological activities, including kidney function, are drastically reduced and, depending on the period of dormancy, large increases in the osmotic concentration of the plasma usually occur. The reptiles sequester themselves in various places that provide thermally benign environments including rock crevices, burrows and vegetation. When the pools of water in which they live dry up, long-necked turtles, *Chelodina rugosa*, which live in tropical northern Australia, dig themselves into the soil (Grigg et al. 1986; Kennett and Christian 1994). They then await the next rains, which may not occur for two or three seasons. Under such conditions the turtles lose weight, which includes fat and, mainly, water. Nevertheless, they can survive. Other reptiles that live in desert conditions undergo less rigorous periods of aestivation. In the Mohave Desert, following the drying out of the vegetation in mid-summer, chuckwalla lizards cease eating (Nagy 1988). They then retire to rock crevices from which they emerge for only an hour or so every 3 days. The desert tortoises under such conditions spend longer periods, on average 6 days, in burrows.

2 Salt Exchange

2.1 Skin

The relative impermeability of the integument considerably impedes the movements of salts across the body surface of freshwater and marine reptiles. Precise values for the permeability to Na^+ are difficult to obtain and assess as they are so low and their measurement is subject to experimental errors. Calculations based on measurements in whole animals may be overestimated in salt water, due to the possibility of drinking and uptake across buccal and cloacal membranes and the eyes. Observations performed on the skin in vitro are also prone to error, due to the leakage through peripheral regions of skin maintained between clamps that separate the solutions on each side of the skin (Dunson 1978). When juvenile caiman are kept in distilled water there is a net loss of Na^+ across their skin amounting to about $1\,\mu mol\ cm^{-2}h^{-1}$ (Bentley and Schmidt-Nielsen 1965). I have also measured such Na^+ loss in softshell turtles, *Trionyx spinifer*, in distilled water and found it to be similar to that of the caiman. Loss (unidirectional efflux) of Na^+ across the integument of marine green turtles, *Chelonia mydas*, has been measured, using isotopic Na^+, in sea-water and fresh water (Kooistra and Evans 1976). In sea-water the efflux amounted to only about 5% of the total sodium loss. When placed in fresh water, the loss of Na^+ decreased from $5.8\,\mu mol\ (100\,g)^{-1}$ h^{-1}, in sea-water, to $0.9\,\mu mol\ (100\,g)^{-1}h^{-1}$. This difference in permeability could be due to a direct effect of the bathing solution on the skin, a change in transcutaneous electrical potential difference or vascular changes. The permeability, in sea-water, of the skin of a variety of terrestrial, freshwater and marine snakes has been investigated (using isotopic Na^+) both in vivo and in vitro (Dunson 1978). The skin of the marine snakes was virtually impermeable to Na^+, but a small influx of this ion, from sea-water, could be detected in the freshwater species. The skin of the estuarine (salt-water) crocodile was also found to be almost impermeable to Na^+ (Taplin 1985). Movements of Na^+ across the skin of aquatic reptiles appear to make little contribution to their salt balance. However, other integumental membranes are more permeable to Na^+ (and water) especially those of the mouth, eye and cloaca. Such routes may be the sites of significant exchanges of salts and, possibly, in crocodiles, even active ion accumulation (Taplin 1988).

2.2 Kidney

The urine of reptiles may contain excess salts, especially sodium and potassium chlorides, which are usually obtained in the diet. In circumstances when there is an excretion of excess water by the kidney, the salt excretion in the urine will be reduced by its reabsorption across the renal tubule. The formation of urine by the kidneys in reptiles, as described earlier, usually involves substantial changes in the GFR, and hence the rate of delivery of water and solutes to the renal tubules. The proportion, or "fraction", of Na^+ that is reabsorbed varies considerably, from more than 98% to as little as 50% (Bradshaw 1997). About 60 to 70% is thought to occur in the proximal tubule and the remainder in the distal tubule and collecting ducts (Braun and Dantzler 1997). Potassium is reabsorbed in the proximal tubule and, when necessary, can be secreted by the distal tubule. Considerable variations have been observed in the osmotic and electrolyte concentrations of the urine in reptiles. The ratio of the urine/plasma osmotic concentrations can vary from 0.1 to 1. The lowest salt concentrations are usually seen in freshwater species, which have a surfeit of water. When the estuarine diamondback terrapin, *Malaclemys centrata*, is maintained in fresh water, the urine concentration is about $60\,mosmol\,l^{-1}$ and its Na^+ concentration is less than $1\,mEq\,l^{-1}$ (Bentley et al. 1967a). I have found that the urine of softshell turtles maintained under such conditions has a similar concentration of sodium. However, Roberts and Schmidt-Nielsen (1966) found that in three species of lizards (gecko, horned toad and Galapagos lizard) the concentration of Na^+ in the ureteral urine was always about $100\,mEq\,l^{-1}$, even when the animals had been hydrated. Solute reabsorption from the renal tubule varied from about 50% in the Galapagos lizard to 85% in the geckos. The experimental administration of various amounts (loads) of sodium chloride to reptiles often results in a decrease of the GFR compared to that of hydrated animals (see Bradshaw 1997). (However, this effect can be variable and it is not, for instance, seen in some snakes.) The administration of sodium chloride loads to the bobtail lizard *Tiliqua (Trachysaurus) rugosus* and the gopher tortoise results in anuria, which appears to reflect a glomerular shutdown (Bentley 1959; Dantzler and Schmidt-Nielsen 1966). In the bobtail, even small loads of sodium chloride are poorly excreted, only about 20% appearing in the urine over a period of 24 h. Renal responses to the administration of potassium chloride are somewhat more rapid, apparently reflecting a renal tubular secretion of the K^+ (Shoemaker et al. 1966). However, the reptile kidney appears to generally have a poor ability to excrete excess salt, especially when water is restricted. This deficiency can be compensated for in two ways. Some reptiles have a remarkable ability to tolerate high concentrations of Na^+ in their body fluids. This ability was observed in the laboratory and, in summer, in free-ranging bobtail lizards (*Tiliqua rugosus*) in Australian coastal sand dunes (Bentley 1959). Bradshaw and Shoemaker (1967) found that plasma sodium chloride concentrations of *Amphibolurus* lizards rise as high as 300 mmol l^{-1} during periods of drought in desert areas in Australia. Following a summer rainstorm, the lizards were observed to dart about and rapidly drink the water as it fell. After 10 h the plasma electrolyte concentrations had returned to normal. The desert tortoise also utilizes such occasions to drink and excrete solutes that have accumulated in its urinary bladder (Nagy 1988). Other reptiles, as will be described later in this chapter, principally utilize cephalic salt glands for the excretion of excess sodium and potassium chlorides.

In mammals the *adrenal cortex*, and its secreted steroidal hormones, has an established role in promoting the reabsorption of Na^+ and the secretion of K^+ by the distal renal tubule-collecting duct system. These effects are principally mediated by aldosterone. Such hormonal effects also occur at extrarenal sites, including the colon. Experimental investigations of the possible roles of the adrenocortical tissue on such processes in reptiles have yielded inconclusive results. In mammals, adrenalectomy results in a rise in the concentration of K^+ in the plasma and a decline in that of Na^+. Adrenalectomy is a difficult procedure in reptiles due to the dispersal of the tissue and its association with large blood vessels. Cauterization of the adrenocortical tissue in the bobtail lizard resulted in an increase in the concentration of K^+ in the plasma but no change in Na^+ (Bentley 1959). A similar response was observed following adrenalectomy in another Australian lizard, *Varanus gouldii* (Rice et al. 1982). The failure to detect an expected decline in the Na^+ concentration in the plasma could be reflecting the tardy metabolism of these cold-blooded animals. Elizondo and LeBrie (1969) induced adrenocortical insufficiency in a water snake, *Nerodia (Natrix) cyclopion*, by occluding the blood supply to the tissue. Sodium, as well as K^+, concentrations in the plasma decreased. The painted turtle, *Chrysemys picta*, responds to adrenalectomy in a more mammal-like manner. The K^+ concentration in the plasma rises and Na^+ declines (Butler and Knox 1970). Study of the renal function of the "adrenalectomized" water snakes suggested that there was a decrease in the fractional reabsorption of Na^+ across the proximal renal tubule.

The administration of aldosterone to the intact water snakes during a water diuresis had no demonstrable effect on renal sodium excretion. However, when they were first given a load of sodium chloride, this hormone increased the fractional reabsorption of Na^+ by the renal tubule (LeBrie and Elizondo 1969). With the benefit of hindsight, we now know that the endogenous levels of aldosterone are high in hydrated reptiles and are decreased by sodium chloride loading (Bradshaw and Grenot 1976; Bradshaw and Rice 1981). Thus, the effects of administered aldosterone would be expected to increase following the administration of the salt load. Administered aldosterone has also been shown to decrease Na^+ and increase K^+ concentrations in the urine of the desert iguana (Templeton et al. 1972a) and three species of freshwater turtles (Brewer and Ensor 1980).

Aldosterone was first measured in the plasma of reptiles in 1976 by Bradshaw and Grenot. It was identified in two lizards; *Uromastix acanthinurus* from Algeria and *Tiliqua rugosa* from Australia. The concentrations were similar to those observed in mammals, and in *Uromastix* were inversely related to the plasma Na^+ levels. Similar observations have since been made in the lizard *Varanus gouldii* and the tortoise *Testudo hermanni* (Bradshaw and Rice 1981; Uva et al. 1982). However, the concentrations of aldosterone in the plasma were lower in the tortoises. The observations of the effects of exogenous aldosterone and its release in response to changes in the Na^+ levels in the plasma are consistent with a physiological role of this hormone on kidney function in reptiles. Corticosterone is also present in the plasma of reptiles, but its release, in contrast to that of aldosterone, is increased in lizards by the administration of sodium chloride (Bradshaw et al. 1972; Bradshaw and Rice 1981). In alligators, this corticosteroid is released when they are placed in dilute sea-water (Laurén 1985). Such release of corticosterone possibly constitutes a stress response.

2.3 Cloaca and Urinary Bladder

The cloacal-colon complex and the urinary bladder, as described earlier, can be sites of substantial fluid absorption. In some reptiles the renal tubular reabsorption of Na^+ is quite low and further conservation of this ion occurs at these extrarenal sites. Various observations have been made in vitro demonstrating the presence of an active Na^+ transport mechanism in the cloaca-colon of the caiman, the Mediterranean tortoise, snakes and several lizards (Bentley 1962a; Bentley and Schmidt-Nielsen 1965; Junqueira et al. 1966; Bentley and Bradshaw 1972; Diaz and Lorenzo 1992). The process of Na^+ reabsorption has also been studied using in vivo perfusion techniques in lizards (Skadhauge and Duvdevani 1977; Bradshaw and Rice 1981). The rate of such Na^+ transport is reduced in *Varanus gouldii* following the administration of sodium chloride (Bradshaw and Rice 1981). However, the change could not be related to the concentrations of vasotocin or aldosterone in the plasma. The Na^+-dependent electrical short-circuit current (in vitro) across the colon of the Mediterranean tortoise and *Amphibolurus* lizards was unaffected by vasotocin (Bentley 1962a; Bentley and Bradshaw 1972). However, aldosterone has been shown to increase such Na^+ transport, in vitro, across the colon of the lizard *Gallotia gallati* (Diaz and Lorenzo 1992). As described earlier, aldosterone has ubiquitous effects in increasing Na^+ reabsorption, including the colons of mammals and birds. Its effect on the lizard colon is reminiscent of these actions.

The urinary bladders of turtles and tortoises are the site of active Na^+ reabsorption from the urine (Brodsky and Schilb 1960; Bentley 1962a). This process can be increased, in vitro, by aldosterone (Bentley 1962a; LeFevre 1973). In contrast to amphibian urinary bladders, vasotocin has no effect on such ion transport in the Mediterranean tortoise or the bobtail lizard (Bentley 1962a; Bentley and Bradshaw 1972).

The urinary bladder and cloaca of turtles may also be involved in the accumulation of sodium from the external environment. *Pseudemys cripta* exchange sodium with bathing fluids at the rate of 0.04 to $10\,\mu mol\ (100\,g)^{-1}h^{-1}$ (Dunson 1967). When the cloaca is blocked, this rate of exchange is reduced by 70%, indicating that the processes of sodium transfer in the cloaca and urinary bladder may be involved. Turtles are known to irrigate their cloacal regions with the fluid that bathes them, so that an active transport of sodium in such regions could mediate a net accumulation of this ion by turtles. In the softshell turtle, Dunson and Weymouth (1965) consider that active sodium transport across the pharynx may be responsible for such solute collection.

The cloaca-colon complex and, when present, the urinary bladder of reptiles, appear to have the capacity to reabsorb, and hence conserve, substantial amounts of Na^+ from the urine. Information regarding possible hormonal control of this process is meagre. However, the fragmentary information that is available suggests that, as in mammals and birds, aldosterone could be contributing to such regulation.

2.4 Cephalic Salt Glands

Like their avian descendants, many reptiles also utilize salt glands for the secretion of concentrated, hyperosmotic, solutions of sodium chloride. Unlike birds, some reptiles can also secrete solutions of potassium chloride. The salt glands of reptiles are more morphologically diverse than the nasal salt glands in birds. There may also be differences in the manner that they regulate the composition of these fluids. Schmidt-Nielsen and Fänge in 1958 first described the functioning of reptile salt glands in the loggerhead turtles and the diamondback terrapin. These salt glands are not nasal glands but modified supraorbital ("tear") glands. However, like birds, the marine iguana, which lives on the shores and reefs of the Galapagos Islands, utilizes a nasal gland for this purpose. Many other lizards utilize such lateral nasal glands for salt secretion. The salt glands that are present in sea snakes are submaxillary glands that lie under the tongue. Somewhat unexpectedly, the estuarine (salt-water) crocodile also has been found to possess salt glands that lie in the tongue (Taplin and Grigg 1981). These lingual salt glands have been found in other crocodiles, but not alligators. Salt glands are also apparently not present in terrestrial snakes and have been described only in a single terrestrial chelonian, *Testudo carbonaria* (Peaker 1978). Salt glands appear to have evolved several times in the reptiles. In marine and intertidal species, they secrete solutions that are hyperosmotic to sea-water and have a Na/K concentration ratio over 30:1. In terrestrial herbivorous lizards, they form equally concentrated solutions but they principally contain potassium chloride and have K/Na ratios as high as 80/1 (Table 5.3). Such K^+ secretion appears to be dictated by the ion content of their plant diet.

Table 5.3. Sodium and potassium secretion by the salt glands of reptiles

	Concentration mEq l⁻¹ Na⁺	K⁺	Ratio Na/K	Rate of secretion µEq kg⁻¹ h⁻¹ Na⁺	K⁺	Habitat
Crocodilia						
Crocodylus porosus[a]	509	11	46	49	–	Estuarine
Chelonia						
Malaclemys centrata[b]	784	–	–	70	–	Estuarine
Caretta caretta[b]	878	31	28	–	–	Marine
Chelonia mydas[c]	685	21	33			
Testudo carbonaria[d]	0.1–6	233–260	0.01	–	–	Terrestrial
Lacertilia						
Sauromalus obesus[e]	44	430	0.1	3	27	Desert
Dipsosaurus dorsalis[f,g]	494	1387	0.35	22	31	Desert
Uromastyx aegyptus[f]	639	1398	0.46	–	–	Desert
Amblyrhynchus cristatus[h]	1434	235	6	2550	510	Marine
Tiliqua rugosa[i]	167	433	0.39*			
Ophidia						
Laticauda semifasciata[k]	686	57	12	730	33	Marine
Pelamis platurus[l]	620	28	22	2180	92	Marine

[a] Taplin and Grigg (1981). [b] Schmidt-Nielsen and Fänge (1958). [c] Holmes and McBean (1964). [d] Peaker (1978). [e] Templeton (1964). [f] Schmidt-Nielsen et al. (1963). [g] Templeton (1966). [h] Dunson (1969). [i] Bradshaw et al. (1984b). [k] Dunson and Taub (1967). [l] Dunson (1968). * Potassium loading.

The process of the initiation of secretion by reptilian salt glands has not been studied in as much detail as that in birds. It may even vary in different species. The administration of an analogue of acetylcholine, methacholine, usually stimulates secretion (Schmidt-Nielsen and Fänge 1958; Templeton 1964; Taylor et al. 1995). This effect is consistent with control by the parasympathetic nervous system, as occurs in birds. The administration of hyperosmotic solutions of sodium chloride initiates secretion. However, in some reptiles, hyperosmotic solutions of sucrose are not an effective stimulus. Hyperosmotic solutions of potassium chloride promote salt gland secretion in herbivorous lizards, but sodium chloride is also, though somewhat less, effective (Templeton 1964; Grenot 1967; Braysher 1971; Shoemaker et al. 1972b; Shuttleworth et al. 1987). The submaxillary glands of sea snakes and the lingual glands of crocodiles respond to the administration of hyperosmotic solutions of sodium chloride (Dunson and Dunson 1974; Taylor et al. 1995). The initiation of the neural processes that apparently control secretion thus appear to reside in receptors responsive to hyperosmotic concentrations of sodium chloride or potassium chloride. Their precise sites have not been identified in reptiles but, as described earlier in birds, they may be associated with the hypothalamus. The presence of a cholinergic nerve supply to the supraorbital gland of the diamondback terrapin has been questioned (Belfry and Cowan 1995). The nerve supply to these glands was found to contain vasoactive intestinal peptide (VIP), but not acetylcholine. Such nerves in birds usually contain both substances. The nerves supplying the lingual salt glands of the estuarine crocodile also contain both acetylcholine and VIP (Franklin et al. 1996). The injection of VIP into these crocodiles evokes a massive increase in salt gland secretion. Methacholine has been observed to *decrease* secretion by the supraorbital salt gland of the green sea turtle (Reina and Cooper 2000). Administered adrenaline usually decreases salt gland secretion in reptiles, as also observed in birds. This response appears to reflect a decreased blood flow to the glands. There are, possibly, differences in the precise mechanisms that are utilized by various reptiles and birds to initiate secretion from their salt glands.

There has been considerable interest in the possible role of adrenocorticosteroids in regulating secretion from salt glands in reptiles. Holmes and McBean in 1964 found that "chemical" adrenalectomy, with injected amphenone, reduced salt gland secretion in green sea turtles. This deficiency could be overcome by the administration of corticosterone. The adrenalectomy also decreased the concentration of Na^+ in the secretion relative to that of K^+. In contrast to this observation, surgical adrenalectomy in the desert iguana (a potassium secretor) resulted in a large increase in secretion from their nasal salt glands and a decrease in the K/Na concentration ratio (Templeton et al. 1968, 1972b). These effects could be reversed by the administration of aldosterone. In the normal lizards, injected aldosterone reduced Na^+ excretion by the salt gland and increased the K/Na concentration ratio. This effect of aldosterone on the secretion of the salt glands of the desert iguana was confirmed by Shoemaker and his colleagues (1972b). They also found that corticosterone and cortisol were effective, but that they were less potent. These corticosteroids may be mimicking the action of aldosterone. The administration of aldosterone to the North African desert lizard *Uromastyx* acanthinurus, a potassium secretor, also decreased Na^+ excretion by their salt glands (Bradshaw et al. 1984a). The administration of K^+ to these lizards increased, while Na^+ decreased, the concentration of aldosterone in the plasma. The rates of K^+ secretion from the salt glands showed a positive correlation and Na^+ excretion a negative correlation with increases in plasma aldosterone. Repeated administration of sodium or potassium chlorides results in appropriate adjustments of the relative amounts of these ions excreted by the salt glands of *Uromastix*. These changes may be mediated by aldosterone. Bradshaw (1997) proposed that the aldosterone may contribute to the process of secretion by influencing both its composition and rate. This action could be a primary one on the metabolism of Na^+ and K^+ by the salt glands. It is reminiscent of the action of aldosterone on the mammalian kidney.

The bobtail lizard *Tiliqua (Trachysaurus) rugosa* can also adjust the composition of its salt gland secretion following the administration of solutions of potassium and sodium chlorides (Braysher 1971). The administration of sodium chloride to this lizard resulted in a modest decline in the concentration of aldosterone in the plasma. However, in contrast to *Uromastix*, corticosterone levels increased (Bradshaw and Grenot 1976; Bradshaw et al. 1984). Binding sites, possibly receptors, for aldosterone and corticosterone were identified in extracts of the nasal salt glands of bobtail lizards. They had a higher binding affinity for corticosterone than for aldosterone. It is possible that corticosterone may be stimulating salt gland secretion in these lizards (Bradshaw 1997). Such an effect would be consistent with the earlier observation of Holmes and McBean (1964) in green turtles. A return to this study of the possible roles of adrenocorticosteroids in turtles would be welcome. Currently, there appear to be no observations on the possible effects of adrenocorticosteroids on the salt glands in crocodiles or sea snakes.

3 Accumulation of Water and Salts

Water and salts are acquired by reptiles from their food and by some species as a result of drinking. Only small accumulations of water and salts can take place across the integument of aquatic species. Carnivorous reptiles generally appear to obtain adequate water from their diets. Some such reptiles, including the Nile crocodile (Balment and Loveridge 1989), the estuarine crocodile (Taplin 1984), sea snakes (Dunson 1969) and the estuarine diamondback terrapin (Bentley et al. 1967a) have been observed to drink fresh water when it is available. Terrestrial herbivores, such as tortoises and lizards, generally consume succulent plants, which usually contain a surfeit of water. However, some conditions, such as life in hot arid environments, evaporative water loss and an excessive intake of salt in the food, may result in drinking behaviour. Drinking in such reptiles may occur periodically following rainstorms, such as observed in gopher tortoises in the Mohave Desert (Nagy 1988) and *Amphibolurus* lizards in arid regions of Australia (Bradshaw and Shoemaker 1967). Schmidt-Nielsen (1964a) described the copious drinking of a desert tortoise that consumed water equivalent to 40% of its body weight in a single such session. Charles Darwin (1839) commented on the prominent drinking habits of the tortoises on the Galapagos Islands. Desert tortoises have even been observed to construct water catchment basins to trap rainwater (Medica et al. 1980).

Amphibians do not drink, so the reptiles appear to be the first tetrapod vertebrates to have developed this custom. As described earlier, mammals and birds may be induced to drink following the administration of angiotensin II (Kobayashi et al. 1979; Kobayashi and Takei 1997). Such a drinking response to injected angiotensin II has also been observed in representatives of the main phyletic groups of reptiles. Most species exhibited such drinking responses. However, the Mediterranean tortoise failed to respond. The tropical lizard *Anolis carolinensis* was relatively unresponsive. The tortoise is adapted to life in dry regions and *Anolis*, which is insectivorous, has little need for drinking water in nature. The relative lack of an angiotensin II-induced drinking response in such reptiles is reminiscent of such insensitivity in some desert mammals and birds (Chaps. 3 and 4). Hibernating reptiles were also found to be unresponsive to angiotensin II.

It is uncertain whether marine reptiles voluntarily drink sea-water. In the estuarine crocodile, an ability to discriminate between salt water and fresh water, and to avoid drinking the former, may be vital for its survival during its marine expeditions (Taplin 1988). Similar conclusions have been drawn with respect to the abilities of related species of water snakes to survive in salt water (Pettus 1958). The freshwater snakes could not survive in salt water, and their demise was attributed to drinking. The salt water species did not drink.

4 Reproduction and Osmoregulation

The need for water may influence reproduction in reptiles. The amniotic membrane, which encloses the fluids in which the embryo develops, facilitates reproduction on dry land. It first appeared in the reptiles. This important innovation is associated with the enclosing of the egg in a shell. Such cleidoic eggs are permeable to gases and water. They also contain the allantoic membrane into which crystals of insoluble uric acid are deposited. Some reptile eggs contain sufficient water for normal embryonic development but others, especially among squamate reptiles, take up additional water. Water may even be lost from the eggs as a result of evaporation or osmosis. Reptiles thus usually deposit their eggs in damp situations where such losses may be limited. Additional water may even be gained during development in such "nests". Reptiles are usually oviparous, but some, such as sea snakes and many lizards, retain their eggs, which develop in the oviduct (ovoviviparous). Some species are even viviparous.

The eggs of reptiles are sometimes guarded by the parents, but the young are usually left to fend for themselves. Their survival depends on many factors, and in hot dry conditions can be influenced by the availability of succulent food and moisture. Such juvenile reptiles, with their relatively high surface area, can be particularly susceptible to evaporative water loss. Thus, reptiles, like other vertebrates, may reproduce when the availability of water and food favours the survival of the eggs and young. The timing of reproductive cycles in reptiles appears to be more related to the environmental temperatures than the photoperiodic cues that are predominant in mammals and birds (see Bentley 1998). Endogenous reproductive rhythms may also be important. In desert reptiles, reproduction may cease in periods of drought, as seen in the chuckwalla lizards in the Mohave Desert (Nagy 1988). In contrast, desert tortoises, which live under similar conditions, lay their eggs even during drought conditions. Some reptiles, such as *Scleroporus* lizards, practice egg retention until the substrate conditions are damp enough to fruitfully deposit their eggs (Andrews and Rose 1994; Mathies and Andrews 1996). A release of vasotocin could be playing a role in such oviposition. The access of water to the eggs of turtles and tortoise can result in the development of larger hatchlings that are better able to survive the vicissitudes of overland migrations to water (Packard 1999). Such adaptations in the patterns of reproduction in reptiles clearly involve changes in the release and actions of hormones. They may influence the onset of breeding behaviour, ovulation and the development of the eggs, oviposition and parturition. Possible adaptations in desert species include differences in the permeability of the eggs to water, the size of the clutch and the times taken for the eggs to hatch. Viviparity has evolved many times in vertebrates, including the reptiles. In cold-blooded animals it is usually considered to provide a more controlled temperature for embryonic development. However, in species living is dry arid conditions, it could also provide a more equitable osmotic situation for the development and survival of the young.

Chapter 6

The Amphibia

The Amphibia bridge the phyletic gap between the fishes and terrestrial tetrapod vertebrates. This transition has involved considerable osmoregulatory, as well as respiratory, change so that water and salt metabolism may be expected to be particularly varied and interesting. For both these reasons, and the relatively free liaison between physiologists and amphibians, the group has received considerable scientific attention.

The Amphibia arose from a crossopterygian fish in Devonian times. This transition is generally thought to have occurred in sea-water. There is only one surviving crossopterygian fish, the coelacanth, *Latimeria chalumnae*, which was not found until 1938. This coelacanth is a marine species, several of which have been caught off the southeast coast of Africa. The contemporary Amphibia are represented by three orders: the Apoda, the Urodela and the Anura. The Apoda, or caecilians, are small, shy worm-like animals, of which there are about 17 genera, living exclusively in tropical equatorial areas of America, Africa and Asia. Unfortunately, very little is known about their physiology and there is little information about their osmoregulation. The Urodela, or the Caudata, contain the newts and salamanders of which there are 43 genera, mostly confined to temperate regions in the Northern Hemisphere and completely absent from Australia. The Anura are the most numerous of the three orders, containing more than 200 genera which are widely distributed around the world. "Nothing – neither climates, nor deserts, nor mountains nor even narrow gaps of salt water – has prevented the spread of dominant frogs over the world in the course of time" (Darlington 1957). Anurans are present on all the major land masses (except Antarctica), where they occupy habitats ranging from freshwater ponds and streams, to tropical forests and arid deserts. A few species even venture into the cold polar regions north of the arctic circle. Many are primarily aquatic, but they are usually more comfortable in fresh than brackish water. At least one species, *Rana cancrivora*, the crab-eating frog, lives in sea-water among mangroves along coastal areas of Southeast Asia. Darlington (1957) concludes that frogs must have crossed narrow gaps of salt water, but notes that they usually have a low tolerance to such solutions, and this presumably accounts, as commented upon by Darwin (1839), for their absence from most oceanic islands. Most amphibians readily lose water through their permeable skin, so that desert regions may also be expected to restrict their movements. Nevertheless, while such areas may slow their migrations, they certainly do not stop them.

The amphibians exhibit a number of well-defined physiological differences from their piscine ancestors and reptilian descendants. Such divergences may have a considerable influence on the osmoregulation of the respective groups so

Table 6.1. Some physiological characters of Amphibia (compared with those of reptiles and bony fishes) that influence their osmoregulation

	Reptilia	Amphibia	Osteichthyes
Poikilothermic	+	+	+
Branchial respiration (gills)	−	− or +	+
Egg amniotic	+	−	−
Kidney:			
1. Hyperosmotic urine	−	−	−
2. Antidiuretic to neurohypophysial peptides	+	− or +	− or diuresis
Urinary bladder derived from cloaca	+	+	−
Nitrogen excretion:			
Uric acid	+	−[a]	−
Urea	+	+	+
Ammonia	+	+	+
Drinks water	+	−	+
Skin: restricted permeability	+	−	+
Extrarenal salt excretion	+	−	+
Extrarenal salt conservation	−	+	+
Neurohypophysis: neural lobe persent	+	+	−
Secretes:			
1. Vasotocin	+	+	+
2. Mesotocin	+	+	−[b]
3. Isotocin	−	−	+
Adrenocortical tissue secretes:			
1. Corticosterone	+	+	+
2. Aldosterone	+	+	−
3. Cortisol	−	some (?)	+

[a] In a few xerophilic anurans.
[b] Present in lungfish.

that I have attempted to summarize some of them (Table 6.1). Perhaps the major factor that affects amphibian osmoregulation is the adoption, by most of the adult forms, of atmospheric, instead of aquatic, respiration. The Amphibia are assisted in this process by gaseous exchanges across their skins (the plethodont salamanders have no lungs and depend solely on their skin for this purpose). As a result, the skin is also more permeable to water. It has been suggested that the crossopterygian ancestors of the Amphibia underwent a reduction of their cutaneous scales in order to facilitate aeriform respiration (Szarski 1962). A relatively impermeable skin was not phyletically restored until the emergence of the reptiles. The other major factor influencing the osmotic life of most amphibians is the necessity to lay their eggs in water and for the larvae to undergo an aquatic period of development before they metamorphose into a form which can live on land. Reptiles have adopted an amniotic and cleidoic egg, which is normally laid on land, and so have achieved a considerable further measure of osmotic independence. There are many other differences in the osmoregulatory pattern of fishes, amphibians and reptiles (Table 6.1) that may further hinder or facilitate their osmotic homeostasis. These will be discussed in subsequent sections.

Water makes up about 80% of the total body weight of most amphibians (see for instance, Schmid 1965) and this is a higher proportion than the 70% seen in most other tetrapods. Nevertheless, this may vary; two Australian tree frogs (family Hylidae), *Hyla moorei* and *H. caerulea*, have water contents equivalent, respectively, to 71 and 79% of their body weights (Main and Bentley 1964). The osmotic concentrations of the body fluids of amphibians are also usually less than those of other tetrapods, about $250 \, \text{mosmol l}^{-1}$; there is also less sodium present in the plasma, about $115 \, \text{mEq l}^{-1}$. This may also vary a great deal depending on the particular species, its habitat and osmotic condition (Table 1.2). The North American spadefoot toad, *Scaphiopus couchi*, can survive in deserts during extended periods of drought. McClanahan (1967) has shown that when this toad aestivates in burrows during such dry periods, the concentrations of solutes in its body fluids rise considerable, from $305 \, \text{mosmol l}^{-1}$ to as high as $630 \, \text{mosmol l}^{-1}$. These frogs emerge after rain when they hydrate and restore the solutes to more normal levels. The crab-eating frog, *Rana cancrivora*, also increases the concentrations of its body fluids to very high levels when living in sea-water (Gordon et al. 1961). Both of these anurans tolerate sodium concentrations up to about $250 \, \text{mEq l}^{-1}$ in their plasma, while large quantities of urea are also accumulated and may reach a level of 500 mM. Considerable decreases in plasma solute levels are also tolerated. Amphibians, even more so than reptiles, thus exhibit a considerable ability to tolerate different solute levels in their body fluids.

Tolerance to changes in the osmotic concentrations of the body fluids may vary in different species and account for their distinctive abilities to withstand differing degrees of dehydration. In 1943, Thorson and Svihla compared the abilities of a number of North American frogs and toads to survive desiccation. Their results indicated that amphibians normally living in more aquatic habitats withstand loss of their body fluids poorly, compared to those which live in drier areas. Thus, the aquatic frog, *Rana grylio*, dies after losing water, by evaporation, equivalent to about 30% of its body weight, while the spadefoot toad, *Scaphiopus hammondi*, can survive a loss of 50% in its body weight. A similar relationship between the habitat and survival of desiccation has been shown by Schmid (1965) in North American anurans, and we have also observed this to be so in some Australian tree frogs (Main and Bentley 1964). Warburg (1967) found that some frogs from dry regions in central Australia withstand dehydration better than those from wetter habitats, but he has emphasized the importance of the speed of water loss in determining the animals' survival. Frogs seem to be able to live longer when the rate of dehydration is slow than when it is fast, and although I have made no precise measurements of this, I would agree that this is probably an important factor. Differences in ability to withstand dehydration are not invariably seen among different groups of anurans. The Australian genera *Neobatrachus* and *Heleioporus* each have four or five species that have adopted habitats ranging from wet forests to dry deserts, and they all survive a water loss equivalent to 40 to 45% of their body weight (Bentley et al. 1958). These Australian frogs (family Leptodactylidae) thus all tolerate water losses as great as those of the best-adapted

North American species or of Australian Hylidae. The Australian deserts are far older than those in North America, and the leptodactylids are an ancient anuran family that probably gave rise to the now more widely distributed Bufonidae. The Australian leptodactylids may thus have evolved into a basically xerophilous group which in the course of long periods of time has adapted more uniformly to dry conditions. As we shall see, these leptodactylids also have a remarkable ability to rehydrate rapidly and can store large volumes of water in their urinary bladders.

Most urodeles live in temperate areas, but some species inhabit more arid regions such as the deserts of the southwest USA. Some, such as the desert sala-mander, *Batrachoseps aridis*, inhabit areas where permanent water seeps exist (Shoemaker 1988). Adult tiger salamanders, *Ambystoma tigrinum*, in New Mexico occupy dry habitats that are similar to those of anurans (Delson and Whitford 1973). Five species of salamanders from temperate areas in North Carolina died after losing water that was equivalent to 18 to 30% of their body weights (Spight 1968). However, tiger salamanders can survive after losing water equal to 45% of their body weight (Alvarado 1972). During dry periods, these urodeles burrow into the damp soil where they apparently aestivate for periods up to 9 months (Delson and Whitford 1973). Under such conditions they accumulate salt and urea in their body fluids, which reaches a concentration of $550\,\mathrm{mosmol\,l^{-1}}$. The Urodela appear to have shared a common ancestry with the Anura. An ability to withstand large increases in the concentrations of their body fluids may have developed early in amphibian evolution. Caecilians (Apoda) are an order of the Amphibia which in evolutionary time predated the Anura and Urodela (Stiffler et al. 1990; Jared et al. 1999). Like other amphibians, they have a permeable skin and are prone to evaporation. Some live an aquatic life or, more usually, they are fossorial and burrow into damp soil and vegetation. Their ability to withstand desiccating conditions is unknown.

The remarkable ability of many amphibians to withstand water loss and con-siderable increases in the concentrations of their body fluids is associated with several aspects of their physiology. They appear to be better able to withstand such conditions when the rates of their dehydration are slower. They reduce this process by seeking refuges, such as burrows, rock crevices and damp soil and vegetation. Urea, instead of being excreted, is retained, and is a substantial com-ponent of the increased concentrations of their body fluids. The haematocrit increases during dehydration and the peripheral circulation declines. However, the water content of the deep visceral organs is maintained (Hillman 1978; Malvin et al. 1992; Churchill and Storey 1994). Amphibians can also store water in capa-cious urinary bladders (see later).

Amphibians may live in three contrasting types of osmotic environment: fresh water, dry land or, very occasionally, in salt water. Larval amphibians, certain urodeles, such as the mudpuppy and the congo eel, anurans like the South African clawed toad and the caecilian *Typhlonectes compressicauda* (Stiffler et al. 1990), are almost entirely aquatic. Many species habitually live near ponds and streams into which they make periodic excursions, and they are thus truly amphibious. The fresh water where such species live, or visit, is hyposmotic to their body fluids so that there is a continual accumulation of water occurring across their skin. This may be equivalent to 30 to 40% of their body weight in a day and is excreted by the kidneys as a dilute (hyposmotic) urine. This urine contains only small, but nevertheless significant, amounts of solutes, and the loss of some of these, especially sodium and chloride, may be physiologically significant. In addition, as shown by the classical studies of Krogh (see Krogh 1939), amphibian skin is permeable to sodium and chloride, so that on simple physico-chemical grounds one would expect further losses to occur across the integument. While in certain unphysiological circumstances, like bathing in distilled water, such losses can be demonstrated, they are usually prevented by an active transport, by the skin, of sodium (and sometimes chloride) from the bathing medium.

Darwin (1839) noted that frogs have a poor tolerance to salt solutions and only two species have been discovered for which this is not true. When leopard frogs, *Rana pipiens*, are placed in sea-water they die in 30 to 60 min. Bentley and Schmidt-Nielsen (1971) found that under these conditions they lose water through the skin, but the main reason for their demise is not water loss, but a massive accumulation of salt that takes place through their skin and as a result of drinking. Even when these frogs are placed in saline solutions similar in concentration to their body fluids (0.7% sodium chloride), they often do not survive for longer than 24 h. In more dilute solutions they live for longer periods, gaining weight due to the accumulation of salt and water (Adolph 1933). Nevertheless, some anurans do habitually enter brackish water and at least one, the crab-eating frog, *Rana cancrivora*, lives in sea-water (Gordon et al. 1961). The European green toad, *Bufo viridis*, has been found in ponds where the salt concentration, while less than that of sea-water, is a substantial 2% (Gislen and Kauri 1959). There are other reports of amphibians being found in brackish water, but it is often difficult to judge how typical such an event is in the life of the animal. The sight or sound of an approaching naturalist can conceivably strike sufficient terror into a frog to prompt it, unwittingly, to leap into a pool of salt water. Gordon (1962) collected *Bufo viridis* in different areas of Europe ranging from Naples to Belgrade and found that these toads, in common with Rumanian ones previously studied by Stoicovici and Pora (1951), could live for extended periods of time in solutions with a sodium chloride content of 2% or sometimes even more. It is, indeed, considered to be euryhaline (Katz 1973; Hoffman and Katz 1997). The most basically important osmotic adjustment *Bufo viridis* and *Rana cancrivora* make towards their life in saline solutions is to initiate and maintain the concentrations of their body fluids at levels that are always, sometimes only slightly, hyperosmotic to the bathing fluids. This elevation in the osmotic pressure of the body's fluids results from an elevation of the sodium levels and, particularly in the instance of *Rana cancrivora*, an accumulation of urea (Table 1.2). The tadpoles of *Rana cancrivora*, however, like all amphibian larvae, cannot form urea, but utilize a completely different mechanism for their life in sea-water. They maintain their body fluids, like those in bony fishes, at a concentration which is hyposmotic to sea-water, possibly excreting excess salt through their gills (Gordon and Tucker 1965).

When amphibians utilize their inherent ability to live on dry land, they are potentially subject to an osmotic stress that is initiated in a manner different from that in sea-water, that of evaporation from the integument. Evaporation from the skin is characteristically rapid in the Amphibia as compared to other tetrapods. Nevertheless, amphibians, especially anurans, have successfully occupied arid areas in which high environmental temperatures and a low content of water vapour in the air considerably facilitate evaporation. Such species utilize a number of mechanisms that somewhat mitigate these conditions. Like all vertebrates living in such places, they have adopted patterns of behaviour that reduce the times of exposure to the extremes of the environment; they seek refuge in cracks in rocks, and burrow, often deeply, into the soil. In addition, a variety of physiological processes affect the osmoregulation of such amphibians.

The amphibians, like other tetrapods, are aided in their osmoregulation by the action of some hormones, the most important being those secreted by the neurohypophysis and the adrenocortical tissues.

The *neurohypophysial hormones* in amphibians are of both phyletic and historic interest. They represent the debut of these hormones in tetrapod vertebrates. They have assumed novel roles concerned with the adaptation to terrestrial life. The neurohypophysis of amphibians and other tetrapods is, compared to that of the fishes, enlarged, and forms the neural lobe (Wingstrand 1966). This morphological change is associated with a greater storage of its peptide hormones. However, some aquatic amphibians, such as the mudpuppy *Necturus maculosus*, retain a more fish-like neurohypophysis. Historically, the study of the amphibian neurohypophysial hormones has played a pivotal role in the emergence of our knowledge of the nature and actions of such hormones in non-mammals. In 1941, Hans Heller observed anomalies in the pharmacological activities of extracts of the neural lobes of English frogs (*Rana temporaria*). Such extracts, like those obtained from mammals, when injected into frogs produce a gain in weight due to the retention of water (Brunn 1921). This action is known as the Brunn effect or water balance response. (It is a composite response and reflects the ability of the hormone to produce an antidiuresis and increased osmotic water transfer across the skin and urinary bladder. The magnitude of the response varies in different species.) Heller standardized his amphibian neurohypophysial extracts by measuring their ability to produce an antidiuresis in rats. Equivalent doses of such standardized preparations from mammals and frogs were then administered to frogs. The neurohypophysial extract from the amphibian was found to be far more active than the mammalian one. Heller suggested that it contained a novel constituent, which he called the amphibian water balance principle. Nearly 20 years later, this "principle" was found to be the peptide hormone 8-arginine-vasotocin (Pickering and Heller 1959; Sawyer et al. 1959; Acher et al. 1960). It is generally referred to as vasotocin or AVT (Fig. 2.5). Subsequently vasotocin was identified in representatives of the main phyletic groups of non-mammalian vertebrates. A second peptide separated chomatographically from such extracts of the amphibian neurohypophysis, based on its pharmacological behaviour, initially, appeared to be oxytocin. However, in 1964, Follett and Heller (1964a) found that it behaved differently to oxytocin in some pharmacological bioassay systems. It behaved more like 8-isoleucine-oxytocin, which had earlier been prepared experimentally by chemical synthesis. The novel structure of the natural hormone was confirmed and it was called mesotocin (Acher et al. 1964). This hormone was subsequently

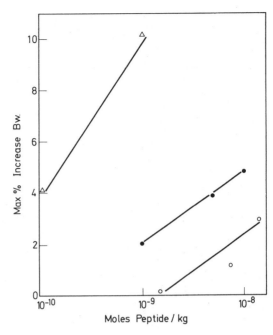

Fig. 6.1. Water balance effects of vasotocin (△), oxytocin (●) and mesotocin (○) in toads (*Bufo marinus*). Note: Log-dose scale. (Bentley 1969a)

found to have a wide phyletic distribution, including birds, reptiles and even lung-fish and kangaroos. It remains possible that oxytocin may also be present in the neurohypophysis of some amphibians (Munsick 1966). Vasotocin, mesotocin and oxytocin can all promote water retention in various amphibians, but vasotocin is by far the most potent of these peptides. In the cane toad *Bufo marinus* it is 200 times more active than mesotocin (Fig. 6.1).

More recently, other vasotocin-like and oxytocin-like peptides have been iden-tified in the amphibian neurohypophysis. Seritocin (5-serine, 8-isoleucine-oxytocin) has been identified in an African toad, *Bufo regularis* (Chauvet et al. 1995). Two natural structural analogues of vasotocin have been found in the amphibian neurohypophysis. They exhibit a similar ability to stimulate water and Na^+ transfer (in vitro) across amphibian skin and urinary bladder. However, in contrast to vasotocin, they have little antidiuretic activity in rats or frogs (Acher et al. 1997). They have been called hydrins, and are formed from the precursor of vasotocin, provasotocin, as a result of differences in the cleavages of amino acids at the C-terminus. Hydrin 1 is a dodecapeptide that is present in the neurohy-pophysis of the aquatic toad *Xenopus laevis*. It is vasotocinyl-Gly-Lys-Arg. Hydrin 2 is a decapeptide that is present in the families Ranidae and Bufonidae. It is vaso-tocinyl-Gly. Such C-terminal-extended natural analogues of vasotocin have, so far, not been identified in other vertebrates. (They are reminiscent of attempts by pharmacologists to produce drug preparations of "impeded" neurohypophysial peptides that have prolonged activities.) The precise physiological roles of the hydrins are unknown. They could be providing more selective effects, such as an increase in water absorption across the skin and urinary bladder without influencing renal function.

The principal *adrenocorticosteroids* that are synthesized by amphibians are aldosterone and corticosterone (Carstensen et al. 1961; Johnston et al. 1967; Vinson et al. 1979; Henderson and Kime 1987). They have been identified in vitro in incubation media containing adrenocortical tissue and also in the plasma of the bullfrog *Rana catesbeiana* and the toad *Bufo marinus*. They have also been identified in the plasma of *Xenopus laevis* (Chan and Edwards 1970; Jolivet-Jaudet and LeLoup-Hatey 1984) and several urodeles including the congo eel and larval tiger salamanders (Stiffler et al. 1986). Cortisol has been identified in some amphibians, but it does not appear to be predominant (Chester Jones et al. 1959; Krug et al. 1983). It is interesting that lungfish possess aldosterone and corticosterone, like the amphibians, but also cortisol, which is predominant in bony fish. Aldosterone, as in other tetrapods, has the greatest mineralocorticoid activity. When assayed, in vitro, for its ability to increase Na^+ transport across frog skin it is 100 times more potent than corticosterone (Yorio and Bentley 1978). Cortisol was about ten times less active than corticosterone. The synthesis and release of corticosteroids in anuran amphibians can be increased by ACTH and the renin-angiotensin system (DuPont et al. 1976). ACTH has also been observed to increase aldosterone secretion in urodeles (DeRuyter and Stiffler 1986). The increased synthesis of aldosterone and corticosterone that occurs in response to ACTH can be inhibited by atrial natriuretic peptide (Lihrmann et al. 1988).

Amphibians have an active *renin-angiotensin system* but lack a discrete juxtaglomerular apparatus. However, renin-secreting cells have been identified in their renal vasculature (Kobayashi and Takei 1997). Amphibian renin is relatively nonspecific as, apart from amphibians, it can also interact with angiotensinogen from birds and reptiles (Nolly and Fasciola 1973). Bullfrog angiotensin I differs from that of other vertebrates by the substitution of asparagine at position 9 (Fig. 2.6). However, the active peptide angiotensin II has the same structure as that of cattle, reptiles and birds, though it differs from that of fish.

Natriuretic peptides (ANP-like, BNP-like and CNP-like) have been identified in the heart and brain of bullfrogs (Sakata et al. 1988; Yoshihara et al. 1990; Fukuzawa et al. 1996).

Prolactin is present in the amphibian pituitary gland and in some species it may influence their osmoregulation. It has a structure that is analogous to that in other tetrapods and it has a similar spectrum of biological activities (for instance stimulation of milk letdown in mammals and crop sac secretion in pigeons). In the bullfrog the homology of its amino acid sequence to that in humans is 65% (see Bentley 1998). It appears to be more closely related to that in lungfishes than any other vertebrates. Two distinct forms of prolactin have been identified in the toad *Xenopus laevis* (Yamashita et al. 1993).

1 Water Exchange

1.1 Skin

Skin plays a major role in the osmoregulation of the Amphibia, due not only to its high permeability to water and salts, but also to the fact that such permeation may vary in different species and be subject at times to physiological regulation.

1.1.1 In Water

Amphibians do not drink, except in certain non-physiological situations as when they are placed in hyperosmotic salt solutions. Instead, water moves across the skin, a process that occurs down a gradient of osmotic concentration. When the solutions on either side of the skin are osmotically equal, little water is transferred. If the external solution is that of an impermeant solute like sucrose, virtually no water transfer occurs. If a permeant solute, such as sodium, is present, small movements of water, consistent with isosmotic solute transfer, may occur in order to maintain osmotic equality. Such transfer of water is normally minor. The movement of water across the skin of frogs and toads is directly proportional to the osmotic gradient between the two sides of the skin (Sawyer 1951). The movement of water is thus considered to be a passive process, not requiring the direct intervention of any metabolic energy. Normally, it is from the outside solution into the animal, but if amphibians inadvertently enter hyperosmotic solutions they lose water; in *Rana pipiens* placed in sea-water, water passes out across their skin at the rate of $55 \mu l \, cm^{-2} h^{-1}$ (P.J. Bentley, unpubl observ).

The rate of water transfer across the skin differs by species. Schmid (1965) has measured the rates of osmotic water transfer across the skins (in vitro) of six different species of North American frogs and toads. He found that the skins of aquatic species like the mink frog, *Rana septentrionalis*, had a lower permeability to water ($6 \mu l \, cm^{-2} h^{-1}$) than those from more terrestrial species like the Dakota toad, *Bufo hemiophrys* ($20 \mu l \, cm^{-2} h^{-1}$). Frogs from habitats judged to be intermediate in their "dryness" showed correspondingly different permeabilities. The skin from anurans with low permeability to water was found to have a higher lipid content than those with a higher permeability (Schmid and Barden 1965).

This osmotic water transfer across the skin of terrestrial toads, *Bufo marinus* and *B. bufo*, is, respectively, 28 and $19 \mu l \, cm^{-2} h^{-1}$, while in the aquatic *Xenopus laevis* it is $7 \mu l \, cm^{-2} h^{-1}$ (Maetz 1963; Bentley 1969a). The skin of adult bullfrogs is about four times more permeable to water than that of their tadpoles (Bentley and Greenwald 1970). Similar differences have been observed among the urodeles. The skin of adult tiger salamanders is about twice as permeable to water as that of their aquatic larvae (Bentley and Baldwin 1980). The skins of aquatic mudpuppies and congo eels are even less permeable to water. Osmotic water uptake in an aquatic caecilian, *Typhlonectes compressicauda*, was found to be about one-third of that in a terrestrial species, *Ichthyophis kohtaoensis* (Stiffler et al. 1990).

When dehydrated amphibians return to water they regain water by absorbing it across their skin. They can also take up water in this way from damp surfaces such as soil. The ability of the Amphibia to gain water from damp earth presumably depends on the moisture content of the soil and the various forces (osmotic, hygroscopic and capillary) that tend to hold water there. The water content of the soil may be quite variable, depending on such things as when rain last fell, the presence of surface vegetation and the proximity of groundwater. Spight (1967a) found that the six species of salamanders studied by him could absorb water from a soil sample that contained about 10% water. When the moisture content was lower, 3 to 4%, the salamanders slowly lost water, showing that the forces controlling water retention by the soil may also have a dehydrating action on the animals. However, it was apparent that these salamanders, which were mostly native to North Carolina, should be able to absorb water from the soil in the agricultural areas of this region at any time of the year. Such an ability to gain water from soil is an important factor in the ability of the fossorial spadefoot toad, *Scaphiopus couchi*, to survive in dry desert conditions (Shoemaker 1988). In 1952, W. T. Stille observed the nocturnal ramblings of three species of frogs (family Ranidae) and a toad, *Bufo woodhousii fowleri*, on a beach in southern Lake Michigan. The toads appeared to search for areas of suitably damp sand to which they apposed the "skin of the groin area" for the replenishment of their body water. The frogs, on the other hand, entered the lake to rehydrate.

When amphibians are dehydrated, they usually absorb water through their skin more rapidly than when they are normally hydrated. There is considerable variability in the magnitude of this process; it may be slow, as seen in aquatic species like *Xenopus laevis*, or fast, as in the terrestrial toad, *Bufo carens*. Such observations on these two species led Ewer (1952b) to suggest that such a response may be better developed in species that are normally terrestrial in their habits as compared to more aquatic species. Parallel with these observations it was found that when neurohypophysial peptides were injected, *Bufo* retained large amounts of water, while *Xenopus* completely failed to respond. These differences are reminiscent of the observations of Steggerda (1937), who found that while terrestrial species of the Amphibia accumulated large quantities of water after being injected with preparations of mammalian neurohypophysial peptides, aquatic ones retained little. An ability to rehydrate rapidly could be important to some amphibians, particularly to those living in areas where water is available only sporadically and quickly disappears either by evaporation or soaking into the soil. It is also possible that entry into open pools of water may increase the risk of attack. Thus, a facilitated rate of absorption would reduce exposure to predators. The rates of rehydration in a large number of North American and Australian amphibians have been measured, and compared in relation to the availability of water in the habitats where they normally live. It is evident that there is considerable divergence in the rates of rehydration, but just how uniformly this may be related to the habitat, thus representing a physiological adaptation, is not clear.

Rehydration in a large variety of North American frogs and toads from the families Ranidae, Bufonidae, Hylidae and Pelobatidae has been measured (Thorson 1955; Claussen 1969), but no relationship between the rate of water accumulation and the "dryness" of the habitat was found. A number or Australian frogs, especially those from the family Leptodactylidae, have also been examined for evidence of such a correlation. We (Bentley et al. 1958) confined our studies to species within the genera *Heleioporus* (five species) and *Neobatrachus* (four species). The habitats of species within these two genera range from areas in which rain falls regularly, on an average of 120 days a year, to desert, or semidesert regions, where the annual rainfall may be spread over only a few days. Frogs of the genus *Heleioporus* all rehydrated at similar rates (about $50\,\mu l\,cm^{-2}h^{-1}$) after being dehydrated to 75% of their normal weight. However, species of *Neobatrachus* showed considerable differences in their abilities to absorb water after such dehydration; *N. pelobatoides*, which lives in areas where rain usually falls on 60–120 days a year, gain water at the rate of $33\,\mu l\,cm^{-2}h^{-1}$, while *N. wilsmorei*, which lives in dry areas towards the interior of the continent, regains water at the rate of nearly $100\,\mu l\,cm^{-2}h^{-1}$. It is notable that such differences in ability to rehydrate were paralleled by the rates at which these frogs accumulated water after being injected with a neurohypophysial peptide, oxytocin. The failure of *Heleioporus* to show any evidence of a correlation between water uptake and the aridity of the habitat could reflect the ability of this group to avoid the excesses of the climate by deep burrowing in friable sandy soils, thus making rehydration ability somewhat redundant. Warburg (1965b) has also found that frogs that live in the arid and semiarid central regions of Australia rehydrate more rapidly than those species that live in more temperate areas. Few urodeles have been examined for their ability to rehydrate following dehydration, but Spight (1967b) measured this in four species of salamanders. He found only relatively small increases in the rates of water absorption by dehydrated animals, the maximum rate of water uptake being observed in *Ambystoma opacum*, and this was only $10\,\mu l\,cm^{-2}h^{-1}$. Urodeles also accumulate water only slowly after being injected with neurohypophysial peptides (see Heller and Bentley 1965).

Whether or not the rate of water absorption through the skin following dehydration can be related to the habitat of the species is not clear. Such a correlation certainly does not exist among all species. Nevertheless, considerable diversity does exist, and this reflects differences in the permeability of the skin that range from about $4\,\mu l\,cm^{-2}h^{-1}$ in dehydrated *Xenopus* to as much as $420\,\mu l\,cm^{-2}h^{-1}$ in the ventral pelvic skin in *Bufo punctatus* (McClanahan and Baldwin 1969). In addition, the permeability of the skin may be augmented during dehydration, in such a manner as to suggest the presence of a regulatory mechanism. In view of these observations, I feel that it is reasonable to suppose that such differences arose and persisted in response to ecological stresses on the Amphibia during their evolutionary history, even though they may not today be precisely related to the life of all contemporary species.

The rate at which amphibians may gain water is directly related to the rate that water passes across their skin. This may depend on several factors. As shown by Schmid's (1965) studies, the permeability to the skin of anurans may differ even in the absence of dehydration, and this will, of course, contribute to the differences observed in the rates of rehydration. The permeability of the skin to water may, as we have seen, be augmented in dehydrated animals, and this could result from several effects. Due to the increased concentration of the body fluids, the osmotic gradient across the skin increases during dehydration, and water would be expected to move along this gradient more rapidly. Such effects must be relatively small, but could contribute to minor increases in water absorption, as Spight observed among urodeles. The increased osmotic pressure of the body fluids probably also exerts a direct action on the skin and increases its permeability to water, such as has been observed in the urinary bladders of toads (Bentley 1964). In 1936, Novelli found that neurohypophysial extracts could increase the permeability of the skin of anurans to water; this action has since been shown to be widespread within this group, especially terrestrial species. The response is absent in *Xenopus*, which is aquatic, and poor in *Rana cancrivora*, which is euryhaline (Maetz 1963; Bentley 1969a; Dicker and Elliott 1969, 1970). The effects of such peptides on the osmotic permeability of the skin of many urodele amphibians, including the tiger salamander, fire salamander (*Salamandra maculosus*) and alpine newt (*Triturus alpestris*), are often small or undetectable (Bentley and Heller 1964, 1965; Bentley and Baldwin 1980). However, cutaneous responses to neurohypophysial peptides have been described in the semiarboreal newt *Aneides lugubris*, the terrestrial, but not the aquatic, phase of *Triturus vitatus* and the California newt *Taricha torosa* (Hillman 1974; Brown and Brown 1980). Mammalian neurohypophysial hormones, like vasopressin, can exert such an action on anuran skin, but the amphibian peptide, vasotocin, is far more active. Bourguet and Maetz (1961) found that it increased the permeability of the skin of *Rana esculenta* even when present at the concentration of only 10^{-10}M. Vasotocin is released into the blood of dehydrated frogs and toads in which it is present at a concentration of 10^{-9} to 10^{-10}M, which should be adequate to mediate increases in the permeability of the skin of dehydrated anurans. Other hormones possibly could also be involved; thus, adrenaline has been shown to increase (in vivo) the osmotic permeability of the skin of toads, *Bufo melanostictus* (Elliott 1968) and frogs in vitro (Jard et al. 1968). This response appears to be a β-adrenergic one (Yokota and Hillman 1984). In *Bufo cognatus* isoproterenol, which is a selective β-adrenergic analogue of adrenaline, increased cutaneous water uptake. Propranolol, a β-adrenergic antagonist drug, blocked 60% of the expected cutaneous water uptake following dehydration. Such an adrenergic effect on water movement across the skin may reflect a sympathetic nerve response and could be facilitating the action of vasotocin.

The entire body surface of amphibians is not always uniformly permeable to water, or Na^+, nor do all areas of the skin respond equally to neurohypophysial hormones. In 1969 McClanahan and Baldwin observed that dehydrated toads, *Bufo punctatus*, could absorb water across a pelvic area of their skin (the pelvic patch) at the rate of $420\,\mu l\,cm^{-2}\,h^{-1}$. Water absorption across the pectoral region of the integument of these toads was too small to measure. The ventral pelvic patch of skin appears to be used for the absorption of water from damp surfaces. In hindsight, it is consistent with the observations of Stille (1952) on the posture of rehydrating toads living on the coastline of Lake Michigan. Comparisons, in vitro, of the permeability, to water and Na^+, of skin from different regions of the integument have been made in several species of amphibians (Bentley and Main 1972; Yorio and Bentley 1977). The permeability to water, and response to vasotocin, were greatest in the pelvic region of the toad *Bufo marinus* and several species of tree frogs (family Hylidae). However, the integument was uniformly permeable in *Xenopus*, which is aquatic, and *Neobatrachus*, which is fossorial. The skin of such pelvic patches is more densely vascularized than that in other cutaneous areas (Roth 1973; Christensen 1974a). The response to vasotocin, in vitro, is reduced when the perfusion of these blood vessels is decreased (Christensen 1974b). This effect appears to reflect localized lowering of the osmotic concentration gradient and the buildup of unstirred layers of fluid in the tissue (Parsons and Schwartz 1991). When the skin is dry and dehydrated there is a reduction in the circulation in pelvic skin, but on contact with water, perfusion is restored (Malvin et al. 1992). The mechanism for this effect is unknown, but it could be due to a decrease in the blood viscosity and changes in neural or hormonal adrenergic activity.

1.1.2 In Air

Amphibians exposed to air can lose considerable amounts of water by evaporation. The rate of water loss may be 50 to 100 times as rapid as in lizards of similar size (Clausen 1969). Evaporation from most amphibians appears to occur at a rate that is similar to that from a free-water surface (Rey 1937; Bentley and Yorio 1979). As demonstrated by Adolph in 1933, it is not changed in frogs by the presence of the skin. However, amphibians can avoid and limit such water loss by utilizing several different strategies. The most common involves their behaviour, nocturnal feeding and adopting postures that reduce their exposed surface area. They often seek cool humid refuges in vegetation, rock crevices and damp soil. Some amphibians, including many Australian frogs, actively burrow into the soil on the onset of desiccating conditions. A number of anurans have developed "waterproofing" mechanisms, which generally involve their exposed dorsal surfaces (Toledo and Jared 1993). Such waterproof frogs include the genus *Chiromantes* (family Rhacophoridae) from Africa, *Phyllomedusa* (Hylidae) from South America and *Litoria* (Hylidae) from Australia. Some amphibians have deposits of relatively impermeable lipids in their skin or "wipe" lipoidal solutions, secreted by skin glands, over their integument. African reed frogs (family Hyperoliidae) (see Chap. 1) accumulate reflective guanine plates in their cutaneous iridophores. The activities of skin glands, which secrete mucus and lipids, in amphibians are controlled by sympathetic adrenergic neural, and possibly hormonal, mechanisms. In the Indian tree frog, *Polypedates maculatus*, which indulges in wiping behaviour, the secretion from skin glands can be stimulated by the injection of adrenaline and blocked by the administration of the β-adrenergic antagonist drug propranolol (Lillywhite et al. 1997).

Many amphibians, especially those living in regions where rainfall is periodic and unpredictable, aestivate during times of drought. This process involves a reduction in their oxygen consumption to levels that may be only 20 to 30% of their standard metabolic rate (Withers 1993). Such aestivation may occur in secluded situations, such as burrows, or exposed areas, as seen in African reed frogs. The osmotic concentrations of their body fluids may gradually rise under such conditions due to water loss and the progressive accumulation of urea. Lee and Mercer in 1967 described cocoons composed of layers of shed skin, the stratum corneum, which envelops several species of Australian lepto-dactylid frogs during their aestivation in burrows. Such cocoons can reduce evaporative water loss by as much as 97% (Withers 1998a). Layers of the stratum corneum are progressively accumulated at the same frequency as the moulting cycle (Withers 1995). Evaporative water loss decreases progressively as the number of layers increases. Such cocoons have also been observed in frogs from Africa and North and South America. They are also formed by some urodeles, including the congo eel (Etheridge 1990). Moulting in amphibians is a rhythmi-cal process that normally occurs every few days and is influenced by endocrine activity arising from the pituitary gland (Budtz 1977). Following the separation of the stratum corneum and the formation of the "slough" shedding occurs. In toads, this shedding can be promoted by the administration of ACTH, corticosterone and aldosterone. The slough is usually eaten. The accumulation of layers of the stratum corneum in aestivating amphibians appears to involve an interruption of normal endocrine activity, but the details of this process are unknown.

1.2 The Kidney

The kidney of adult amphibians, like that of fishes, is a mesonephros. In the early stages of larval development, such as in tadpoles, it is a pronephros. The nephrons usually consist of a glomerulus and Bowmans capsule, a ciliated neck segment, proximal tubule, intermediate segment and distal tubule. The latter is connected to the collecting ducts by a connecting segment. There is no loop of Henle. Arte-rial blood vessels supply the glomeruli and there is a venous, peritubular, renal portal system. This morphological arrangement of the nephron and its blood supply is similar to that in the metanephric kidney of reptiles. The amphibians also lack a capacity to form hyperosmotic urine. The glomeruli are usually quite large. However, two species of Australian desert frogs have been found to only have "rudimentary or degenerate" glomeruli (Dawson 1951). Nevertheless, their kidneys are not completely aglomerular.

Amphibians provided the first experimental preparations for direct observa-tions, by A. N. Richards, of the processes of the formation of the glomerular filtrate and renal tubular absorption in vertebrates (see Smith 1951). The rela-tively large size and accessibility of the nephron allowed sampling of its fluids with glass micropipettes. These studies were made using North American leopard frogs and mudpuppies.

Amphibians in fresh water excrete the water absorbed through their skin as a dilute urine. An anuran like the European frog, *Rana esculenta*, which weighs 100 g, forms about 60 ml of urine during a single day in fresh water (Jard and Morel 1963). Urodeles, in the same situation, secrete similar amounts of urine each day; the mudpuppy (200 g) forms about 50 ml, while the alpine newt (10 g) pro-duces 16 ml, 160% of the body weight (Bentley and Heller 1964). It can be seen that the daily volume of urine that an amphibian in fresh water may secrete is considerable, especially in small animals that have a relatively large surface area in relation to their body weight. When living in sea-water, *Rana cancrivora* forms far less urine; this frog, weighing 50 g, forms only about 2 ml of urine each day (Schmidt-Nielsen and Lee 1962). When amphibians are exposed to a drying atmosphere, urine formation is too small to measure, and in leopard frogs amounts to an anuria (Adolph 1933).

The relationship of the processes of glomerular filtration and tubular water reabsorption to the eventual urine volume has been examined in a number of amphibians. Both of these mechanisms play a prominent role in the regulation of the urine volume. The *Rana esculenta*, referred to above, have a GFR of 125 ml per day, corresponding to the 60 ml of urine per day, so that about 50% of the filtered water is reabsorbed under these conditions. In other circumstances tubular water reabsorption may be greater. The urine volume in *Rana esculenta* is almost directly related to the GFR, and in some individual frogs that form urine at twice the average rate, the GFR is also doubled (Jard 1966). Crab-eating frogs (50 g) in sea-water filter 30 ml of water per day across their glomeruli but produce only 2 ml of urine, so that more than 90% of the filtrate is reabsorbed. The cessation of urine secretion that is observed in dehydrated frogs probably results from the almost complete failure to form a glomerular filtrate. Fewer such observations are available in urodeles and they are rare in caecilians (Pruett et al. 1991). However, both glomerular filtration and tubular reabsorption of water and solutes appear to contribute to the regulation of urine flow in the Amphibia.

Amphibians, in common with reptiles and fishes, cannot secrete urine that is osmotically more concentrated than their body fluids, so that the ability to prevent water loss in this way is limited. The urine can attain isosmoticity with the plasma, as seen, for instance, in aestivating spadefoot toads (McClanahan 1967). In fresh water the urine is markedly hyposmotic, less than $30 \, \text{mosmol} \, l^{-1}$ (see for instance Gordon 1962; Jard and Morel 1963). In sea-water the urine of crab-eating frogs is intermediate in concentration, being slightly hyposmotic to the body fluids, $600 \, \text{mosmol} \, l^{-1}$ compared to $830 \, \text{mosmol} \, l^{-1}$ in the plasma (Gordon et al. 1961).

Neurohypophysial peptides when injected have been shown to reduce the urine volume in many species of anurans and urodeles (Morel and Jard 1968). The Brunn effect, as described earlier, partly reflects this renal response. The antidiuretic effect of the administration of mammalian neurohypophysial extracts (pituitrin, containing both vasopressin and oxytocin) was initially observed to be entirely due to a decrease in the GFR (Burgess et al. 1933). No effect on water reabsorption by the renal tubule was observed. However, in 1952 Sawyer and Sawyer showed that when such hormone preparations were injected into the toad *Bufo marinus* the observed antidiuresis reflected *both* a decline in GFR and an increased reabsorption of water from the renal tubule. When it became available for experimental use, vasotocin was also shown to have such a dual effect during antidiuresis, not only in the toads but also in bullfrogs and European frogs (Sawyer 1957; Uranga and Sawyer 1960; Jard and Morel 1963). Vasotocin was also shown to have an antidiuretic effect in several urodele amphibians (Bentley and Heller 1964; Warburg 1971). However, the contributions of the GFR and renal tubular water reabsorption to this effect were not investigated. In the California newt, *Taricha torosa*, such an antidiuresis was accompanied by an 84% reduction in the GFR (Brown and Brown 1980). The magnitude of the antidiuretic response to administered neurohypophysial hormone preparations varies considerably in different species of amphibians. It appears to be smaller in aquatic species, such as mudpuppies and *Xenopus laevis*. It is absent or poorly developed in larval forms of urodeles and anurans (Alvarado and Johnson 1965,1966; Bentley and Greenwald 1970; Warburg and Goldenberg 1978).

Mesotocin has a *diuretic* effect when injected into frogs (Jard 1966; Pang and Sawyer 1978) and several species of urodeles (Stiffler et al. 1984). This unexpected effect appears to reflect a vasodilatatory action on the afferent glomerular arteriole, which results in an increased GFR (Pang et al. 1982). Mesotocin has also been observed to antagonize the antidiuretic effect of vasotocin in frogs (Morel and Jard 1963). A possible role for mesotocin as a diuretic hormone is still open to speculation.

Although there are sporadic reports of antidiuretic effects of vasotocin in fish (see Chap. 7), such a physiological role is generally considered to be unlikely. The Amphibia appear to be the first vertebrates to have utilized vasotocin as an antidiuretic hormone. However, vasotocin is present in all the phyletic groups of vertebrates and has usually been found to be able to elicit vascular responses that result in an increase in blood pressure. It has been suggested that amphibians first utilized this vascular response of vasotocin to elicit an antidiuretic effect (Pang et al. 1982, 1983). The ability of vasotocin to increase the permeability of epithelial membranes, such as the renal tubule, to water was a subsequent evolutionary event. Administered vasotocin can have a diuretic action in lungfish that is thought to be due to a general vasoconstriction and an increased blood pressure (called a type-I response). In the mudpuppy there is also a rise in blood pressure and a small antidiuresis due to a decreased GFR. This type-II effect is thought to be due to an increase in the vasoconstrictor responsiveness of the afferent glomerular arteriole. The vasoconstrictor effects of vasotocin are mediated by V_1 type receptors. In other amphibians the antidiuretic response to vasotocin involves both a decline in GFR and an increased permeability of the distal renal tubule to water (type-III response). The latter is a phyletically novel response to vasotocin (and vasopressin) that is mediated by V_2-type receptors. In mammals, it provides the principal mechanism for the antidiuretic action of vasopressin (type-IV response).

Neurohypophysectomized amphibians still exhibit an antidiuresis in response to dehydration, despite the absence of vasotocin (Bakker and Bradshaw 1977). The amphibian kidney has a sympathetic-adrenergic innervation and access to circulating catecholamine hormones. This system may provide an alternate mechanism for the control of urine volume. In 1924, Richards and Schmidt showed that adrenaline could influence the renal circulation of frogs. Administered noradrenaline has an antidiuretic effect in bullfrogs (Gallardo et al. 1980; Pang et al. 1982). The α-adrenergic antagonist drug phenoxybenzamine can inhibit this response. It can also promote a diuresis in dehydrated frogs. The antidiuretic response to noradrenaline, like that of vasotocin, is due to a decreased GFR and an increased reabsorption of water across the renal tubule. Exposure of the skin of toads, *Bufo arenarum,* to salt solutions results in a rapid antidiuresis (Petriella et al. 1989). This effect can be inhibited by blockade of sympathetic nerve activity. Water excretion by the amphibian kidney appears not only to utilize vasotocin for the control of urine flow but also the sympathetic-adrenergic system.

1.3 The Urinary Bladder

As a tetrapod innovation, a urinary bladder derived embryologically from the cloaca, makes its phyletic debut in the Amphibia. It is a distensable sac into which the urine passes from the cloaca and, in some species, especially those that live in arid conditions, like the Australian leptodactylid frogs, it can hold fluid equivalent to 50% of the body weight. Townson in 1799 described how frogs "have power of absorbing the fluids necessary for their support . . . through the external skin . . . a large part of them appearing to be retained in the so-called urinary bladder, though gradually thrown off again by the skin." The later experimental observations of Ewer (1952a) and Sawyer and Schisgall (1956) have shown that toads and frogs reabsorb water from their urinary bladders when they are dehydrated (see also Ruibal 1962; Shoemaker 1964). The urinary bladder of many terrestrial amphibians holds large quantities of water compared with more aquatic species and thus affords them a useful store of water, that can be used to replace body water at times when other supplies are limited (Bentley 1966). Such storage of water may be important for the normal daily activities of more terrestrial species away from sources of water. It may allow rehydration to occur at intervals of several days. Stille (1952) observed that toads, *Bufo w. fowleri*, rehydrated, on damp beach sand, only every 5 to 6 days. Such stores of water also appear to be important for the survival of aestivating amphibians.

Before the discovery of vasotocin, mammalian neurohypophysial peptides were shown, when injected, to increase water reabsorption from the bladders of toads and frogs (Ewer 1952a; Sawyer and Schisgall 1956). Vasotocin has been shown to be particularly active in promoting water reabsorption from this organ, being effective, in vitro, at concentrations of 10^{-11} to 10^{-12}M (Jard et al. 1960; Sawyer 1960). Mesotocin and oxytocin also increase the permeability of the anuran bladder to water but, in *Bufo marinus*, they are about 200 times less potent than vasotocin (Fig. 6.2). This reabsorption of urinary water can supplement the role of the renal tubule in water conservation. It may also contribute to the observed antidiuretic effect of vasotocin. In *Bufo marinus*, the urinary bladder contributes about 50% of the total urinary water retained in response to injected vasotocin (Bentley and Ferguson 1967). In bullfrogs, it is about 20% of the total. In the fire salamander, water reabsorption from the urinary bladder has been observed to nearly completely account for the observed "antidiuretic" effect of vasotocin (Bentley and Heller 1965). Neurohypophysial peptides increase the permeability of the urinary bladder to water in many anurans but not that of the aquatic toad *Xenopus* (Bentley 1966). The response of the urinary bladder of the euryhaline crab-eating frog is poor (Dicker and Elliott 1966, 1970). It is also not very prominent in many urodeles, especially aquatic species (Bentley and Heller 1964; Bentley 1971b). However, urinary bladders of the newt *Triturus vitatus*, the salamander *Aneides lugubris*, the eastern spotted newt *Notophthalmus*

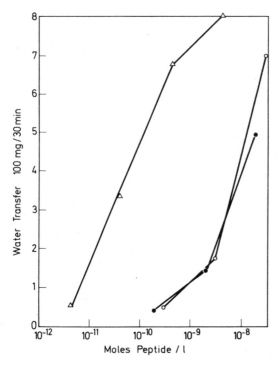

Fig. 6.2. Water transfer across the urinary bladder (in vitro) of the toad, *Bufo marinus*, in the presence of vasotocin ($\triangle$), oxytocin ($\bullet$) and mesotocin ($\bigcirc$). Note: Log-dose scale. (Bentley 1969a)

viridescens, and the California newt *Taricha torosa* have been observed to respond, by increasing water reabsorption, to neurohypophysial peptides (Warburg 1971; Hillman 1974; Brown and Brown 1980). Large urinary bladders have been described in caecilians. They probably act as sites for water storage. However, it is unknown whether they are responsive to vasotocin.

1.4 The Large Intestine (Colon)

The Amphibia have a well-developed large intestine across which sodium is actively transported (Cooperstein and Hogben 1959). Such transport can result in the production of osmotic gradients along which water can move. Mammalian antidiuretic hormone (vasopressin) can increase the osmotic flow of water across this organ in vitro (Cofré and Crabbé 1965, 1967), but this effect is very small, an increase from $3.9\,\mu l\,cm^{-2}\,h^{-1}$ to $5\,\mu l\,cm^{-2}\,h^{-1}$, and is probably not physiologically significant.

While the neurohypophysis stores and secretes vasotocin, which has multiple actions related to water metabolism, its precise physiological role is controversial

(Jørgenson 1993, 1994, 1997a). Vasotocin is released into the circulation of amphibians in response to dehydration, exposure to hyperosmotic solutions and haemorrhage (Bentley 1969b; Rosenbloom and Fisher 1974; Eggena 1986). Its concentrations in the plasma are similar to those in other tetrapods. Neurohypophysectomy is a difficult surgical procedure in amphibians as it is usually accompanied by damage to the adenohypophysis and it can be incomplete. Possibly as a result of such problems, clear deficiencies related to water metabolism have not generally been observed following neurohypophysectomy. In the toad *Bufo marinus*, surgical hypothalamic lesions calculated to block the activity of the neurohypophysis resulted in some deficiencies in their response to dehydration (Bakker and Bradshaw 1977). Water uptake across the skin in response to dehydration and administered hyperosmotic saline were diminished, but not abolished. An antidiuresis still occurred in the toads with lesions following their removal from access to water. However, the response was slower in onset than that of intact toads and it involved only a decrease in GFR. The results were considered to be consistent with the observations of others, using *Rana pipiens* and *B. marinus*, who attempted to ablate the neurohypophysis (Levinsky and Sawyer 1953; Middler et al. 1967; Shoemaker and Waring 1968). However, neurohypophysectomy in European toads, *Bufo bufo*, did not result in any detectable deficiencies in their ability to rehydrate (Jørgenson et al. 1969; Jørgenson 1993). Possibly sympathetic-adrenergic effects on the kidney and skin are predominant (Jørgenson 1993). They may also be able to compensate for a deficiency of vasotocin in such circumstances.

2 Salt Exchange

Most amphibians spend extended periods of their lives in fresh water. At least one species, the crab-eating frog *Rana cancrivora* is truly euryhaline. Some are primarily aquatic, including tadpoles and larval and neotenous urodeles. Others live on the banks of rivers, ponds and lakes and are amphibious, entering the water regularly. Some amphibians lead a more terrestrial existence and only enter water irregularly, such as to breed and rehydrate. While cavorting in aqueous environments, exchanges of water and solutes, especially Na^+ and Cl^-, occur continually across their integument. As described in the previous section, water gained by osmosis is excreted as dilute urine containing low, but significant, amounts of such ions. Solute concentration gradients between the body fluids and the external bathing medium, where the salt concentration may be less than 1 mM, favour a net loss of salt by diffusion. However, as observed by A. Krogh (Krogh 1937, 1939), European frogs, *Rana esculenta,* can survive for many weeks in distilled water. Only small changes in the salt concentrations of their body fluids were found to occur. When they are restored to normal pond water, the frogs then accumulate Na^+ and Cl^- from such solutions by taking it up across their skin. This process can take place from sodium chloride solutions as dilute as 10^{-5} M. It involves active, metabolically dependent, transport of Na^+ and Cl^-. Such active Na^+ transport was demonstrated in vitro in the skin of frogs by Ussing and Zerahn in 1951. It has subsequently also been demonstrated in the skin of many other amphibians. Active Cl⁻ transport, which was demonstrated in vivo (it appears to involve Cl^-/HCO_3^- exchange) is not as readily demonstrated in vitro but has been observed in the skin of some species, such as the South American frog *Leptodactylus ocellatus* (Zadunaisky and Candia 1962).

The magnitude of the losses of salts by aquatic amphibians depends on the permeability of their skin both to ions, and, indirectly, to water. (Excess accumulations of water are excreted as dilute urine with an associated loss of salt.) Net losses of Na^+ and Cl^-, as a result of diffusion follow their electrochemical concentration gradients across the skin. The magnitude of such efflux mainly depends on the special permeability properties of the skin with respect to these ions. It has been observed that amphibians that habitually live in fresh water have an integument that has a lower permeability to water and salts than those species leading amphibious or terrestrial lives (Greenwald 1972; Yorio and Bentley 1978). In addition, most amphibians can utilize cutaneous ion "pumps" to take up salts from bathing solutions. Krogh (1939) demonstrated that after keeping frogs in distilled water for prolonged periods of time, the activity of such ion accumulation mechanisms was enhanced. Maetz and his colleagues (Maetz 1959) also found such an increase in the activity of Na^+ transport. They also found that when the frogs were kept in 0.7% sodium chloride solutions the rate of Na^+ transport by their skin was greatly reduced. Regulatory processes that are sensitive to the amount of salt in the frogs' environment appear to be present. Evidence for such a control of salt accumulation across the integument has been observed in many species of anurans and urodeles and even two species of caecilians (Stiffler et al. 1990). Only sporadic information is available about the nature of such regulation in vivo but active Na^+ transport across amphibian skin can be stimulated in vitro by aldosterone and vasotocin.

2.1 Skin

Transcutaneous active transport of Na^+ has been observed to occur across the skin of anurans from diverse families, including Ranidae, Bufonidae, Hylidae, Leptodactylidae and Pipidae. It has also been demonstrated in many urodeles, including the genera *Triturus, Salamandra, Ambystoma, Siren* and *Amphiuma*. Such ion transport also occurs across the integument of the caecilians *Typhlonectes compressicauda* and *Ichthyophis kahtaoensin* (Stiffler et al. 1990). Transcutaneous salt transport does not appear to be present in all amphibians, suggesting that it is not necessarily essential for their survival. For instance, it is not present in premetamorphic bullfrog tadpoles (Alvarado and Johnson 1966) or larval tiger salamanders (Bentley and Baldwin 1980). It is also absent in the mudpuppy (Bentley and Yorio 1977). However, transcutaneous active Na^+ transport can be promoted in this neotenous urodele by exposing the skin to the amphotericin B. This drug appears to increase the permeability of the apical plasma membranes of the cutaneous epithelial cells. Normally in this species, and probably in other aquatic larval forms, there appears to be a deficiency of sodium channels (ENaC) in the apical plasma membranes of the cutaneous epithelial cells. The promotion of metamorphic climax in amphibian larvae by thyroid hormone appears to be associated with the generation or activation of such sodium channels and cutaneous Na-K-ATPase. Such larval amphibians presumably maintain their salt balance by minerals obtained from their diets and by restricting their losses across the integument and in urine. However, it is also possible that some larval amphibians utilize their gills for salt accumulation (Alvarado and Moody 1970).

Various zones of the integument of amphibians may exhibit different abilities to transport salt. Electrical measurements (in vitro) of the "short-circuit current", which reflects active transcutaneous ion transport (usually of Na^+), indicated that the skin of *Xenopus laevis*, *Rana pipiens* and *Bufo marinus* exhibited uniform such activity, but in the tree frog Litoria (*Hyla*) *moorei* it was much greater in the pelvic area (Bentley and Main 1972). Such differences in regional ion transport have been observed in four other species of tree frogs (Hylidae) (Yorio and Bentley 1977). This enhanced ion transport appears to be associated with the pelvic patch area in such species. The pelvic skin of the toad *Bufo woodhouseii* displays a similarly enhanced ability to actively transport Na^+ (Baker and Hillyard 1992). The possible physiological significance of this activity of the pelvic skin is unknown, but it may reflect morphological differences, such as skin thickness, which are associated with increased osmotic permeability at such sites.

Neurohypophysial peptides not only increase the permeability of amphibian skin to water, but they can also promote an increase in active transcutaneous Na^+ transport. Fuhrman and Ussing (1951) demonstrated that mammalian neurohypophysial peptides increase Na^+ transport across frog skin (in vitro) and vasotocin has since been shown to be even more effective (Jard et al. 1960; Maetz 1963). This action of these peptides on Na^+ transport (the natriferic effect) has been demonstrated in a number of anurans including *Rana esculenta*, *R. catesbeiana*, *Bufo bufo*, *B. marinus* and *Xenopus laevis* (Maetz 1963; Bentley 1969a). It does not occur in *R. catesbeiana* tadpoles (Alvarado and Johnson 1966). A natriferic response has also been observed in several urodeles, including *Triturus alpestris*, *T. cristatus* (Bentley and Heller 1964), *Ambystoma mexicanum* (Aceves et al. 1968), *Aneides lugubris* (Hillman 1974) and adult *Ambystoma tigrinum* (Bentley and Baldwin 1980). It has not been observed in larval *A. tigrinum* (Bentley and Baldwin 1980) or the neotenous urodeles *Amphiuma means*, *Siren lacertina* (Bentley 1975) or *Necturus maculosus* (Bentley and Yorio 1977). The natriferic effect of neurohypophysial peptides is not invariably associated with the increase in water permeability (the hydrosmotic response) that these peptides can also induce. Thus, in the mudpuppy there is a small hydrosmotic response but no natriferic response (Bentley and Yorio 1977). The newt *Triturus alpestris* exhibits a natriferic response to vasotocin but there is no detectable hydrosmotic effect (Bentley and Heller 1964). The two responses involve different effector mechanisms; ENaC channels for the natriferic response and, presumably, an aquaporin for the hydrosmotic response (see Chap. 2). It is unknown if the natriferic effect of vasotocin on amphibian skin has any physiological significance. Compared to the actions of aldosterone (see below), this effect of vasotocin is quite transitory. The conditions for the release of vasotocin are also inappropriate for a role in sodium homeostasis. (Release is inhibited by hyposmotic conditions such as would be expected to occur during sodium deficiency.) Possibly, it makes a localized contribution to the functioning of the skin such as could involve maintenance of its permeability and sensory processes that utilize electrical signals. The latter could be important for the detection of salts in the external medium and the selection, during dehydration, of a suitable solution for "cutaneous drinking" (see later).

Aldosterone is secreted by amphibians, as in other tetrapods, in response to a depletion of sodium in the body. This effect has been observed in toads (Crabbé 1961a; Garland and Henderson 1975) and larval tiger salamanders (Stiffler et al. 1986). Conversely, elevated concentrations of Na^+ in their bathing media depress the levels of aldosterone in the plasma. Adrenocorticosteroids, including aldosterone, have well-defined and persistent effects in stimulating active Na^+ transport, in vitro, across the skin of frogs and toads (Taubenhaus et al. 1956; Maetz 1959; McAfee and Locke 1961; Crabbé 1964; Crabbé and DeWeer 1964; Yorio and Bentley 1978). Injected aldosterone increases the uptake of Na^+ by larval tiger salamanders (Alvarado and Kirschner 1963). A physiological role for aldosterone in promoting Na^+ uptake across the skin of amphibians does not appear to be in doubt, though its precise quantitative contribution is uncertain.

Amphibian skin has a sympathetic-adrenergic innervation that supplies mucous glands and blood vessels, and may influence the permeability and ion transport activities of its epithelial cells (Castillo and Orci 1997). Adrenaline promotes a loss of Cl^- from frog skin by stimulating secretion by mucous glands (Koefoed-Johnsen et al. 1952). This response has been observed in a variety of anurans (Castillo and Orci 1997). Noradrenaline has also been found to stimulate Na^+ transport across frog skin in vitro (Bastide and Jard 1968; Watlington 1968). These effects of catecholamines appear to be principally β-adrenergic ones. The secretory response of the mucous glands may be important in maintaining the hydration of the skin in terrestrial environments. However, the consequences of the various cutaneous actions of catecholamines on the amphibian's salt metabolism are unknown.

Insulin has been shown, in vitro, to increase active Na^+ transport across the skin of *Bufo marinus* (André and Crabbé 1966). Its action appears to be synergistic with that of aldosterone. It is unknown whether this effect reflects a normal physiological interaction between the two hormones in vivo.

Prolactin has been observed to influence the permeability of the skin of urodeles, which is reminiscent of its action on the gills of teleost fish (see Chap. 7). Its general effect, which is seen in larval and neotenous amphibians, appears to involve a reduction in permeability to water and salt (Brown and Brown 1987). Pang and Sawyer (1974) observed that hypophysectomy results in large losses of Na^+ across the integument of mudpuppies. This deficiency could be corrected by the administration of prolactin. The injection of prolactin into the newt *Triturus cristatus* during its terrestrial phase results in a decline in osmotic permeability and Na^+ transport across their skin (Lodi et al. 1982). These conditions were normally observed in the newts when they were in their aquatic phase. In the newt *Notophthalmus (Triturus) viridescens*, long-term treatment with prolactin has been observed to increase the thickness of the skin and decrease its permeability to water and Na^+ (Brown and Brown 1973). An electrical analysis of the permeability of the skin of the Japanese newt *Cynops pyrrhogaster* demonstrated that prolactin could block Na^+ transport channels (ENaC ?) (Takada and Shomazaki 1988). Prolactin may have a special role in the adaptation of the skin of urodeles to aquatic life.

2.2 The Kidney

The urine of amphibians kept in fresh water contains little salt. Thus, the ureteral urine from *Rana esculenta* living in fresh water contains only about $5\,mEql^{-1}$ sodium (Jard and Morel 1963) while in urodeles, such as *Triturus*, the sodium levels in the bladder urine are similar to this. Analysis of the glomerular filtrate obtained by micropuncture of the renal tubules indicates that sodium and chloride are progressively reabsorbed as they pass down the nephron, this occurring mainly in the proximal, but also in the distal, segment (Walker et al. 1937). When euryhaline green toads, *Bufo viridis*, are acclimated to hyperosmotic bathing media containing 115 and $250\,mEql^{-1}$ sodium chloride, they initially accumulate salt (ShPun and Katz 1995, 1999). The subsequent urine flow and renal clearances of Na^+ and Cl^- were related to the concentration of salt present in the external solution. The GFR was similar in the toads in either solution. Salt excretion was regulated by changes in the renal tubular reabsorption of salt. The kidneys of the toads contributed "efficiently" to regulation of the ions in such media.

The role of hormones in regulating these processes in amphibians is not clear. While from the mammalian evidence it seems likely that corticosteroids influence sodium reabsorption and potassium excretion, this has not been directly shown in frogs and toads. Injections of aldosterone into *Bufo marinus* have no consistent effect on renal sodium losses (Middler et al. 1969). Mayer (1963, 1969) also found this to be so in *Rana esculenta*. When larval tiger salamanders are adapted to an external medium of distilled water or $150\,mEql^{-1}$ sodium chloride, the concentrations of salt in the urine, respectively, decline or increase (Stiffler et al. 1986). The concentration of aldosterone in the plasma was found to be six-fold higher and corticosterone was 45% greater when the salamanders were kept in the distilled water as compared to normal pond water. In the salt solution the plasma aldosterone concentration was depressed by 47%. The correlation between urine Na^+ levels and the aldosterone in the plasma of the salamanders suggested that the hormone was increasing renal tubular reabsorption of salt. It has also been observed that administration of the drug aminoglutethimide, which reduces synthesis of adrenocorticosteroids (chemical adrenalectomy), decreased the renal tubular reabsorption of Na^+ (Heney and Stiffler 1983). The failure to demonstrate direct effects of aldosterone on renal tubular Na^+ reabsorption in anurans is vexing.

Vasotocin increases sodium reabsorption by the renal tubule of *Rana esculenta* (Jard and Morel 1963; Jard 1966), an observation similar to that of the action of ADH in the mammalian renal tubule. As described earlier (Chap. 1), this response in mammals may contribute to the maintenance of the renal medullary solute concentration gradient that is necessary for the formation of hyperosmotic urine. Amphibians lack the necessary renal morphology for such a process, but the natriferic tubular response is present. Whether it normally contributes to renal Na^+ conservation in amphibians or is merely a "preadaptation" is unknown.

Natriuretic peptides, ANP, BNP and CNP, have been identified in the Amphibia. Their actions have been studied in bullfrogs, *Rana catesbeiana*, toads, *Bufo marinus* and *Xenopus laevis*. Their effects are similar to those that have been observed in mammals, but the precise mechanisms mediating the responses may differ. In bullfrogs, several natriuretic peptides had a vasodilatatory action, reduced the force of contraction of the heart and decreased blood pressure (Uchiyama et al. 1997). Frog ANP was found, in vitro, to induce a diuresis and natriuresis in the kidney of *Bufo marinus* (Meier and Donald 1997). Specific binding sites, probably receptors, for natriuretic peptides have been identified in the glomeruli of *Xenopus laevis* and *Bufo marinus* (Kloas and Hanke 1992; Meier and Donald 1997). (Such binding sites were also identified in the toad urinary bladder.) Cyclic GMP mediates the effects of natriuretic peptides. Increases in its concentration in response to the presence of natriuretic peptides have been observed in preparations of the glomeruli, but not the renal tubules, of bullfrogs (Uchiyama et al. 1997). The action of natriuretic peptides on the amphibian kidney appears to involve actions on the glomerulus. At this time it would appear that a renal tubular response, which is seen in mammals, is absent.

2.3 The Urinary Bladder

The urinary bladder of the Amphibia, in addition to performing the function of a water reservoir, can decrease the sodium content of the urine. Active sodium transport has been shown (in vitro) to take place across the urinary bladders of frogs, *Rana esculenta*, and toads, *Bufo marinus* (Leaf et al. 1958).

Sodium transport across the urinary bladders of *Bufo marinus* and *Rana ridibunda* is increased by aldosterone, either when the animals are injected 24 h prior to measuring sodium transport in vitro, or after the tissue is directly exposed to the steroid's action (Crabbé, 1961b,a; 1963). Other steroids, including corticosterone, are not as active, and at very high concentrations (100 times as great) may inhibit its action. Changes in the sodium concentration of the media bathing the toads influences the rate of sodium transport across the bladder in a manner that parallels that of aldosterone in the blood; a low external sodium level increases aldosterone in the blood and stimulates sodium transport.

Leaf et al. (1958) found that oxytocin increases sodium transport across the toad urinary bladder and vasotocin has been found to be very active in this process in frogs (Jard et al. 1960). In *Bufo marinus* vasotocin is about 100 times more active in increasing sodium transfer across the bladder than either oxytocin or mesotocin (Fig. 6.3).

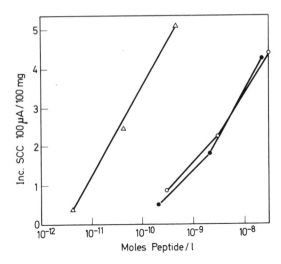

Fig. 6.3. Sodium transfer (as short-circuit current) across the urinary bladder (in vitro) of the toad *Bufo marinus* in the presence of vasotocin (△), oxytocin (●) and mesotocin (○). Note: log-dose scale. (Bentley 1969a)

As in the anuran skin, several other hormone preparations have been shown to promote sodium transport across the bladder. These include insulin (Herrera 1965) and adrenaline (Jard et al. 1968). Various endocrine factors may interact in their effects on the bladder. Aldosterone can facilitate the effects of neurohypophysial peptides (Fanestil et al. 1967; Handler et al. 1969). Aldosterone and insulin, similarly, together can produce a far greater increase in sodium transport in vitro than either can produce alone (Crabbé and Francois 1967). Such synergistic endocrine interactions probably assist such ussues to attain optimal levels of sodium transport.

The urinary bladder is the site of Na^+ reabsorption in several urodele amphibians, including *Necturus maculosus*, *Ambystoma tigrinum*, *Amphiuma means* and *Siren lacertina* (Bentley and Heller 1964; Bentley 1973a). This process, measured in vitro as the electrical short-circuit current, which could be inhibited by amloride, was usually unresponsive to vasotocin and aldosterone. However, these amphibians were all maintained in dilute solutions, which are conditions when the endogenous concentrations of aldosterone in the plasma, and Na^+ transport would already be expected to be high (Stiffler et al. 1986). Responses to exogenous aldosterone may then be suppressed. Aldosterone, when preinjected into *Necturus*, increased the short-circuit current across its urinary bladder (measured in vitro) (Bentley 1971b).

Sodium reabsorption from urine stored in the urinary bladders of amphibians may contribute to the conservation of this ion in many species. It appears to be acting as a functional extension of the distal renal tubule and, as in the kidney of many tetrapods, can utilize aldosterone to control this process.

2.4 The Large Intestine (Colon)

Sodium is actively transported from the mucosal to serosal side of the colon in anurans (Cooperstein and Hogben 1959). In the toad, *Bufo marinus*, this process can be increased by aldosterone and neurohypophysial peptides (Cofré and Crabbé 1965, 1967).

Aldosterone has been identified in the plasma of several amphibians and its release occurs under conditions of a sodium deficiency. This adrenocorticosteroid exhibits a typical mineralocorticoid action, which is absent in fishes, but present in other tetrapod vertebrates. In amphibians an important target tissue is their permeable skin. The colon and urinary bladder are additional sites of its action, and these tissues are similarly responsive in other tetrapods. However, it is not clear whether a renal response to aldosterone is widespread in amphibians. The overall contribution of aldosterone to sodium and potassium metabolism in amphibians is not clear. Adrenalectomy is a difficult operation in such animals and reported attempts to perform this procedure are sparse. The results are complicated by the simultaneous loss of the glucocorticoid action of corticosterone. The toad *Bufo arenarum* has been shown to suffer an excessive loss of salt following adrenalectomy (Marenzi and Fustinoni 1938). Fowler and Chester Jones (1955) destroyed the adrenocortical tissue in the frog *Rana temporaria*. In summer, the frogs died within 2 days, although in winter they survived for prolonged periods of time. The summer frogs lost large amounts of Na^+ and accumulated K^+. This response is what occurs in mammals suffering from adrenocortical deficiency. When the summer frogs were placed in isosmotic saline they survived longer, which suggests that their demise was due to faulty electrolyte balance (Chester Jones et al. 1959). The site, or sites, of the excessive Na^+ loss in the adrenalectomized frogs was not determined. In larval tiger salamanders, chemical adrenalectomy with the drug aminoglutethimide resulted in a decline in the concentration of Na^+ in the plasma and increased its excretion in urine (Heney and Stiffler 1983). This deficiency could be corrected by the administration of aldosterone or, in larger doses, corticosterone.

The synthesis of aldosterone in amphibians is under the control of both ACTH and the renin-angiotensin system (Du Pont et al. 1976; De Ruyter and Stiffler 1986). Adenohypophysectomy in *Bufo marinus* results in an excessive loss of Na^+ that did not appear to involve kidney function (Middler et al. 1969). This salt loss could be prevented by the administration of ACTH. Hypophysectomy in the newt *Triturus cristatus* also resulted in an excessive loss of Na^+ (Peyrot et al. 1963). The synthesis of corticosterone is also regulated by ACTH and its loss following hypophysectomy complicates the interpretation of such experiments. Nevertheless, the available evidence supports the view that the adrenocorticosteroids, especially aldosterone, play an important role in the osmoregulation of the Amphibia.

3 Nitrogen Metabolism

The main catabolic end products resulting from the deamination of amino acids can be ammonia, urea or uric acid (see Chap. 1). They all require water for their excretion. Ammonia, which is quite toxic, is the principal such excretory product of ammonotelic animals, including aquatic amphibians such as *Xenopus laevis* and the mudpuppy. Ammonotelic amphibians can often also form urea, but in their normal aquatic conditions ammonia is predominant. The ammonia is principally excreted across the skin into the surrounding water. Terrestrial amphibians, with limited access to water, usually utilize urea as the main end product of nitrogen metabolism. It is mainly excreted in the urine. A few arboreal frogs can convert catabolic nitrogen to uric acid, which is also excreted in the urine. However, it requires less water for this process than urea.

Urea has a relatively low toxicity and can, under various circumstances, be accumulated in high concentrations in the tissue fluids of amphibians (see Wright 1995; Jørgenson 1997b; Withers 1998b). In euryhaline species, such as *Rana cancrivora* and *Bufo viridis*, urea can be utilized as a "balancing osmolyte". It helps maintain the concentrations of the body fluids at levels that are slightly hyperosmotic to those of the external saline media (Gordon et al. 1961; Gordon 1962; Katz 1973). This osmotic strategy is also used by marine elasmobranchs. These fish also accumulate other solutes, such as trimethylamine oxide, which reduce the potential toxic effects of high concentrations of urea. However, amphibians appear to be even more tolerant to urea than elasmobranchs, and do not appear to resort to this strategy (Withers and Guppy 1996). Aestivating amphibians, such as the desert spadefoot toad *Scaphiopus couchi* (McClanahan 1967) and the urodele *Siren lacertina* (Etheridge 1990), also accumulate urea. Concentrations of urea as high as 900 mM have been observed in the plasma of the toads. In these circumstances it acts as a "storage osmolyte" which is subsequently excreted in the urine when adequate water becomes available.

Urea synthesis and excretion may be regulated in response to the amphibian's current needs. *Xenopus laevis* in their normal aquatic environment are ammonotelic. They periodically aestivate, in the mud, when the pools where they live dry up. The aestivating toads accumulate urea at high concentrations in their body fluids (Balinsky et al. 1967). This metabolic change is associated with an increase in the levels of the liver enzyme carbamoylphosphate synthetase that directs ammonia into the ornithine-urea cycle (see Fig. 1.5). This response may be a direct result of the toxicity of rising levels of ammonia in the body or a consequence of dehydration. Such increases in the ureogenic activity of the hepatic ornithine-urea cycle occur in other amphibians, including *Scaphiopus couchi*, *Bufo viridis* and *Rana cancrivora*. Such adaptation is not associated with changes in oxygen consumption. However, increases in urea synthesis may also be favoured by a shift to protein catabolism, as observed in *Scaphiopus* (Jones 1980) and a stimulation of gluconeogenesis, as seen in *Bufo viridis* (Hoffman and Katz 1998). Adrenocorticosteroids promote gluconeogenesis and mobilization of tissue proteins in vertebrates, but there appears to be no information about this hormonal response in such amphibians (Jungreis 1976).

Renal retention of urea provides an important mechanism for its conservation in *Rana cancrivora* (Schmidt-Nielsen and Lee 1962) and *Bufo viridis* (ShPun and Katz 1995). Urea excretion by the kidneys in many amphibians is influenced by the GFR and its secretion by renal tubule (Forster 1954). However, in anurans which utilize urea as a balancing osmolyte, urea appears to be mainly conserved by its tubular reabsorption (Carlisky et al. 1968; Schmidt-Nielsen and Lee 1962). Urea can also be reabsorbed across the urinary bladder of the toad *Bufo marinus* and this process can be increased, in vitro, by neurohypophysial peptides (Leaf and Hays 1962). Vasotocin has been observed to increase urea absorption, in vitro, across the urinary bladder of the crab-eating frog (Dicker and Elliott 1973). It may be contributing to urea conservation by this euryhaline frog. The administration of ACTH, corticosterone and aldosterone increases plasma urea concentration in neotenic Mexican axolotls (*Ambystoma mexicanum*) (Schultheiss 1977). It was suggested that a decreased excretion of urea could be contributing to this effect. Increased protein utilization due to a gluconeogenic effect of the adrenocorticosteroids could also be involved. Amphibian skin is permeable to urea, and neurohypophysial peptides can increase this process (Andersen and Ussing 1957). The magnitude of this response to vasotocin varies in different species (Yorio and Bentley 1978). It could provide an extrarenal avenue for urea excretion in aquatic conditions such as could be utilized during rehydration following aestivation.

Uricotelism has been observed in frogs of the genus *Chiromantis* (family Rhacophoridae) that live in Africa (Loveridge 1970; Drewes et al. 1977) and *Phyllomedusa* (Hylidae) from South America (Shoemaker et al. 1972a). These tree frogs live, without access to water, in dry exposed situations for prolonged periods of time. They reduce their evaporative water loss by waterproofing their skin. Their uricotelism would be expected to further reduce their need for water. Until the discovery of these amphibians, it had been generally assumed that the use of uric acid as a method for disposing of catabolic nitrogen had first evolved in reptiles. Campbell and his colleagues (1987) have suggested that uricotelism probably arose in an amniotic amphibian ancestor of the reptiles. However, it seems doubtful that these extant frogs are relicts of such an ancestry. They probably reflect separate evolutionary origins of this important biochemical process.

4 Reproduction

The Amphibia, in contrast to other tetrapods, produce anamniotic eggs which lack a shell. Such non-cleidoic, fish-like eggs generally must be deposited in water. Fertilization is usually external. Thus, even terrestrial amphibians generally need to return to water to breed. However, a few frogs and urodeles, and most caecilians, can retain eggs in the oviduct, where they undergo a period of gestation (Wake 1993). In anurans and urodeles such ovoviviparous and viviparous species usually live in cool alpine areas. This reproductive strategy can thus be viewed as an adaptation to cold. However, as will be described below, at least one such frog lives in hot deserts, and all the caecilians inhabit hot tropical regions.

The aquatic larvae of amphibians, on hatching, undergo a period of growth and differentiation before undergoing metamorphosis to a juvenile adult form. Some frogs, such as the Australian leptodacylids and myobatrachids, deposit their eggs in damp soil at the bottom of deep burrows (Main et al. 1959; Roberts 1981, 1984). They emerge from this seclusion, as juveniles, following rain. The eggs of some such frogs are enclosed in a gelatinous capsule. Other amphibians are ovoviviparous or are even considered to be viviparous. Species that live in dry desert conditions utilize various such methods of reproduction, but most are conventional and, following rain, seek out ephemeral ponds and pools.

The amphibian reproductive cycle, like that in other vertebrates, is controlled by the hypothalamic-pituitary-gonadal axis of hormones (Lofts 1984). Environmental temperature and rainfall, rather than light, appear to be the principal external stimuli that influence their reproduction. In areas with predictable weather, reproduction is seasonal, and may also involve internal physiological rhythms. However, in dry regions, such as occur over much of Australia, the trigger to reproduce may be the occurrence of rain. Such "opportunistic" breeding also occurs in Australian birds. The frogs appear to achieve a "tonic" reproductive condition so that ovulation, oviposition and spermiation can occur rapidly, sometimes within hours of rain falling. It has been observed that the oviduct of amphibians can contract in response to the presence of vasotocin (Heller 1972), and this hormone could be involved in such oviposition.

Preparative reproductive behaviour and migration to water to breed (water drive) can be promoted in the eastern spotted newt *Notophthalmus viridescens* by the administration of prolactin (Reinke and Chadwick 1939; Chadwick 1940, 1941). This effect has been observed in other newts including *Triturus cristatus*, *T. alpestris* (Tuchmann-Duplessis 1948; Giorgio et al. 1982) and the Japanese newt *Cynops pyrrhogaster* (Toyoda et al. 1996). The actions of prolactin, as described earlier, also include morphological changes in the structure and osmotic properties of the skin, which are consistent with the adoption of aquatic life. Sexual behaviour in male rough-skinned newts, *Taricha granulosa*, can be stimulated by the administration of vasotocin (Zoeller and Moore 1988). However, this effect appears to be distinct from its actions on osmoregulation and involves local actions in areas of the brain concerned with sexual behaviour.

Premetamorphic larval development in amphibians is assisted by the activities of the hypothalamo-pituitary-thyroid axis, prolactin and, possibly, adrenocorticosteroids. Thyroid activity is low in early premetamorphic life, during which growth is promoted and metamorphosis is inhibited by prolactin (Dickoff 1993). Metamorphic climax is precipitated by rising levels of thyroid hormone in association with thyroid-stimulating hormone and hypothalamic TRH. The nature of the involvement of adrenocorticosteroids is not clear. They may contribute to the activity of thyroid hormone by enhancing the formation of triiodothyronine from thyroxine and decreasing its inactivation (Galton 1990). The period of larval life in anurans can vary from about 7 days in spadefoot toads living in the desert to 3 years in bullfrogs from colder regions in North America. A more usual period of premetamorphic development in spadefoot toads is 30 to 40 days (Warburg 1997). A longer premetamorphic period usually results in larger juveniles, which are better able to withstand life stresses, including desiccating conditions.

However, metamorphosis sometimes appears to be "facultative" and may depend on conditions in the pond, including the food supply, pH and solute concentrations. Such limnological circumstances may impinge on the hormonal processes regulating metamorphosis. There is little information about this possibility.

The West African viviparous frog *Nectophrynoides occidentalis* aestivates in burrows during periods of seasonal drought. Fertilization usually occurs in October and drought conditions commence in November. The development of the larvae in the oviduct is suppressed during aestivation by the secretion of progesterone from the ovary (Xavier and Ozon 1971; Xavier 1974). The frogs emerge the following April and larval development continues until June. This interesting strategy of an amphibian to adapt its reproduction to drought conditions appears to be unique.

The larvae of some urodeles, such as the tiger salamander, achieve a size that is similar to that of the adults. A number of other urodeles are neotenous and retain the larval form, including external gills. They are paedomorphic and can reproduce. These amphibians include mudpuppies, mud eels, congo eels and the Mexican axolotl. The latter can even be artificially induced to metamorphose by administering thyroid hormone. The integument of such urodeles, as described earlier, generally exhibits a lower osmotic and ion permeability than adults and a reduced or lack of response to vasotocin. Such properties of their integument are similar to those of anuran tadpoles. As metamorphosis becomes imminent, the skin of tadpoles begins to behave more like that of adults and displays active transcutaneous Na^+ transport (Taylor and Barker 1965; Cox and Alvarado 1983). One of the changes that takes place at metamorphosis is an increase in the levels of Na-K-ATPase (Taylor and Barker 1965) and the appearance of epithelial sodium channels (ENaC). An ability to respond to vasotocin also develops.

5 Accumulation of Water and Salt

Most amphibians are carnivorous, principally consuming invertebrates, though some, such as large cane toads, sometimes even eat small mammals. Such a diet can be expected to meet their normal need for salt, probably even when they lead an exclusively aquatic existence. However, the water contained in such a diet appears to be inadequate for the total needs of most terrestrial amphibians. Feeding may be sporadic in some amphibians, and in aquatic species and those that hibernate under water it is possible that their ability to accumulate Na^+ and Cl^- across their skin could be a physiological necessity.

It is generally accepted that amphibians do not drink in order to rehydrate (Adolph 1933; Bentley and Yorio 1979). Instead, they absorb water osmotically across their permeable skin. The tadpoles of the crab-eating frog may be an exception, as they drink the brackish water in which they live and, as occurs in teleost fish, apparently excrete the salt across their gills (Gordon and Tucker 1965). Aquatic amphibians, such as the urodeles *Amphiuma means* and *Siren lacertina*, normally imbibe small amounts of the water in which they live (Bentley 1973a). Frogs may be induced to swallow even unpalatable salt solutions when exposed to severe stress (Krogh 1939; Bentley and Schmidt-Nielsen 1971). However, normal rehydration occurs by absorption in what has been called cutaneous drinking. As described above, the rate of such water absorption in response to dehydration is a process that is increased by vasotocin. There are differences in the speed of such rehydration in various amphibians. It often occurs across a specialized area of the integument called the pelvic patch that is well vascularized and very responsive to vasotocin (see above). Amphibians have been observed to periodically migrate and search for suitably damp surfaces from which to absorb water (Stille 1952; Brekke et al. 1991). Some even appear to exhibit "anticipatory cutaneous drinking" prior to any measurable dehydration (Jørgensen 1994).

Mammals, birds and reptiles utilize the renin-angiotensin system to induce thirst in response to decreases in the volume of the body fluids. As described above, a resulting formation of angiotensin II stimulates thirst receptors associated with the hypothalamus and induces drinking behaviour. Attempts to induce such drinking by the administration of angiotensin II has not been successful in any amphibians (Kobayashi et al. 1979; Kobayashi and Takei 1997). However, von Sechendorff Hoff and Hillyard in 1991 made the fascinating observation that when red-spotted toads, *Bufo punctatus*, were injected with angiotensin II they exhibited prolonged "water-seeking behaviour." This involved persistent apposition of their pelvic patch region to proffered damp surfaces of soil. There was a resulting water uptake. This water absorption response (WR) could be inhibited by the prior injection of the angiotensin II antagonist drug saralasin. The injection of angiotensin II into the brain also promoted the water-absorption response (Propper et al. 1995). Amphibians appear to utilize a hormonal mechanism to promote cutaneous rehydration behaviour that parallels the effects of angiotensin II on thirst and oral drinking in other tetrapods (Hillyard et al. 1998).

Angiotensin II also has a vasoconstrictor effect and can increase blood pressure and the release of aldosterone. Optimal osmotic water transfer across the pelvic patch of amphibians depends an adequate blood flow in the region of the pelvic patch. Captopril, a drug which blocks the conversion of angiotensin I to II, can inhibit water absorption across the pelvic patch of dehydrated toads and decreases the blood pressure (McDevitt et al. 1995). It was suggested that angiotensin II may facilitate water absorption across the pelvic patch by maintaining blood pressure during dehydration. Aldosterone has been observed to reduce the threshold when dehydration evokes water-seeking behaviour in green toads (Hoffman and Katz 1999). The authors suggested that angiotensin II and aldosterone increase the frequency of water-seeking behaviour while vasotocin subsequently promotes water uptake across the pelvic patch. However, as just described, angiotensin II may also contribute to the latter response. Aldosterone appears to act only in the initial water-seeking phase. The nature of this adrenocorticosteroids action is unknown, but it clearly merits further investigation.

The search for "potable" water by toads is a discriminating one (Stille 1952; Brekke et al. 1991; Hillyard et al. 1998). They appear to utilize the skin on their feet and ventral surface to "taste" the water. When they locate a damp surface, toads have been observed to exhibit "seat-patch-down" behaviour when they briefly explore its suitability for cutaneous absorption. They appear to be able to discriminate between the concentrations of such solutions as sodium chloride, potassium chloride and urea. Their sodium taste appears to be related to the rate of Na^+ transport across the skin (von Seckendorff Hoff and Hillyard 1993). When this process is blocked by the drug amiloride, they lose the ability to taste solutions of sodium chloride. The activity of taste bud cells in the mammalian tongue can also be blocked by amiloride. The enhanced rate of Na^+ transport, which has been observed in the pelvic patch of anurans and which can be further stimulated by vasotocin, may be contributing to such a cutaneous sense of taste.

Apart from accumulating salt from their diets, amphibians can take up Na^+ and Cl^+ across their skin from solutions as dilute as 10^{-5} M. The ability of the skin to limit losses of such ions from the body fluids is not in doubt. However, the physiological importance of skin as a conduit for salt accumulation is not clear. Krogh in 1939 remarked that "it (cutaneous salt uptake) may be of vital importance to the frogs in winter when about 7 months are spent under water without food". He (Krogh 1937) maintained fasting European frogs in distilled water, which was constantly changed, for periods of up to 12 months. Most survived, but they displayed a slow decline in serum Cl^- concentrations. In 1972, McAfee repeated these experiments using North American leopard frogs (Rana pipiens) and over a period of 60 days could not detect any change in the concentration of Na^+ in the plasma or the whole body content of this ion. I maintained three species of anurans (Rana pipiens, Bufo marinus and Xenopus laevis) for 15 days in tap water containing 0.25 mEql^{-1} Na^+ and amiloride (10^{-5} M) (Bentley 1973b). The latter drug blocks the transcutaneous uptake of Na^+. There was a small decline in the serum Na^+ concentration in the B. marinus, but no significant change in the levels of this ion in the other two species. The skin of B. marinus subsequently displayed what appeared to be a compensatory increase in transcutaneous Na-dependent short-circuit current. Such a change was not observed in the behaviour of skin from the other two species. It should be recalled that B. marinus normally lives a mainly terrestrial life, while R. pipiens is amphibious and X. laevis is completely aquatic. Prolonged exposure to distilled water can result in a rise in the concentrations of aldosterone in the plasma in amphibians. This hormone is probably mediating the increased salt transport in B. marinus. The affinity constant (K_m) for the transcutaneous Na^+ transport has been observed to be lower in aquatic as compared to more terrestrial species of amphibians (Greenwald 1972). Possibly this difference reflects an adaptation of this process to life in a salt-poor environment. However, on a short-term basis, the importance of such a mechanism for the accumulation of salt is questionable, except in amphibians hibernating under water. Active transport of Na^+ occurs across many epithelial membranes. Not all of these are primarily concerned with the accumulation of salt. The process may have other roles, including the maintenance of the integrity of epithelium itself, the cotransport or exchange of other solutes, acid-base balance and even participation in sensory processes. The active transport of Na^+ across the skin appears to be important for the accumulation of salt in some, but probably not all, amphibians.

Chapter 7

The Fishes

The fishes, or Pisces, are aquatic vertebrates that breathe with the aid of gills. They are thought to have originated 500 million years ago, a time that predates the origin of tetrapods by some 200 million years. It is uncertain whether the transformation from an invertebrate ancestor took place in fresh water or in the sea, but subsequently, fish have adopted both of these environments, as well as inland lakes and springs where the solute concentrations of the waters may even exceed those in the sea. The Pisces contains about 23 000 species, more than all of the tetrapods, which occur in several widely divergent evolutionary classes. The principal of these are: (see Table 1.1) the Agnatha (the jawless lampreys and the hagfishes), the Chondrichthyes (sharks, rays and chimaeras) and the Osteichthyes (bony fishes). The teleostean fishes, which are a group of the latter bony fish, are today the predominant group in both fresh water and the sea; there is more information about the physiology of these than for the other classes of fish. From the biological viewpoint, however, further information about osmoregulation in the sparser relict fishes would probably be even more interesting, especially as it could aid our understanding of the evolution of osmoregulation in vertebrates. Such archaic fish include the lungfishes, the coelacanth, the sturgeons, the bowfin and garpike, and the lampreys and hagfishes. At present, physiological information about these fishes is somewhat limited.

The sharks and rays and the bony fishes probably originated in the sea (see Halstead 1985), but they have subsequently migrated into fresh water. While contemporary species of some of the main orders within these groups of fish may today live solely in fresh water (the lungfishes and bichir) or the sea (the coelacanth), in the course of geological time some representatives of all such evolutionary divisions have occupied both habitats (see Darlington 1957). The dominant teleosts are thought to have originated in the sea, but some subsequently moved into fresh water and a number of these have even returned to the sea. Darlington (1957) gives a fascinating description of the dispersal of freshwater fishes throughout the world, and this has often involved several successive migrations between the sea and fresh water. The freshwater catfishes of Australia, for instance, moved into the antipodean rivers from the sea, but their immediate marine ancestors previously originated from freshwater ancestors that probably lived in Asia. When one considers the complexities of such transformations, it is not surprising to find considerable diversity in the detailed pattern of osmoregulation of fishes, though the basic blueprint remains remarkably uniform.

Most contemporary species of fish have only a limited ability to move between fresh water and salt solutions like sea-water. Such osmotically conservative fishes are stenohaline, while others, which can adapt more readily to either type of solution are euryhaline. As emphasized by Fontaine and Koch (1950), such a classification is not rigid, some fish being "more or less" stenohaline while others are "more or less" euryhaline. Thus, the freshwater goldfish and the marine perch can both live in salt solutions about one-third the concentration of the sea (Lahlou et al. 1969a; Motais et al. 1966). Some euryhaline fishes can withstand direct transfer from fresh water to sea-water, while others, such as the lamprey, once having migrated into rivers, will die if they are then replaced in the sea (Hardisty 1956; Morris 1960). Euryhaline fishes are often migratory breeders; some, like lampreys and salmon, move from the sea into rivers, where they produce their eggs, while others, like the eel, return to the sea to breed. Many euryhaline fishes occupy estuaries and certain inland lakes and streams where the salinity may vary considerably during different seasons. The osmotic adjustments which must be made by such fishes would seem to offer a situation for endocrine coordination. In natural conditions, adaptation to solutions of differing osmotic concentration probably can take place gradually, as when such water is evaporated from an inland lake or fish migrate through estuarine areas between rivers and the sea (Fontaine and Koch 1950). Such situations would seem to be particularly suitable to the relatively long-term actions that are characteristic of many hormones.

Fish can live in solutions with a wide range of osmotic concentration. As described above, fish that normally live in the sea or fresh water can often tolerate solutions of intermediate concentration. Others can live in waters with a solute concentration considerably higher than that of sea-water; *Oreochromis mossambica* may survive in a salinity of 6.9%, while the cyprinodont *Cyprinodon variegatus* has been found in saline waters with a concentration of 14.2% (see Parry 1966). Such osmotic tolerance presumably constitutes a rather special physiological adaptability about which we have little knowledge.

Fishes contain an amount of water that is equivalent to 70 to 75% of their body weight, which is similar to the water content of most other vertebrates. The distribution of this water in the body varies somewhat in different species (Thorson 1961). The plasma volume of agnathans and chondrichthyeans constitutes about 5% of their body weight, which is similar to that of the tetrapods, but in osteichthyeans it is, according to Thorson (1961), about half as large. However, as the number of species of bony fishes is vast and the measurements have been made on comparatively few, it remains to be seen whether such a difference is a general characteristic of these animals.

The solute content of the plasma in representatives of the main groups of fishes has been measured (Table 7.1). The bony fishes that live in the sea generally have higher solute levels than those of fish in fresh water. Euryhaline fish, like the flounder, salmon and eel, have a plasma osmolarity that is about 20% greater in the sea than in fresh water, and this is the result of an elevated sodium chloride concentration. Such marine fish are considerably hyposmotic, two- to threefold less, to the sea-water that bathes them. The only exception among the Osteichthyes appears to be the relict crossopterygian *Latimeria* (arguably a group that once also may have lived in fresh water) whose plasma is hyperosmotic to sea-water, due mainly to a retention of urea. It is interesting that the closest living phyletic relatives of this fish are the lungfishes, and one of these, *Protopterus*, can also tol-

Table 7.1. Osmotic constituents in the plasma of various fishes

	Habitat	Sodium mEq l^{-1}	Potassium mEq l^{-1}	Urea mM	Osmolarity mosmol l^{-1}
Sea-water		470	12		1070
Osteichthyes					
Latimeria chalumnae[a] (Coelacanth)	SW*	181	–	355	1181
Protopterus aethiopicus[b] (African lungfish)	FW*	99	8		238
Carassius auratus[c] (Goldfish)	FW	115	3.6		259
Opsanus tau[d] (Toadfish)	SW	160	5.2	–	392
Platichthys flesus[e] (Flounder)	FW	157	5		304
	SW	194	5		364
Anguilla anguilla[f] (European eel)	FW	155	2.7	–	323
	SW	177	2.8	–	371
Salmo salar[g] (Atlantic salmon)	FW	181	1.9	–	340
	SW	212	3.2	–	400
Chondrichthyes					
Raja clavata[h] (Thornback ray)	SW	289	4	444	1050
Squalus acanthias[i,j] (Spiny dogfish)	SW	287	5.4	354	1000
Chimaera montrosa[i] (Rabbitfish)	SW	360	10	265	
Potamotrygon[k] (Freshwater stingray)	FW	146	–	1.1	308
Agnatha					
Myxine glutinosa[l] (Hagfish)	SW	549	11	–	1152
Lampreta fluviatilis[m] (River lamprey)	FW	120	3	–	270

* SW = sea-water; FW = fresh water.
References: [a] Pickford and Grant (1967). [b] Smith (1930a). [c] Maetz (1963). [d] Lahlou et al. (1969b). [e] Lange and Fugelli (1965). [f] Sharratt et al. (1964a). [g] Parry (1961). [h] Murray and Potts (1961). [i] Fänge and Fugelli (1962). [j] Burger (1965). [k] Gerst and Thorson (1977). [l] Bellamy and Chester Jones (1961). [m] Robertson (1954).

erate large amounts of urea in its body fluids. The marine chondrichthyean fishes, like the relict crossopterygian *Latimeria*, which lives in sea-water, also maintain their body fluids slightly hyperosmotic to sea-water by retaining urea as well as some sodium chloride (Smith 1936). Chondrichthyeans that live in fresh water, like the stingray, *Potamotrygon*, have a much lower plasma concentration due to a reduction in the levels of both urea and sodium (Table 7.1). The agnathans exhibit both types of osmotic constitution, the lampreys are hyposmotic to sea-water, while the hagfishes are slightly hyperosmotic. The latter group, however, does not retain urea for this purpose, but has high salt concentrations in its extracellular fluids.

We can predict the osmotic stresses on the various fishes from the osmotic concentrations in their body fluids. In fresh water, all fishes tend to gain water by osmosis while at the same time they may be expected to lose solutes by diffusion. In the sea, chondrichthyeans, *Latimeria* and the hagfishes, all gain water by osmosis, and (except for the latter) may be expected to accumulate sodium by diffusion. The marine bony fishes and lampreys, which are hyposmotic to sea-water, lose water osmotically and gain sodium. Measurements of water and solute balance in the different groups of fishes indicate that such osmotic changes are occurring, but they differ widely and are adequately compensated for by physiological adjustments.

1 The Piscine Endocrines

The vertebrate endocrine glands and their secreted hormones make their phyletic debut in the fishes (see Chap. 2). They possess hormones similar to those present in tetrapod vertebrates. Notable differences in the fishes are the absence of parathyroid glands but the presence of what appear to be two additional contributors to the endocrine armoury: the corpuscles of Stannius and the urophysis. The secreted hormones in the fishes have structures that are homologous to those in other vertebrates. However, growth hormone and prolactin have not been identified in the pituitaries of hagfishes and lampreys (superclass Agnatha), but a novel pituitary hormone, somatolactin, is present in many bony fish. The precise chemical structures of their hormones, especially peptides, polypeptides and proteins, often display considerable differences from those in other species of fish and the tetrapods. These differences appear to reflect the evolution of the hormones among the vast number of piscine species during the long periods of time that have elapsed since their primaeval origin. The receptors for such hormones in fish display many differences in their interactions with homologous hormones from other species. Such differences apparently reflect an evolution of their chemical structures that parallels that of the hormones. Although the piscine hormones and their receptors may make contributions to osmoregulation similar to those in other vertebrates, there are also some remarkable differences. Information about the particular roles of hormones in the osmoregulation of fishes is often meagre, and generalizations are often not possible. The fishes are a very diverse group of vertebrates containing vast numbers of species. They can be difficult to obtain and maintain under experimental conditions. They are also often highly prone to stress in the laboratory. However, observations on the actions and physiological roles of hormones in fish can provide important insights into the evolution of osmoregulation in vertebrates.

1.1 Neurohypophysial Hormones

The neurohypophysis of fish lacks the distinct neural lobe that is present in terrestrial vertebrates. However, it contains homologous peptide hormones, though usually they are stored in smaller quantities. As in non-mammalian tetrapods, vasotocin is present in all the main phyletic groups extending from the Agnatha to the Chondrichthyes and Osteichthyes (Fig. 2.5). In lampreys and hagfish, vasotocin appears to be the only such hormone present, suggesting that it is the primaeval hormone in this family of peptides. Chemically related peptides have been identified in invertebrates, including molluscs (Van Kesteren et al. 1992a,b). The Agnatha may have acquired this gene from such an ancestor. The biosynthetic precursor of vasotocin, provasotocin, in lampreys is more similar to that in other vertebrates than to the prohormone in hagfish (Suzuki et al. 1995). This observation supports the suggestion that the lampreys are part of the main phyletic line of vertebrates. Most other fish possess at least two different, but homologous, such peptide hormones (Acher 1996). Lungfish (Dipnoi) have vasotocin and, like many tetrapods, also mesotocin. However, most other bony fish possess a novel such peptide, which differs from mesotocin by the presence of a serine residue, instead of glutamine, at position 4 in the molecule (Heller et al. 1961). It is called isotocin. A plethora of such "neutral" oxytocin-like peptides have been identified in the cartilaginous fish. As described in earlier chapters of this book, vasotocin contributes to osmoregulation in non-mammalian tetrapods by an ability to exert an antidiuretic effect. The evidence for such an effect is equivocal in fishes, but it could have other roles. It has been suggested that the considerable diversity of neurohypophysial peptides in cartilaginous fish may reflect the lack of a vital specific effect in these fish (Acher et al. 1999). However, Acher has made the interesting suggestion that they could be contributing to the regulation of urea levels in such ureosmotic fish.

1.2 Adrenocorticosteroids

Adrenocortical tissues have been identified in all the phyletic groups of the fishes (Henderson 1997). These tissues usually lie adjacent to the kidneys and may be present on their ventral surfaces. In teleost and agnathan fish, presumptive adrenocortical tissue may also be situated along the posterior cardinal veins. In these fish it can be intermixed with chromaffin tissue that secretes adrenaline and noradrenaline. In lungfish, aldosterone is a major secretion of adrenocortical tissue. It also produces cortisol and corticosterone in these fish (Idler et al. 1972; Joss et al. 1994). This mixture of corticosteroids is similar to that seen in amphibians. It is possible that small amounts of aldosterone are produced in some teleost fish. However, the principal such steroid in the plasma of the Teleostei, Holostei and Chondrostei is cortisol. Corticosterone is also usually present, and sometimes small amounts of cortisone. While the primary physiological role of cortisol is that of a glucocorticoid in most vertebrates, this hormone also makes an important contribution to the salt metabolism in fishes. Sharks and rays

(Elasmobranchii) secrete a novel corticosteroid, 1α-hydroxycorticosterone, from their interrenal glands (Table 2.3; Idler and Truscott 1972). This steroid has not been found in any other group of vertebrates. It is not even present in other cartilaginous fish, such as the Holocephali (chimaeroids). The latter fish secrete cortisol. Corticosterone and cortisol have been identified in the plasma of a hagfish, *Myxine glutinosa* (Phillips et al. 1962b; Idler and Truscott 1972). However, adrenocorticosteroids could not be identified in another agnathan fish, the lamprey *Lampetra fluviatilis* (Buus and Larsen 1975). The synthesis of adreno-corticosteroid hormones in fishes can be increased by ACTH and angiotensin II (Henderson 1997). Urotensin I, which has structural similarities to mammalian corticotropin-releasing hormone, also promotes cortisol synthesis in teleosts (Arnold-Reed and Balment 1989, 1994). In tetrapods, the natriuretic peptides usually inhibit the synthesis of corticosteroids but in flounder a *stimulation* has been observed (Arnold-Reed and Balment 1994).

1.3 The Renin-Angiotensin System

A distinct renin-secreting juxtaglomerular apparatus, such as that in mammals, has not been identified in the kidneys of most fish. However, a comparable structure has been observed in elasmobranchs (Lacy and Reale 1990). Granulated cells, like those containing renin in other species, have been found in many bony fish, where they are associated with the afferent glomerular arteriole. They include the Teleostei, Dipnoi, Chondrostei and Holostei (Nishimura 1978, 1987). In 1942, Friedman and his associates found renin activity to be present in the kidneys of freshwater but not in marine teleosts. However, it has subsequently also been identified in marine species, though its concentrations are lower than in the freshwater fish (Chester Jones et al. 1966; Sokabe et al. 1966; Capelli et al. 1970). Teleost renin can interact with the plasma substrate angiotensinogen from all non-mammalian tetrapods, to produce angiotensin (Nolly and Fasciola 1973). Elasmobranch renin can even interact with mammalian angiotensinogen (Henderson et al. 1981). Renin activity has also been found in the corpuscles of Stannius, which abut the kidney, in eels (Chester Jones et al. 1966). The chemical structures of angiotensins in fish usually differ in a characteristic way from those of tetrapods (Fig. 2.6). "Fish angiotensin" typically contains asparagine, instead of aspartic acid, at position 1 in the molecule. An interesting exception is the bowfin (Holostei), which has the "tetrapod-type" peptide (Takei et al. 1998). Elasmobranch angiotensin is unique as it contains proline, instead of valine, at position 3. This amino acid substitution could be influencing its tertiary structure and activity (Takei et al. 1993). The Agnatha appear to lack the renin-angiotensin system.

1.4 Natriuretic Peptides

The fishes possess at least four different types of natriuretic peptides. Two of these are homologous to tetrapod ANP and CNP (Fig. 2.9). However, they lack a homologue to BNP. A different type of natriuretic peptide has been found in the ventricular heart muscle of teleosts and is called ventricular natriuretic peptide (VNP) (Takei et al. 1994a, b, c). A fourth distinct type of such peptide has been identified in salmon, *Salmo salar* (Tervonen et al. 1998). It is formed only in the heart, and its structure suggests that it could be an ancestor of ANP and BNP. This cardiac peptide (CP) contains 29 amino acids residues and has been identified in the plasma of 9 genera and 15 species of teleosts (Tervonen 2000). CNP has been identified in the heart and brain of several elasmobranchs (Schofield et al. 1991; Suzuki et al. 1992). It is, apparently, the sole such natriuretic peptide present in these fish, and it has been suggested that it also may be a primordial form of these hormones. Immunohistochemical observations indicate that natriuretic peptides are present in the heart and brain of hagfish (Reinecke 1989; Donald et al. 1992). However, at this time, their precise chemical nature does not appear to have been described.

1.5 Catecholamines

The hormones adrenaline and noradrenaline are described chemically as catecholamines (Fig. 2.8). Both may be present in the plasma of fish, where they also function as neurotransmitters in sympathetic-adrenergic neurons (Randall and Perry 1992). They are secreted by chromaffin tissue. Some of the catecholamines in the plasma may originate as an overflow from adrenergic nerves. The chromaffin tissue in fish may be dispersed, as observed along the posterior cardinal veins in the "head" kidney of teleosts. In elasmobranchs, this tissue forms paired clusters along the surface of the kidneys. Chromaffin tissue is present in the heart of agnathans. In bony fishes, chromaffin tissue often intermingles with adrenocortical tissue. In elasmobranchs, noradrenaline quantitatively predominates over adrenaline, while in teleosts the latter tends to be more plentiful. The secretion of these hormones occurs in response to such stimuli as hypoxia, hypercapnia, metabolic acidosis, intense exercise and severe stress. Release may be mediated by autonomic preganglionic sympathetic nerves, which secrete acetylcholine. It may also occur in response to direct stimulation of the endocrine tissue. Circulating concentrations of these hormones vary in different species of fish (Reid and Perry 1994) and at different times of the day and year (Le Bras 1984). There are several types and subtypes of adrenergic receptors (Chap. 2). For instance, α- and β-adrenergic receptors, α_1-, α_2-, β_1- and so on.

1.6 Thyroid Hormone

Thyroxine (T_4) and triiodothyronine (T_3) are secreted by the thyroid gland (Fig. 2.8). This tissue has been identified in members of all the main phyletic groups of fishes, including hagfish. Homologous tissues and thyroid hormone-like activity have even been identified in protochordates, such as the lancelet (amphioxus). Calorigenic effects of thyroid hormone do not appear to be manifested in fish. However, they may contribute to the ability of some fish to adapt to solutions of differing osmotic concentration (see Fontaine 1956). They also can exhibit morphogenetic actions, such as may be involved in the proliferation of salt-secreting chloride cells (Subash Peter et al. 2000).

1.7 Urotensins I and II

Urotensins I and II (Fig. 2.10) are peptides that were first identified in spinal cord neurons in teleosts. However, they have since been found in most of the other groups of bony fish and elasmobranchs but not hagfish or, apparently, lungfish (Onstott and Elde 1986). In teleosts, these neurons aggregate in the caudal region to form the urophysis. This tissue has a neurohaemal junction with the caudal veins, which lead to the kidney and urinary bladder. The possibility that the urophysis may be involved in osmoregulation is largely based on observations of changes in its structure following transfers between fresh water and sea-water (see Larson and Bern 1987). Changes in the concentrations of the biologically active peptides stored in the urophysis have also been observed in such circumstances. Urotensin I, like its homologue, corticotropin-releasing hormone, has been observed to promote secretion, in vitro, of ACTH from the pituitary of a teleost, the white sucker, *Catastomus commersoni* (Fryer et al. 1983). It has also been shown (in vitro) to directly stimulate the secretion of cortisol from adrenocortical tissue in trout (Arnold-Reed and Balment 1994). Urotensin II can stimulate Na^+ transport across the isolated intestine and urinary bladder of the marine goby, *Gillichthys mirabilis* (Loretz and Bern 1981; Loretz et al. 1985). It has also been found to stimulate Cl^- transport across the opercular skin of this fish (Marshall and Bern 1981). Both urotensin I and II are vasoactive, which has contributed to the suggestion that the latter may be involved in cardiovascular regulation in fish (Conlon et al. 1996; Platzack et al. 1998). Such responses could influence osmoregulation. A homologous radioimmunoassay for urotensin II has recently been developed (Winter et al. 1999). It was found that in flounder the plasma concentrations of this peptide were generally higher when the fish were adapted to sea-water than to fresh water. When flounder adapted to sea-water are transferred to fresh water there is an transitory decline, for at least 72 h, in its concentrations in the plasma (Winter et al. 2000). This technique is an important advance, as it may provide a basis for establishing a role for this putative hormone in piscine osmoregulation.

1.8 The Growth Hormone/Prolactin Family

The adenohypophysis of fishes can secrete growth hormone and prolactin, which are structurally homologous to these hormones in tetrapods. They have not been directly identified in the Agnatha, though immunocytochemical observations suggest that they may be present in lampreys (Wright 1984). Prolactin and growth hormone can contribute to the adaptation of some euryhaline teleosts to their transition between life in fresh water and the sea. The chemical structures of the hormones exhibit marked differences in their amino acid sequences to those in other vertebrates. Thus, salmon growth hormone and prolactin have a homology of only about 35% to the hormones that are present in humans. However, in lungfish, the homology to the human hormones is about 60%.

Prolactins in teleost and holostean (the bowfin) fish contain only two disulphide bridges, compared to three in the tetrapods. However, prolactin in lungfish (Dipnoi) and sturgeon (Chondrostei) have three such bonds (Noso et al. 1993a,b). The Teleostei appear to have followed a separate line of evolution with respect to the structure of prolactin. They have also utilized this hormone in their osmoregulation (see later). Whether these two events are related is unknown. The teleost *Oreochromis niloticus* (tilapia) possesses two different prolactins, which have a homology in their amino acid sequences, of only 69% (Specker et al. 1993). Little is known about prolactin in elasmobranchs. Its presence was originally identified by its biological activity. Like other vertebrate prolactins, it can promote a "water drive" response in newts (see Chap. 6).

Most of the actions of growth hormone in vertebrates are mediated by insulin-like growth factor-I (IGF-I). This polypeptide (see Chap. 2) is induced, mainly in the liver, by growth hormone. Several IGF-Is have been identified in teleosts (Chen et al. 1994; Duan 1997). Their structures are highly conserved and remarkably similar to those in other vertebrates. Although IGF-I has not been identified in the plasma of agnathans, an IGF-I gene has been found in hagfish (Nagamatsu et al. 1991).

Somatolactin is a novel member of the growth hormone/prolactin family which was first identified in the pituitary glands of flounder (Ono et al. 1990). It is formed in the pars intermedia and has since been identified in many other bony fish (Rand-Weaver et al. 1992; Dores et al. 1996; Amemiya et al. 1999). It is released into the circulation in response to a variety of potentially stressful stimuli such as external acid conditions, low environmental calcium levels and changes in osmotic concentrations of the bathing media (Rand-Weaver et al. 1993; Kakizawa et al. 1993, 1996). Its physiological function is unknown.

Many fish can adapt to life in either fresh water or the sea. Migrations, such as are often associated with breeding, involve the movement of the fish from the sea into rivers and back into the sea again. Salmon, trout and lampreys breed in fresh water, where their young undergo initial growth and development before returning to the sea. Others, like eels, breed in the ocean but subsequently spend a large part of their life in fresh water. Some species, like the rainbow trout, *Oncorhynchus mykiss*, and the marine lamprey, *Petromyzon marinus*, have landlocked populations that have discarded such migratory habits, and undergo their entire life cycle in fresh water. Certain other species like the flounder, killifish and toadfish occupy coastal waters near the mouths of rivers and streams, where they experience variations in the salinity of the water in which they live. There is not only a variety of euryhaline fishes, but they exhibit diverse morphological development and physiological conditions associated with different stages of their life period. These include the juvenile freshwater salmonid parr and premigatory smolt, and silver eels, as well as the adult fishes in various stages of their breeding cycle. Many such fish cannot withstand direct transfers from fresh water to sea-water during all stages of their development, or even at any season of the year. When such a migration is made, several days are required to adjust completely to the new osmotic conditions.

Conte, Wagner and their collaborators (Conte and Wagner 1965; Conte et al. 1966) made some interesting observations of the migratory behaviour, and ability to adapt to sea-water, of steelhead trout and coho salmon. Populations of these fish breed and undergo their initial development in the rivers of the northwest of the United States. When very young juveniles of these salmonids are placed in sea-water they die, but subsequently, as they grow larger, they can adapt to such solutions. It is well known that this often occurs at about the time when they metamorphose from a parr to a smolt, and this also corresponds to their seaward migration. It was found that it is not necessarily this metamorphosis per se that results in their ability to adapt to sea-water, but their size also contributes. When they attain a length of about 15 cm, they can adapt to sea-water and indeed in the juvenile parr coho salmon studied, this was usually seen 6 or 7 months *before* they metamorphosed into smolts. This suggests that surface area relative to the body weight may be important, and that physiological and morphological features may develop that are associated with the general growth rate, rather than any sudden metamorphic transformation. Hoar (1951) found that the chloride-secreting cells in the gills of Pacific salmon underwent a period of rapid development before migration. This is consistent with the observations of Conte and Lin (1967), who found an increased rate of branchial cell renewal during salt water adaptation in young salmonids.

When fish are transferred from fresh water to sea-water, or from sea- to fresh water, a number of changes take place in the osmotic composition of their body fluids. Initially, the changes are usually more pronounced than are subsequently maintained. This period of equilibration lasts for about 48 h in eels and flounder (Keys 1933; Lahlou 1967) transferred from fresh water to sea-water, but may be as long as 170 h in the steelhead trout (Houston 1959). Some of the changes which occur in the steelhead trout during this time are shown in Fig. 7.1. Initially, there is a marked increase in the concentration of chloride in the plasma and the tissues, while the overall chloride space increases. After the rather abrupt initial increases, these levels start to decline after about 24 h, and eventually reach equilibrium 4 or 5 days later. The salt concentration in the plasma of such sea-water-adapted fish, however, remains somewhat elevated, compared to fish in fresh water.

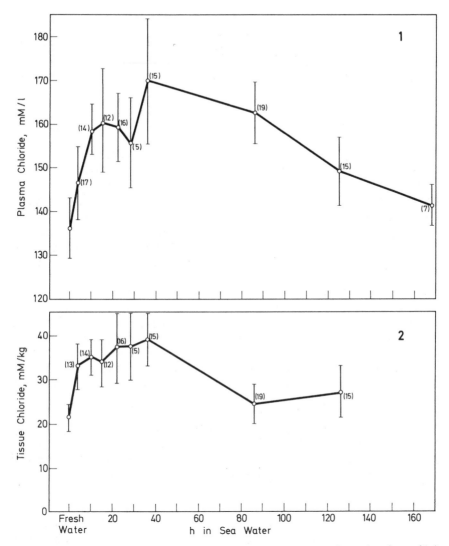

Fig. 7.1. Changes in plasma and muscle chloride concentrations in trout (*Oncorhynchus mykiss*) after transfer from fresh water to sea-water. (Houston 1959, by permission of the National Research Council of Canada)

2 Water Exchanges

The skin and gills of fishes are continuously bathed by the environmental solutions and are avenues for movements of water in or out of the animal. The direction of any net transfer depends on the direction of the osmotic gradient.

The relative surface areas of the skin and gills vary considerably in different fishes; in the toadfish, *Opsanus tau*, the gills have an area ten times larger than

197

that of the skin, while in the mackerel, *Scomber scombrus*, it is 60 times greater (see Parry 1966). The toadfish has been shown to have a very low permeability to water and sodium (Lahlou and Sawyer 1969), and this may partly reflect its relatively small branchial surface.

2.1 The Skin and Gills

The skin of fishes is a thick, complex, poorly vascularized structure, which is liberally supplied with mucous glands and often covered with scales. It has the appearance of being a substantial barrier to diffusion, but there are few direct measurements of its permeability. In 1939, Krogh, after considering the available evidence, concluded that the gills, rather than the skin, are the principal avenue for exchanges of water and solutes across the integument of fish. Parry (1966) agreed with this conclusion. Motais and his colleagues (1969) measured the movements of tritiated water across the gills of several freshwater and marine fish (*Anguilla, Carassius, Platichthys* and *Serranus*) and found that virtually all the water that was exchanged occurred across the gills. However, the permeability of the gills of these teleosts was still found to be low compared to other epithelia, such as amphibian skin and urinary bladder. The permeability coefficient for water of the gills of rainbow trout, in sea-water, is at least 30 times less than that of the toad urinary bladder (Isaia 1984). Water movements across the gills of teleosts have been found to be much less when they are bathed by sea-water instead of fresh water. The change in permeability occurs very rapidly following transfer to sea-water. It has been suggested that it could be a direct effect on the epithelial cells covering the gills (Motais et al.1969; Isaia 1984), such as occurs when the urinary bladders of toads are bathed (in vitro) with hyperosmotic solutions (Bentley 1961, 1965). The gills of fish are well vascularized. The magnitude of the flow and the pathways of distribution of blood in the gills can be regulated by adrenaline and noradrenaline originating both from sympathetic nerves and endocrine chromaffin tissue (Nilsson 1984). Changes in the distribution of blood within the gills can result in the recruitment of additional respiratory lamellae. It is possible that other vasoactive hormones, such as angiotensin II, natriuretic peptides and vasotocin, could be contributing to changes in the distribution of blood. The vascular actions of adrenaline on the gills appear to be insufficient to account for the observed changes in permeability to water. A more direct effect of this catecholamine on the cell membranes of the gill epithelial cells appears to be occurring. Such a response results in an increased permeability to water (Pic et al. 1974; Isaia 1979). Other hormones, such as prolactin, may also contribute to changes in the permeability of the gills to water.

Information about the permeability of the skin and gills of cartilaginous and agnathan fish is sparse compared to bony fish. The skin of dogfish, *Scyliorhinus canicula*, is about three-times more permeable to water than that of the mudpuppy, *Necturus maculosus*, which is an aquatic amphibian (Payan and Maetz 1970). However, it is less permeable than that of the leopard frog *Rana pipiens*. The gills of dogfish have a similar permeability to water as those of rainbow trout.

Cartilaginous fish, in contrast to marine teleosts, maintain their body fluids at concentrations that are nearly isosmotic to sea-water. Thus, limiting the permeability of the gills to water would not appear to be of paramount physiological importance in such fish. Marine myxinoid agnathans also maintain their body fluids at near-isosmotic concentrations with sea-water. The integument of Pacific hagfish has been found to be very permeable to water (McFarland and Munz 1965). However, as the osmotic gradient between the body fluids and sea-water is low, little net transfer of water occurs. Hagfish appear to gain water across their integument that is equivalent to less than 0.5% of their body weight each day (Morris 1965). Agnathan fish living in fresh water accumulate greater amounts of water across their integument. The migratory river lamprey, *Lampetra fluviatilis*, takes up water at about the same rate as goldfish. It is equivalent to about 33% of their body weight each day (Bentley and Follett 1963). Measurements of osmotic water movement across the skin, in vitro, of the river lamprey (osmotic gradient $210\,\mathrm{mosmol\,l^{-1}}$) indicate that it occurs at the rate of about $2.5\,\mu\mathrm{l\,cm^{-2}\,h^{-1}}$ (Bentley 1962a). This value is similar to the estimate, in vivo, of the rate of water uptake that occurs across the entire integument, i.e. skin and gills (Wikgren 1953). The relative osmotic permeability, in fresh water, of the entire integuments of eels: river lampreys: frogs have been estimated to be $1:20:60$ (Wiikgren 1953).

2.2 The Gut

Fishes living in fresh water drink little water, but teleosts in the sea must swallow sea-water in order to compensate for their osmotic losses (Smith 1930b). Indeed, if marine teleost fish are prevented from drinking, they rapidly die from dehydration. The rate of drinking appears to relate directly to the fish's needs, for, as the concentration of the external media increases, so does the amount that the fish drinks; eels in sea-water drink $325\,\mu\mathrm{l}\,(100\,\mathrm{g})^{-1}\mathrm{h}^{-1}$, but if they are placed in a solution twice as concentrated as the sea, they drink $800\,\mu\mathrm{l}\,(100\,\mathrm{g})^{-1}\mathrm{h}^{-1}$ (Maetz and Skadhauge 1968). The quantity of sea-water swallowed by marine teleosts each day varies from volumes equivalent to 5% of the body weight in the flounder to 12% in the sea-water perch (Table 7.2). These quantities are presumably largely dictated by the osmotic water loss of the fishes, a factor influenced by their surface area, as well as the precise properties of the skin and gills. There are few determinations of drinking rates in non-teleost fishes; the lamprey drinks when placed in 50% sea-water and myxinoids have been observed to drink, but this water does not appear to be absorbed (McFarland and Munz 1965; Morris 1965). Marine chondrichthyeans do not usually appear to drink (Smith 1931).

Drinking by fish principally depends on their osmotic circumstances. However, some water may be imbibed during feeding and, as observed in amphibians, be induced by stressful conditions. Hirano (1974) observed that euryhaline Japanese eels, *Anguilla japonica*, adapted to fresh water, started drinking immediately when they were transferred to sea-water. They then ceased drinking as soon as they were returned to fresh water. Haemorrhage resulted in drinking in

Table 7.2. Rates of drinking of various fish in fresh water (FW) and sea-water (SW)

		ml $(100\,g)^{-1}$ body weight day^{-1}	
Osteichthyes			
Carassius auratus (Goldfish)	FW	2	Lahlou et al. (1969a)
Tilapia mossambica	FW	6	Potts et al. (1967)
	SW	27	
Anguilla anguilla (European eel)	FW	3	Maetz and Skadhauge (1968)
	SW	8	
Anguilla japonica (Japanese eel)	FW	Not detectable	Oide and Utida (1968)
	SW	8	
Platichthys flesus (Flounder)	SW	5	Motais et al. (1969)
Salmo gairdneri (Trout)	SW	8	Oide and Utida (1968)
Serranus scriba (Sea-water perch)	SW	12	Motais and Maetz (1965)
Chondrichthyes	SW	nil.	Smith (1931)
Agnatha			
Lampetra fluviatilis (River lamprey)	SW (50%)	15	Morris (1960)
Myxine glutinosa (Hagfish)	SW	6.5[a]	Morris (1965)

[a] Drinking was irregular; most fish did not drink. There was no evidence of absorption from the gut.

the freshwater eels. As observed in many tetrapod vertebrates, the injection of angiotensin I or II can induce drinking in many fish. However, this response is not seen in all fish. The differences have been related to the usual osmotic conditions in the environments that the particular fish occupies (Kobayashi et al. 1983; Kobayashi and Takei 1997). Fish that are habitually stenohaline and live exclusively in either fresh water or sea-water generally do not respond to administered angiotensin. The drinking response of fish to this hormone has been viewed as an emergency one which is not present in species that live in a predictable osmotic environment. However, there are exceptions, and some teleosts that have been described as "more or less" stenohaline have been found to respond to angiotensin. Such fish may be exposed to changes in external osmotic concentrations, such as may occur near the mouth of rivers or when they become trapped in tidal pools. Euryhaline species, such as eels, killifish and flounder, all respond to the injection of angiotensin by drinking. Two agnathans, the freshwater Arctic lamprey and the hagfish, *Eptatretus burgei*, which is marine, and two species of elasmobranchs were found to be unresponsive. Somewhat unexpectedly, it has been found that in another elasmobranch, the dogfish *Scyliorhinus canicula*, angiotensin II induces drinking (Hazon et al. 1989). Acute decreases in the blood pressure of these dogfish also resulted in drinking, apparently reflecting an activation of the renin-angiotensin system. The latter response has also been observed in European eels (Tierney et al.1995). The drinking responses to reduction in blood pressure can be inhibited by the drug captopril, which blocks ACE and the conversion of angiotensin I to II. The infusion of atrial natriuretic peptide to Japanese eels has been observed to inhibit drinking (Tsuchida and Takei 1998). The response, which did not result in changes in blood pressure, was accompanied by a decrease in the plasma levels of angiotensin. It is possible that the ANP may modulate the drinking response, especially during the early adaptation of such euryhaline fish to sea-water.

About 75% of the sea-water that is imbibed by teleost fishes is absorbed into the body fluids. The intestine absorbs water, but as this is a consequence of the absorption of sodium, a large amount of salt is also absorbed. Most of the divalent ions appear to be retained in the gut, though some are absorbed and later excreted by the kidney. The bulk of the salt is sodium chloride, which is excreted through the gills (Keys 1931) leaving osmotically free water in the body. The amounts of sodium (and chloride) absorbed across the gut of fishes kept in seawater amounts to $100 \mu Eq$ $(100 g)^{-1} h^{-1}$ in the flounder, *Platichthys flesus* (Maetz 1969), $280 \mu Eq$ $(100 g)^{-1} h^{-1}$ in the sea perch, *Serranus scriba* (Motais and Maetz 1965) and $170 \mu Eq$ $(100 g)^{-1} h^{-1}$ in the eel, *Anguilla anguilla* (Maetz and Skadhauge 1968). This is equivalent in a day to 100 to 200% of the total sodium present in the fish.

Water absorption from imbibed sea-water takes place from the anterior and posterior intestine of marine teleosts. The fluid that enters the intestine approaches isosmoticity with the plasma. Water absorption then occurs as a result of a favourable concentration gradient that is established by the absorption of sodium chloride (Loretz 1995). It was once considered likely that the dilution of imbibed sea-water occurred solely in the stomach as a result of the movement of body water by osmosis and secretion (Smith 1930b). This water was assumed to be subsequently reabsorbed in the intestine. This recycling process would be expected to occur at a metabolic cost. Such a dilution of sea-water has been observed in vitro in the stomach of eels (Sharratt et al. 1964b). There was a simultaneous movement of Na^+ from the stomach to the plasma. However, the process of dilution of such sea-water has since been shown to commence in the oesophagus (Hirano and Mayer-Gostan 1976). The oesophagus of freshwater eels has a very low osmotic permeability to water but in the sea-water-adapted fish, sodium chloride and water readily cross the mucosa into the plasma. The absorption of sodium chloride from the oesophagus involves its diffusion down a concentration gradient. There is, in addition, a linked Na-Cl cotransport (Hirano and Mayer-Gostan 1976; Kirsch and Meister 1982; Parmalee and Renfro 1983; Nagashina and Ando 1994). The precise nature of this salt transport has not been defined. However, the process can be partly inhibited by the sodium channel-blocking drug amiloride and also ouabain, which inhibits Na-K-ATPase. This oesophageal desalination process has been observed in several teleosts, including flounder, plaice and cod (Loretz 1995). When freshwater-adapted eels were treated over a period of 7 days with cortisol, the permeability of the oesophagus to Na^+ increased (Hirano 1980). When sea-water-adapted eels were treated with prolactin, the permeability to Na^+ decreased. Such responses would be physiologically apt for such fish adapting to, respectively, a life in sea-water or fresh water.

The absorption of water from the intestine of eels is dependent on the active absorption of Na^+ (Skadhauge and Maetz 1967; Skadhauge 1969). This process in teleost fish depends on the activity of Na-K-ATPase (Smith 1964, 1967; Oide 1967). This enzyme mediates the extrusion of Na^+ across the basolateral surfaces of the intestinal mucosal cells. Its activity increases in fish adapted to sea-water. If the adenohypophysis is removed, from eels and killifish, this increase in enzyme activity fails to occur. However, it can be restored by the administration of ACTH (Hirano 1967; Hirano and Utida 1968; Pickford et al. 1970b). The absorption of Na^+ from the intestine of teleosts commences with its entry into the mucosal cells, across their apical plasma membranes (Loretz 1995). The latter process involves the activity of at least two cotransport proteins that mediate a coupled ion transport of Na^+-K^+-$2Cl^-$ and Na^+-Cl^-. The Na^+ is extruded across the basolateral surface by the Na-K-ATPase, and the Cl- by a Cl^-/HCO_3^- exchange mechanism and a linked K^+-Cl^- cotransport. Atrial natriuretic peptide has been shown to inhibit the Na^+-K^+-$2Cl^-$ cotransport across the intestine of winter flounder (O'Grady et al. 1985). It has also been shown to inhibit such salt transport across the intestine of sea-water-adapted eels (Ando et al. 1992) and the goby *Gillichthys mirabilis* (Loretz 1996). Thus, at least three different hormones, angiotensin II, cortisol and ANP, may contribute to the process by which teleost fish regulate salt, and ultimately water, absorption from the gut.

2.3 The Kidney

The urine produced by the kidneys of fish living in fresh water is the primary route for the excretion of the excesses of water that are gained across the integument. In sea-water the kidneys are an important avenue for excretion of divalent ions that are gained as a result of drinking.

The kidneys of fish (Dantzler 1989) are mesonephric, though non-functional remnants of the pronephros are sometimes still present. A venous renal portal system, which supplies the tubules, is usually present. There is an arterial blood supply to the glomeruli. The nephrons of different species of fish have quite diverse structures. In freshwater and euryhaline teleost fish, the glomeruli are usually quite large and numerous, but in marine stenohaline teleosts they are reduced in size and number. About 30 species of teleosts are aglomerular, including a few freshwater species, where the condition is apparently secondary to a marine ancestry. The renal tubules have multiple separate segments. Freshwater teleosts have a distinct distal segment that is almost impermeable to water but it is an important site for the reabsorption of Na^+. This segment is not present in marine stenohaline teleosts. The proximal renal tubules may be subdivided into segments, usually two. The second segment can be the site of a tubular secretion of a solution of sodium chloride, which may equal the volume of the glomerular filtrate. In aglomerular species it is the primary site of formation of the renal tubular fluid. The proximal tubules are separated from the distal ones by a short intermediate segment. These tubules lead into the collecting duct system and the archinephric duct. Over 90% of the sodium chloride and as little as 25% of the water in the tubular fluid is reabsorbed in the tubular-collecting duct system. The kidney structures of marine elasmobranchs and hagfish (Agnatha) have special features, which will be described later.

Teleost fish living in fresh water form a copious dilute urine containing low concentrations of sodium chloride (Table 7.3). In sea-water the volume is much less and the concentration approaches isosmoticity with the plasma. A reabsorption of sodium chloride still occurs from the renal tubules of marine teleosts, the excess salt being excreted by the chloride cells in the gills. Tubular secretory processes are also important for the excretion of divalent ions in marine teleosts. The initial determinant of the urine volume, except in aglomerular fish, is the glomerular filtration rate (GFR). The basal filtration is dictated by the blood pressure and the opposing osmotic pressure of the plasma proteins. Changes in urine volume mainly reflect the GFR. In lampreys (Agnatha) the rate of filtration across individual glomeruli can be adjusted (Brown and Rankin 1999). However, in teleosts, increases and decreases in the GFR are due to changes in the numbers of functioning glomeruli. This mechanism is described as glomerular recruitment or glomerular intermittency. Non-filtering glomeruli may still be perfused with blood, but the pressure is insufficient for filtration to occur. Such changes in the GFR appear to be due to constriction or dilatation of the afferent and efferent glomerular arterioles. However, local shunting of blood to different regions of the kidneys could still be occurring. There are several hormones that are vasoactive and could be contributing to changes in glomerular activity. They include vasotocin, catecholamines, angiotensin II, natriuretic peptides and possibly urotensins. Adrenergic nerve stimulation may also be contributing to changes in GFR. These possibilities will be discussed later. None of them has established physiological roles in this process in fish. However, most of them have been shown, when administered pharmacologically, to be able to induce changes in GFR.

There is little evidence to indicate that the reabsorption and secretion of water and solutes in the renal tubules of fish are directly influenced by their hormones. It has been suggested that in fresh water prolactin may decrease the permeability of the renal tubule to water (Nishimura and Imai 1982). It is possible that renal tubular permeability and transport processes may respond directly to the local concentrations of water and solutes. Such concentrations could be influenced by the rates of delivery of fluid to the tubular mechanisms and reflect the GFR (glomerular-tubular balance).

Neurohypophysial hormones, especially vasotocin, when administered to fish, can change the GFR (Sawyer et al. 1982; Pang et al. 1983). In high doses, such peptides invariably produce a diuresis (Maetz et al. 1964; Chester Jones et al. 1969; Sawyer et al. 1976; Logan et al. 1980; Uchiyama and Murakami 1994). The fish include goldfish, eels, lungfish and lampreys. The response appears to reflect an increased blood pressure and glomerular recruitment. However, in lower doses, vasotocin may result in an antidiuresis, due to a decrease of the GFR (Babiker and Rankin 1973, 1978; Henderson and Wales 1974; Amer and Brown 1995). It is generally accepted that vasotocin lacks any action on the permeability of the fish renal tubule to water. However, indirect observations on the ability of vasotocin to induce the formation of cAMP in nephrons of rainbow trout suggest that it has some action at this site (Perrott et al. 1993).

Table 7.3. Composition of the urine of fishes

		vol. (ml kg⁻¹ h⁻¹)	GFR	Tubular H₂O % reabsorption	Sodium conc. (mEq l⁻¹)	Total (μEq kg⁻¹ h⁻¹)	Osmolarity (mosmol l⁻¹)
Osteichthyes							
Protopterus aethiopicus[a] (African lungfish)	FW	4.9	14	69	5.5	27	17
Amia calva[b] (Bowfin)	FW	5.3	8.3	36	9.6	51	31
Carassius auratus[c] (Goldfish)	FW	13.7	20.4	33	11.5	158	36
Esox lucius[d] (Pike)	FW	0.6	3.1	77	<0.5	≈0.5	37
Opsanus tau[e] (Toadfish)	SW	0.18	aglom.	–	73	13	356
Platichthys flesus[f] (Flounder)	FW	1.8	4.2	57	30	43	90
	SW	0.6	2.4	75	60	36	275
Anguilla anguilla[g] (Eel)	FW	3.5	4.7	25	19	56	
	SW	0.6	1	40	6.5	4	
Salmo gairdneri[h,i,j] (Trout)	FW	4.0	6.5	38	9.3	37	
	SW	0.03	0.4	93	220 (Cl)	–	–
Chondrichthyes							
Squalus acanthias[k,l] (Spiny dogfish)	SW	0.18	3.0	94	339	61	780
Pristis microdon[m] (Freshwater sawfish)	FW	10.4	19	45	"Traces"		54
Agnatha							
Lampetra fluviatilis[n] (River lamprey)	FW	13.7	21	34	15	106	–
Myxine glutinosa[o] (Atlantic hagfish)	SW	0.22	–	–	480	72	1000

FW = fresh water; SW = sea-water.

[a] Sawyer (1966). [b] Butler and Youson (1988). [c] Maetz (1963). [d] Hickman (1965). [e] Lahlou et al. (1969b). [f] Lahlou (1967). [g] Sharratt et al. (1964a). [h] Holmes and McBean (1963). [i] Fromm (1963). [j] R.M. Holmes in W.N. Holmes and McBean (1963). [k] see B. Schmidt-Nielsen (1964). [l] Burger and Hess (1960). [m] Smith (1936). [n] Bentley and Follett (1963). [o] Morris (1965).

The acute transfer of fish from fresh to brackish water has often been found to result in a depletion of the storage of vasotocin in the neurohypophysis. This effect has been observed in many species, including flounder, rainbow trout, medaka and Japanese lampreys (Haruta et al. 1991; Hyodo and Hirano 1991; Perrott et al. 1991; Uchiyama et al. 1994). Such observations suggest that changes in the release and / or synthesis of vasotocin are occurring. However, until the development of a specific radioimmunoassay for vasotocin became available, it was not possible to measure the concentrations of this peptide in the plasma. Such an assay was introduced in 1994 (Warne et al. 1994). The concentrations of vasotocin the plasma of eels, flounder and trout were found to be between 10^{-12} and 2×10^{-11} M, which is a little lower than those usually observed in tetrapods. No consistent differences were observed in the concentrations of this peptide in the plasma in such fish following adaptation to fresh or sea-water. However, the infusion of hyperosmotic sodium chloride solutions into sea-water-adapted flounder resulted in a 60% increase in the concentration of vasotocin in the plasma (Warne and Balment 1995). No change in the concentration of the hormone was observed following haemorrhage. A transient increase in the release of vasotocin was also observed in rainbow trout transferred from fresh to brackish water (Kulczykowska 1997). It is possible that vasotocin contributes to the initial adjustment of the GFR and urine flow in some fish, following sudden exposure to hyperosmotic conditions. However, vasotocin may have other functions in fish.

The transfer of euryhaline teleosts from fresh to sea-water usually results in a decline in blood pressure. In European eels there is then a rise in the concentration of renin in the plasma (Henderson et al. 1976). Such a response would appear also to initiate a rise in angiotensin II. The administration of large doses of angiotensin II to American eels results in an increase in blood pressure (Nishimura and Sawyer 1976). In fresh water there is an associated rise in GFR and urine volume. However, in rainbow trout maintained under experimental conditions where normal blood pressure was maintained, angiotensin II was found to induce an antidiuretic effect. This response reflects a decrease in the number of filtering glomeruli (Brown et al. 1980; Gray and Brown 1985; Kenyon et al. 1985). Such an effect of the renin-angiotensin system could be contributing to the decreased urine flow that occurs when euryhaline teleosts enter sea-water.

Administered natriuretic peptides can influence the urine flow in fish. Such effects may reflect their vascular action, changes in the GFR and an increased excretion of sodium (natriuresis). The latter effect could be a primary effect of such hormones and will be discussed later. Natriuretic peptides can induce an increase in blood pressure and a diuresis in freshwater and sea-water-adapted rainbow trout (Duff and Olson 1986; Olson and Duff 1992). This effect appears to be an indirect one due to the activation of α-adrenergic neural mechanisms. A diuretic and natriuretic response has been observed in marine aglomerular toad-fish (Lee and Malvin 1987). Eel atrial natriuretic peptide induces an *antidiuresis* in Japanese eels adapted to sea-water but not fresh water (Takei and Kaiya 1998). The natriuretic peptides have multiple interrelated actions and their possible contributions to changes in the urine flow of fish are not clear. Their concentrations are usually higher in the plasma of fish adapted to sea-water than those in fresh water (Westenfelder et al. 1988; Evans et al. 1989; Freeman and Bernard 1990; Smith et al. 1991). The effects of natriuretic peptides on the kidney would appear to be more apt in a marine than freshwater environment.

3 Salt Exchanges

Fishes may gain salts from, or lose them into, the fluids that bathe them. The principal exchanges involve sodium and chloride and, to a much lesser extent, potassium.

On simple physico-chemical grounds, a net loss of sodium and chloride would be expected to occur across the integument of fishes into fresh water, while a gain of these ions may be expected to take place from sea-water. In addition, accumulation of salts occurs as a result of drinking and feeding. Earlier experiments, especially those of Krogh (1939), indicated that such movements of sodium chloride do indeed take place, but the technical methods then available did not allow a prompt assessment of the relative magnitude of the various exchanges to be made. The subsequent ready availability of radioisotopes considerably facilitated such measurements. In 1950 Mullins used ^{24}Na to measure the total rate of salt exchange in sticklebacks, *Gasterosteus aculeatus*. The isotopes were placed in the external bathing fluids (alternatively, they can be injected into the fish), and the amounts accumulated by the fish were measured at intervals. If the fish are in salt balance, the total influx and efflux of the ion will be similar and can be taken to indicate the total rate of its turnover by the fish. This has been called the rate constant (K) for the ion and is often expressed as the percentage of the total (exchangeable) amount of that ion in the body that moves in (K_i) or out (K_o) of the fish each hour. This provides a relatively simple and accurate method for comparing the permeability of different fish to salt in various circumstances. Sticklebacks in sea-water were found to exchange 20% of their total exchangeable sodium each hour, while in fresh water only about 1% was moved. Potassium was exchanged far less rapidly; in sea-water only $0.4\,mEq\,kg^{-1}\,h^{-1}$ was transferred (compared to $11\,mEq\,kg^{-1}\,h^{-1}$ for sodium), but this increased about three-fold in fresh water.

When the rate constants for sodium were measured in other fishes in sea-water they were found to vary greatly, though the movement of this ion in fresh water is always small (Table 7.4). In sea-water, *Tilapia mossambica* exchange 66% of their total sodium each hour, but in the toadfish, *Opsanus tau*, only 16% is moved in this time. Marine chondrichthyeans and hagfish exchange sodium far less rapidly than teleosts: less than 1% each hour.

Differences in the total ion exchange of different fish are due to several factors, probably the most prominent being their relative surface areas. Nevertheless, other factors also can result in differences in the rates of exchange, and these include the specific permeability of the gills and skin, the amounts of salt that may be accumulated through the gut as a result of feeding and drinking, and that which is unavoidably lost in the urine. As will be seen the latter is usually of minor importance.

Table 7.4. Total sodium fluxes* in fish bathed in various media

		Total Na flux μequiv $(100 \, g)^{-1} h^{-1}$	K (% of total exchangeable Na h^{-1})	Weight/g
Osteichthyes				
Carassius auratus[a,b]	FW	27	<1	90–220
(Goldfish)	190 mM Na Cl	553	8	
Opsanus tau[c]	10% SW	52	<1	200–700
(Toadfish)	SW	805	16	
Blennius pholis[d]	10% SW	50	8	2.5–7
(Blenny)	SW	2700	45	
Platichthys flesus[e]	FW	43	<1	60–390
(Flounder)	SW	2600	45	
Anguilla anguilla[f]	FW	4	<0.1	60–120
(Eel)	SW	1321	33	
Fundulus heteroclitus[g,h]	FW	60	<0.1	9–20
(Killifish)	SW	2020	35	
Gasterosteus aculeatus[i]	FW	60	1	1–2
(Stickleback)	SW	1000	20	
Tilapia mossambica[j]	FW	200	3	0.5–3
	SW	6000	66	
Serranus scriba[k]	SW	3100	Approx. 60	30–80
(Sea-water perch)				
Chondrichthyes				
Scyliorhinus caniculus[l]	SW	59	0.5	40–380
(Spotted dogfish)				
Squalus acanthias[m]	SW	90	0.9	4000
(Spiny dogfish)				
Hemiscyllium plagiosum[n]	SW	75	0.74	155–1100
(Lip shark)				
Agnatha				
Myxine glutinosa[o]	SW	40	<0.1	Approx. 25–50
(Hagfish)				

* Given from measurements of outflux.
[a] Bourguet et al. (1964). [b] Lahlou et al. (1969a). [c] Lahlou and Sawyer (1969). [d] House (1963).
[e] Motais and Maetz (1965); Motais et al. (1966). [f] Maetz et al. (1967a). [g] Maetz et al. (1967b).
[h] Potts and Evans (1967). [i] Mullins (1950). [j] Potts et al. (1967). [k] Motais and Maetz (1964).
[l] Maetz and Lahlou (1966). [m] Burger and Tosteson (1966). [n] Chan et al. (1967b). [o] Evans and Hooks (1983).

3.1 The Skin and Gills

As described in the previous section, the skin is generally considered to be relatively impermeable to the movements of solutes, as well as water, the principal exchanges taking place through the gills. However, in some teleost fish there are areas of the skin on the head, including the lining of the operculum, that are the site of cells which can contribute to ion exchanges (Burns and Copeland 1950; Karnaky and Kintner 1977; Marshall 1977). We must, however, await precise measurements of cutaneous permeability in different species of fish.

Gills in Sea-Water. The flounder, *Platichthys flesus*, in sea-water exchanges about $2600\,\mu Eq$ sodium $(100\,g)^{-1}$ body weight each hour (Table 7.4). This represents 40% of its total exchangeable sodium. About 25% of this sodium is absorbed through the gut (Motais and Maetz 1965) so that 75% of the total sodium accumulated by these fish takes place through the gills. In the sea perch, *Serranus scriba*, 90% of the sodium taken up from the sea-water takes place through the gills. The extrabranchial gains of sodium, which occur as a result of drinking, are extruded through the gills, urinary losses making up less than 0.1% of the total. The gills of such fish in sea-water are thus the site of considerable sodium exchange, with the efflux exceeding the influx by an amount which is about equivalent to that gained through the gut.

Keys (1931) used a perfused heart-gill preparation of the eel to demonstrate an active extrusion of chloride by the branchae into the external sea-water. Isolated gills of eels have also been shown to secrete chloride actively into the sea-water that bathes them (Bellamy 1961). Such an extrarenal mechanism for excretion of sodium chloride provided the channel that Homer Smith concluded must be present in marine teleost fish in order that they can drink sea-water and maintain a positive water balance. Such active salt excretion appears to be characteristic of marine teleost fish and probably the lampreys (see Morris 1960), but has not been clearly demonstrated in chondrichthyeans. Some of the latter, however, possess an alternative channel for extrarenal salt excretion, the rectal gland.

Immediately following the original demonstration, by Keys, of chloride secretion by fish gills, a histological search was made for a structural element which could be involved in this process. Keys and Willmer (1932) found some large and prominent epithelial cells at the base of the gill leaflets in the eel (Fig. 7.2). As these were not initially seen in freshwater teleosts or marine chondrichthyeans, it was concluded that they may be concerned with the extrusion of chloride and were termed *chloride-secreting cells*. They are now usually called chloride cells or, sometimes, ionocytes (see Chap. 1). These cells are rich with mitochondria and have a complex plasma membrane and intracellular tubular system. Na-K-ATPase is present on their basolateral borders. This enzyme maintains a low concentration of Na^+ in the cells, thus allowing a linked diffusion of Na^+ and Cl^- to occur into the cell across its basal surface (Silva et al. 1977). This process is now known to be mediated by the Na^+-K^+-$2Cl^-$ cotransport protein. The accumulated Cl^- diffuses out of the cell, down its electrochemical gradient, across the apical surface of the chloride cell. The Na^+ crosses the gills in the same direction, also following its electrochemical gradient, but through paracellular pathways (Marshall and Bryson 1998; Fig 1.2).

Chloride cells have been identified in most fishes, including freshwater teleosts, lampreys, elasmobranchs and even hagfish. However, their number and morphological development varies. They appear to have reached the epitomy of their frequency in marine teleosts. In euryhaline fish they may proliferate, differentiate and hypertrophy prior to their migration into sea-water. Conte and Lin in 1967 made the first such observations on the proliferation and renewal of chloride cells in the gills of young salmon in sea-water. The enzyme Na-K-ATPase has been identified in the gills of killifish (Epstein et al. 1967) and Japanese, European and North American eels (Utida et al. 1966; Jampol and Epstein 1970; Motais 1970).

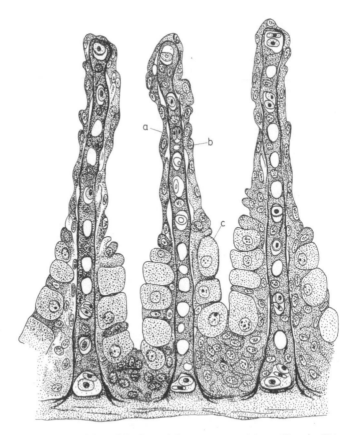

Fig. 7.2. Diagramatic representation of the gill leaflets of the European eel (*Anguilla anguilla*) showing the chloride-secretory cells. *a* Respiratory epithelium; *b* blood vessels; *c* chloride-secretory cell. (After Keys and Willmer 1932)

The activity of this enzyme increases considerably when these fish are adapted to sea-water. When euryhaline teleosts are transferred from fresh water to sea-water, this increase in the levels of Na-K-ATPase in the gills usually occurs over a period of 3 to 7 days. However, in some species, such as killifish, *Fundulus heteroclitus*, there may also be a more rapid response, taking 0.5 to 3 h (Towle et al. 1977; Mancera and McCormick 2000). The increase in the enzyme's activity, as well as salt extrusion from the gills, can be inhibited in European eels by the administration of actinomycin D (Maetz et al. 1969; Motais 1970). The effect was also observed in the killifish. This drug blocks the formation of mRNA. This mRNA could be mediating the synthesis of new subunits of Na-K-ATPase or factors involved in its activation (Mancera and McCormick 2000). The activity of Na-K-ATPase is four to five times greater in teleosts that live in sea-water as compared to fresh water (Kamiya and Utida 1969; Jampol and Epstein 1970). Its concentration in the gills of marine elasmobranchs is about the same as that in freshwater teleosts.

While chloride cells are clearly involved in the extrusion of Cl⁻ from the gills of marine teleosts (Foskett and Schaffey 1982), their possible role in the absorption of salt in fresh water is not clear. In fresh water the branchial chloride cells of euryhaline teleosts "dedifferentiate" and become reduced in size and possibly number. Sodium chloride is accumulated across the gills of such fish from very dilute solutions. However, the quantity of salt is very much less than that which is extruded by the fish when they are in sea-water. The lower levels of Na-K-ATPase in the gills of freshwater fish is consistent with this difference in salt transport. In sea-water-adapted teleosts, the Cl⁻ follows its electrochemical gradient through the chloride channels that are present in the apical plasma membrane of the chloride cells (Fig. 1.2). In fresh water, Na^+ and/or Cl⁻ are accumulated across this membrane. Several processes may be involved in this uptake, including Cl⁻/HCO_3^- and Na^+/H^+ or NH_4^+ exchange transport proteins. However, it is currently considered more likely that the Na^+ enters the cell through specific sodium channels and follows an electrochemical gradient established by a vacuolar (V)-type H^+-ATPase. This enzyme extrudes protons across the apical plasma membrane (Fig. 1.2). Conclusive evidence that Na^+ and Cl⁻ uptake across the gills of freshwater teleosts involves chloride cells is lacking. Pavement epithelial cells, which predominate on the lamellar and interlamellar surfaces of the gills, could also be involved.

Several hormones have been observed to influence the movements of salt across the gills of fish.

Cortisol. In 1967, Ian Chester Jones and his collaborators removed the adrenocortical tissue from European eels (Chan et al. 1967a; Henderson and Chester Jones 1967; Mayer et al. 1967). (This surgical operation does not appear to be possible in other teleosts.) In the sea-water-adapted eels, an excessive accumulation of Na^+ followed but the administration of cortisol restored normal concentrations of this ion. In freshwater eels, the Na^+ levels became depleted, but could also be restored by the injection of cortisol. This adrenocorticosteroid thus appeared to be contributing to the salt balance of the eels in both fresh water and sea-water. Hypophysectomy results in a loss of salt by many teleosts living in fresh water. This absence of the pituitary glands results in the lack of several hormones that can contribute to osmoregulation. In freshwater eels there is a loss of Na^+, which can be reduced by the administration of ACTH (Chan et al. 1968). This effect appears to reflect the ability of this hormone to promote the secretion of cortisol. Hypophysectomy in euryhaline killifish reduces the activity of Na-K-ATPase in the gills (Epstein et al. 1967). This deficiency can be corrected by the administration of cortisol (Pickford et al. 1970b).

It has been observed that when North American eels are transferred from fresh water to sea-water there is a rise in plasma cortisol concentration (Fig. 7.3) (Forrest et al. 1973a,b). Over a period of several days there is also a closely related increase in branchial Na-K-ATPase. The administration of cortisol to freshwater eels has been shown to increase the levels of Na-K-ATPase in the gills (Epstein et al; 1971; Kamiya 1972). This enzyme was found to be present in the chloride cells. Cortisol has also been shown to increase Na-K-ATPase activity in branchial chloride cells in tilapia in fresh water (Dang et al. 2000). In brown trout, cortisol

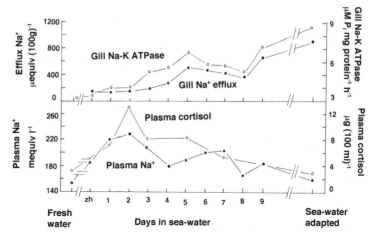

Fig. 7.3. A diagram showing the effects of adaptation of North American eels (*Anguilla rostrata*) to sea-water following transfer from fresh water. *Upper section* Changes in the concentrations of Na-K-ATPase in the gills and the efflux of Na⁺; *lower section* changes in plasma concentration of cortisol and the plasma Na⁺ concentration. (After Forrest et al. 1973a,b)

increases the concentrations of mRNA for the α-subunit of Na-K activated ATPase (Madsen et al. 1995).

Cortisol receptors have been identified in the gills of North American eels (Sandor et al. 1984) as well as in brook trout, steelhead trout and rainbow trout (Chakkraborti et al. 1987; McLeese et al. 1994; Shrimpton and McCormick 1999). They have been localized in the chloride cells of chum salmon fry (Uchida et al. 1998).

Cortisol has been observed to increase the numbers and differentiation of branchial chloride cells (Thompson and Sargent 1977; Dang et al. 2000). It has been suggested that the primary effect of cortisol in sea-water is to promote the proliferation and differentiation of chloride cells (Foskett et al. 1983; Foskett 1987).

In fresh water, cortisol appears to be necessary to maintain the integrity of the branchial sodium chloride transport system. In Japanese eels in fresh water the administration of cortisol has been found to stimulate the differentiation and pro-liferation of two freshwater types of chloride cells (Wong and Chan 2001). These changes were associated with increases in Na-K-ATPase activity. Cortisol also appears to contribute to maintaining levels of the vacuolar H⁺-ATPase that is present in the apical plasma membranes of the branchial epithelial cells. The chronic administration of cortisol has been found to increase the activity of this enzyme in freshwater- but not sea-water-adapted rainbow trout (Lin and Randall 1993).

Growth Hormone and IGF-I. In 1956, D. C. Smith observed that the adminis-tration of growth hormone to brown trout enhanced their ability to adapt and survive when they were transferred from fresh water to sea-water. This important observation was confirmed, much later, in other salmonids (Komourdjian et al. 1976; Clarke et al. 1977; Bolton et al. 1987). The concentrations of growth hormone in the plasma of young Atlantic and coho salmon have been observed to increase at about the time of their transformation from the juvenile parr stage to the subadult premigratory smolt (Prunet et al. 1989; Young et al. 1989). This hormone is released into the plasma of rainbow trout and coho salmon when they are transferred from fresh water to sea-water (Sweeting and McKeown 1987; Sakamoto et al. 1990). The administration of growth hormone to brown trout and Atlantic salmon has been shown to increase the numbers of chloride cells (Madsen 1990b; Prunet et al. 1994). There is also an increase in the activity of branchial Na-K-ATPase. Such effects are not confined to salmonids. In euryhaline tilapia, *Oreochromis mossambica*, the injection of growth hormone also increases the levels of this enzyme in the gills (Sakamoto et al. 1997). These effects of growth hormone in fish are reminiscent of those of cortisol. In brown trout, the administration of both of these hormones enhanced the salinity tolerance in a manner that may be synergistic (Madsen 1990b). Growth hormone has been observed to increase the numbers of cortisol receptors in the gills of juvenile Atlantic salmon (Shrimpton and McCormick 1998). This effect could be the basis for the interaction of the two hormones.

Many of the effects of growth hormone in tetrapods are mediated by IGF-I. This growth factor is present in the plasma of fish. Administration of IGF-I to rainbow trout improves their adaptation to sea-water (McCormick et al. 1991). When these fish are adapted to sea-water there is an increased expression of the IGF-I gene in the gills and kidney but not the liver (Sakamoto and Hirano 1993). Administered IGF-I can increase Na-K-ATPase in the gills of coho salmon (Madsen and Bern 1993), Atlantic salmon (McCormick 1996) and brown trout (Madsen et al. 1995; Seidelin et al. 1999). In the latter study, the effects were found to be additive to those of cortisol, which suggests that they may have separate sites of actions.

Prolactin. In 1956, C. E. Burden found that removal of the pituitary gland of the euryhaline killifish, *Fundulus heteroclitus*, resulted in their death within 10 days if they were kept in fresh water. However, when they were in sea-water they survived. Their death in fresh water is accompanied by an excessive loss of salt from the body. A similar observation has been made in many, though not all, teleosts in fresh water (Schreibman and Kallman 1969). Pickford and Phillips in 1959 found that the fatal effects of hypophysectomy in killifish, in fresh water, could be overcome by injecting them with prolactin. No other pituitary hormones were effective. Ball and Ensor (1965; 1969) demonstrated that in mollys, *Poecilia latipinna*, the loss of Na^+ which followed hypophysectomy was prevented by administered prolactin. Sticklebacks, *Gasterosteus aculeatus*, normally migrate from the sea into rivers in the spring (Lam and Hoar 1967; Lam and Leatherland 1969). If these fish are taken from the sea in winter and placed in fresh water, they die. However, if they are first injected with prolactin, they survive. This hormone is normally released from the pituitary gland during the spring migration. The effect of prolactin is due to an inhibition of salt loss from the gills in fresh water (Maetz et al. 1967a,b; Ensor and Ball 1972; Lam 1972). The precise mechanism of this effect is still unknown, but it appears to involve a decrease (dedifferentiation) in the size, but not the number, of chloride cells in gills and cutaneous epithelium (Foskett et al. 1983; Herndon et al. 1991). Prolactin is present in the plasma in higher concentrations in teleosts in fresh water than in salt water (Young et al. 1989; Auperin et al. 1994a; Yada et al. 1994). Receptors for this hormone have also been identified in the gills (Auperin et al. 1994b). In tilapia in *sea-water* the branchial extrusion of Na^+ is reduced by 75% following the injection of prolactin (Dharmamba et al. 1973; Dharmamba and Maetz 1976). This effect suggested that an inhibition of the activity of Na-K-ATPase could be occurring. Prolactin has been shown to have such an effect on this enzyme not only in tilapia but also in killifish and mullet (Pickford et al. 1970a; Gallis et al. 1979; Sakamoto et al. 1997). Prolactin has also been shown to inhibit the effect of IGF-I on the induction of mRNA for the α-subunit of Na-K-ATPase (Seidelin and Madsen 1999). It is also possible that prolactin reduces the diffusion permeability of the gills to Na^+, but the nature of such a process is unknown.

Natriuretic Peptides. In tetrapods, the primary nature of the effects of the natriuretic peptides are to compensate for the expansion of the volume of the extracellular fluids. Such adjustments involve cardiovascular changes and salt excretion by the kidneys. Indeed, as described earlier, it has been suggested that the primaeval role of these hormones was to limit the expansion of the extracellular fluids in marine fish. As such extant fish principally utilize the gills for salt excretion, there has been considerable interest regarding possible effects of these hormones at this site. Receptors for natriuretic peptides have been identified in both the vasculature and epithelial cells of the gills of teleosts. The receptors in the branchial blood vessels of toadfish were found to be predominantly of the C-type, which are not linked to guanylate cyclase (Olson and Duff 1993; Donald et al. 1994). These receptors are associated with the afferent and efferent gill arteries and the blood vessels of the respiratory lamellae. Receptors for natriuretic peptides have been studied in vitro in preparations of gill epithelial cells from European eels (Broadhead et al. 1992). These receptors appeared to be linked to guanylate cyclase and were presumably of the A or B type. It has been suggested that natriuretic peptides may increase the blood flow in the gills (Evans et al. 1989). Such an effect could influence the exchange of Na^+. However, a more direct effect on the chloride cells could also be occurring. Under in vitro conditions atrial natriuretic peptide was found to stimulate Cl^- secretion from chloride cells in opercular epithelia of killifish (Scheide and Zadunaisky 1988). This natriuretic peptide has also been shown to stimulate the branchial efflux of Na^+ in sea-water-adapted flounder, plaice and dab (Arnold-Reed et al. 1991).

Catecholamines. Adrenaline and noradrenaline can influence the blood flow and its distribution, and permeability of fish gills (Nilsson and Sundin 1998). α-Adrenergic receptors can mediate vascular responses that decrease the blood flow to the interlamellar regions, which contain most of the chloride cells, and divert it to the respiratory lamellae. On the other hand, β-adrenergic receptors can mediate vasodilatatory effects that may facilitate blood flow to the interlamellar region. Keys and Bateman (1932) in their seminal observations on chloride cells in the gills of eels observed that adrenaline inhibited Cl^- secretion. This effect has been observed in other teleosts (Mayer-Gostan et al. 1987). It could be due to vascular effects, but direct actions also occur. Catecholamines have been observed in vitro to variously decrease or increase Cl^- secretion from chloride cells in preparations of the opercular epithelium of killifish (Marshall and Bern 1980; Zadunaisky and Degnan 1980). Specific antagonists for the α- and β-adrenergic receptor-mediated effects of catecholamines were used to study the nature of the responses. An inhibition of Cl^- secretion by catecholamines was found to be the result of stimulation of α-adrenergic receptors. In contrast, stimulation of β-adrenergic receptors resulted in increased Cl^- secretion. In the isolated perfused head of trout adapted to fresh water, α-adrenergic stimulation increased the influx of Na^+ while β-adrenergic stimulation inhibited it (Perry et al. 1984). These effects were not related to haemodynamic changes in gill perfusion. Intracellular measurements of the Cl^- concentration have been made on chloride cells from the gills of freshwater-adapted brown trout (Morgan and Potts 1995). The β-adrenergic drug isoproterenol reduced the concentration of Cl^-. This observation is also consistent with a direct action on the cells. The effects of catecholamines on salt exchanges across the gills are unlikely to be persistent ones. Apart from their actions on CFTR-like chloride channels, the catecholamines may also influence the activity of Na-K-ATPase. An inhibitory effect appears to be mediated by an adenylate-cyclase-induced phosphorylation of the enzyme (Tipsmark and Madsen 2001). The responses are rapid and may occur acutely, such as in response to stress. It has been suggested that they may mediate the rapid decline in Cl^- secretion that occurs when fish move from sea-water to fresh water (Foskett 1987).

Neurohypophysial Hormones. The administration of neurohypophysial peptides, including vasotocin, has been observed to promote the exchange of Na^+ across the gills of teleost fish (Bourguet et al. 1964; Motais and Maetz 1964, 1967; Maetz et al. 1964). Such effects could be reflecting their vasoactive effects and the redistribution of blood in the gills (Maetz and Rankin 1969; Rankin and Maetz 1971). In contrast to the effects of catecholamines, vasotocin appears to constrict the arterioles supplying the lamellae so that there is an increased blood flow to the interlamellar regions, which are rich with chloride cells. Such a vascular response may occur following the administration of quite low doses of vasotocin (Bennett and Rankin 1986). Whether it could be a normal physiological response is unknown. It is also possible that the vasotocin is acting directly on the branchial epithelial cells. Receptors for vasotocin have been identified in such cells of European eels (Guibbolini et al. 1988). Adaptation of these fish to sea-water resulted in an increase in the amount of vasotocin binding to these receptors. The increase was paralleled by changes in the number of chloride cells present. The vasotocin receptors appeared to be of the V_1 type, that stimulate the phospholipase C-phoshatidylinositol mechanism. In teleost gills, somewhat unexpectedly, it induces a reduction in the activity of adenylate cyclase (Guibbolini and Lahlou 1987). Specific vasotocin V_1 type receptors have been cloned and found to be present in the gills of several teleosts (Mahlmann et al. 1994). Such receptors are also associated with the pituitary gland in fishes, where, like the V_1 type for vasopressin in mammals, they could be mediating a release of ACTH (Pierson et al. 1996).

Thyroid Hormone. The earliest observations on the possible roles of hormones in piscine osmoregulation were made on the actions of thyroid hormones (see Fontaine 1956). They were mainly concerned with the process of osmotic adaptation of teleosts following their transfer from salt water to fresh water. H. J. Koch and M. J. Heuts in 1942 (quoted by Fontaine 1956) found that feeding thyroid hormone preparations to three-spined sticklebacks reduced their ability to adapt to fresh water. However, as related by M. Fontaine, the results of the administration of thyroid hormone and antithyroid drugs to other species were quite variable. In some fish, osmotic adaptation was enhanced. Fontaine concluded "as to the mode of action of the thyroid hormone, it is possible that it includes an effect on the osmoregulatory cells of the gills, . . .". D. C. Smith (1956) also observed that the injection of thyroxine into brown trout "generally raises the salinity tolerance of trout, (but) the doses required seem above the physiological level". Such observations have since often been confirmed in other teleosts, especially salmonids (see, for instance, Fagerlund et al. 1980; McCormick 1995; Saunders et al. 1985). The administration of thyroid hormone to freshwater tilapia has been observed to increase the Na^+ and CI^- concentrations in the plasma and branchial Na-K-ATPase activity (Subash Peter et al. 2000). However, the results in different species are quite variable and the relationship to salinity tolerance is considered to be not proven. The physiological basis for such an effect of thyroid hormone on salinity tolerance is not clear, but has tended to focus on its possible actions on salt-extrusion mechanisms in the gills. Thyroid hormone receptors have been identified in the gills of many teleost fishes (Bres and Eales 1988; Lebel and Leloup 1989; Marchand et al. 2001).

Increases in the concentrations of thyroid hormone in the plasma have been observed in Atlantic salmon and coho salmon during their transformation from parr to smolts, which migrate to the sea (Prunet et al. 1989; Young et al. 1989). Smoltification has also been related to increases in the concentrations of cortisol and growth hormone in the plasma and a proliferation and differentiation of branchial chloride cells. Thyroid hormone replacement in hypophysectomized coho salmon resulted in an *inhibition* of gill Na-K-ATPase activity (Björnsson et al. 1987). A possible synergistic effect between thyroxine and cortisol on the activity of branchial Na-K-ATPase has been investigated in rainbow trout (Madsen 1990a). However, the thyroxine was found to be without an effect. In contrast, in a non-salmonid fish, tilapia, thyroxine "augmented" the action of cortisol in increasing branchial Na-K-ATPase (Dangé 1986). A synergistic relationship between thyroxine and growth hormone has been observed in the sea-water adaptation and smoltification of amago salmon (Miwa and Inui 1985).

Fish can undergo various types of developmental metamorphoses and in some, including flounder, thyroid hormone is involved in the transformation (Miwa et al. 1988; Inui et al. 1994). In summer flounder, *Paralichthys dentatus*, this process is accompanied by the development of branchial chloride cells (Schreiber and Specker 1999). There is also an increase in salinity tolerance and Na-K-ATPase in the gills. These changes could be inhibited by an antithyroid drug, thiourea, and enhanced by thyroxine (Schreiber and Specker 2000). Unfortunately, at this time, there seems to be no consensus regarding the physiological significance and role of the thyroid in osmoregulation. However, there are a number of observations suggesting that thyroid hormone could be contributing to this process, especially during metamorphic changes.

A diagrammatic summary of the possible roles of hormones in regulating salt movements across choride cells in the gills of teleost fishes is shown in Fig. 7.4.

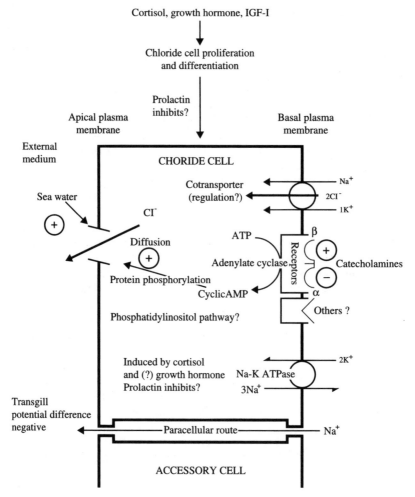

Fig. 7.4. A summary of the possible roles of hormones on Cl⁻ transport across the gills of teleost fish. For a more detailed explanation see the text. (Bentley 1998)

3.2 The Kidney

The fish kidney (see Dantzler 1989; Braun and Dantzler 1997) functions as a sodium chloride-conserving organ in fresh water fish. The GFR is high in such fish and most of the salt in the filtrate is reabsorbed by the renal tubule. Marine teleosts have a surfeit of salt, but most of this is secreted across the gills. The GFR in these fish is low. Sodium chloride may be secreted into the renal tubules across the second segment of the proximal tubule. This process is especially important in the formation of urine by aglomerular teleosts. The filtered Na^+ and Cl^- is mainly reabsorbed in marine teleosts and is subsequently excreted across the gills. The kidneys are the principal avenue for the excretion of divalent ions (Ca^{2+}, Mg^{2+}, SO_4^{2-} and PO_4^{2-}) and K^+ in these fish.

In sea-water, the flounder and sea perch excrete only about 0.1% of the total accumulated sodium in their urine (Motais and Maetz 1965). When euryhaline fish such as the flounder and eel are transferred from fresh water to sea-water, the total renal sodium excretion changes little; if anything, it may decrease in the latter medium (Table 7.3). In fresh water, sodium and potassium are lost in the copious dilute urine that is formed. In goldfish, this daily sodium loss equals about 8% of the total sodium in the body, an amount similar in magnitude to the total accumulated by the fish (Maetz 1963). Lampreys lose a similar proportion of their sodium in this manner. The pike, *Esox lucius*, loses as little as 0.02% of its body sodium each day (Hickman 1965). These renal losses can be replaced by active branchial uptake of sodium, but in some feeding fish this may not be necessary. The kidneys of fish do not appear to respond dramatically to excesses of sodium chloride, but this situation is unlikely to occur in fresh water. Holmes (1959) observed that sodium loads were mainly excreted extrarenally in rainbow trout. The urinary losses of sodium are little affected when sodium chloride solutions are injected into goldfish; indeed, if these are hypertonic, there is a decreased renal loss (Bourguet et al. 1964). In lampreys, we (Bentley and Follett 1963) found that only 10% of a dose of injected sodium chloride was excreted in the urine after 24 h.

Sodium is absorbed from the glomerular filtrate of fish. In the goldfish and lamprey, 93% of this is reabsorbed (Bentley and Follett 1963; Maetz 1963), while in the pike 99.95% may pass back into the plasma (Hickman 1965). Potassium is also reabsorbed from the glomerular filtrate of fish and, as observed in the pike and the white sucker, *Catostomus commersoni*, it may also be secreted into the urine across the wall of renal tubules (Hickman 1965).

Information regarding possible hormonal control of renal salt conservation and excretion in fish is meagre. The administration of vasotocin, usually in large amounts, has been observed to result in a natriuresis in several species, including goldfish, European eels, African lungfish and river lampreys (Bentley and Follett 1963; Maetz et al. 1964; Sawyer 1966; Chester Jones et al. 1969). These effects appear to reflect the vasoactive properties of this peptide, which produces an increased GFR. Angiotensin II evokes a similar natriuretic response in North American eels (Nishimura and Sawyer 1976). These effects are unlikely to be of physiological significance.

Atrial natriuretic peptide has a natriuretic effect in the toadfish, *Opsanus tau*, but as this teleost is aglomerular, the response would appear to involve the renal tubule (Lee and Malvin 1987). However, Evans (1995) was unable to detect ANP receptors in the renal tubules of a close relative, *Opsanus tau*, and has suggested that the response may be a "secondary" one. Receptors for natriuretic peptides have been identified in the kidneys of fish, but they are usually associated with the glomeruli (Perrott et al. 1993; Sakaguchi et al. 1996). However, Perrott and his collaborators also demonstrated that ANP stimulated cGMP formation in isolated nephrons of rainbow trout. This observation suggests that ANP receptors may also be present in the renal tubules of these fish.

Adrenocorticosteroids regulate the processes of Na^+ reabsorption and K^+ secretion in the renal tubules of tetrapod vertebrates (with the possible exception of some amphibians). However, despite numerous attempts, a direct effect of such steroid hormones has not been observed in fish. The renal tubular transport of ions ultimately depends on the integrity of Na-K-ATPase in the renal tubular cells. The activity of this enzyme has been observed to change in the kidneys of teleosts exposed to different salinities. However, the observations have not been consistent, possibly reflecting species variability and the prevailing environmental conditions. In the thick-lipped mullet, *Chelon labrosus*, renal Na-K-ATPase activity increases when they are adapted to fresh water (Gallis et al. 1979). On the other hand, sea-water acclimation has been observed to increase renal mRNA for Na-K-ATPase in the migrating silver phase of European eels (Cutler et al. 2000). In killifish, the administration of cortisol results in a rise in the activity of renal Na-K-ATPase in the hypophysectomized fish maintained in sea-water (Pickford et al. 1970b). In fresh water, prolactin inhibits branchial Na-K-ATPase in these fish but stimulates the renal enzyme (Pickford et al. 1970a). The latter effect was interpreted as a response, which could, appropriately, be mediating Na^+ retention. In intact brown trout adapted to fresh water, cortisol had no effect on the genetic expression of Na-K-ATPase (Seidelin et al. 1999). Cortisol also had little effect on renal Na-K-ATPase in thick-lipped mullet adapted to sea-water (Gallis et al. 1979). However, it *decreased* the activity of this enzyme in the fish adapted to fresh water. Due to the variability of such observations, the possible role of cortisol in regulating renal Na-K-ATPase is difficult to assess.

Receptors for angiotensin II have been identified in the renal tubules of teleosts (Cobb and Brown 1992; Marsigliante et al. 1997). It has been suggested that this peptide may modulate the activity of Na-K-ATPase in the renal tubules of European eels (Marsigliante et al. 2000). Perfusion of the kidney of the freshwater-adapted fish with angiotensin II resulted in a 1.8-fold increase in the activity of this enzyme. In sea-water-adapted eels, where the activity of the Na-K-ATPase was already high, there was no change. The renin-angiotensin system can thus be considered as another candidate for the hormonal regulation of Na-K-ATPase activity in the teleost kidney.

3.3 The Urinary Bladder

The urinary bladder of fishes is derived from the mesonephric ducts, and thus represents an extension of the renal tubules. It is present in many teleost fish, but it is not as capacious as the tetrapod urinary bladder. (The latter is derived from the cloaca.) Brahim Lahlou in 1967 observed that the urine stored in the urinary bladder of flounder had lower concentrations of sodium chloride than that collected directly from the ureters. This difference was observed when the fish were in both fresh water or sea-water. It was suggested that salt was being reabsorbed from the urinary bladder. Evidence of a similar reabsorption of salt was also found in the bladder of toadfish (Lahlou et al. 1969b). Studies, in vitro, on rainbow trout confirmed that an active transport of Na^+ and Cl^- was occurring (Lahlou and Fossat 1971; Fossat and Lahlou 1979). The transport process involved a partial coupling of the movements of these two ions. Such in vitro experiments have been extended to many other species of freshwater and marine teleosts (Marshall 1995). The nature of the Na^+-Cl^- coupling varies in different species. The urinary bladders of teleosts are also osmotically permeable to water. It is usually less in fish living in fresh water than in sea-water. The active salt transport results in a further conservation of urinary Na^+ and Cl^- and appears to be a physiologically useful saving in fresh water (Curtis and Wood 1991). In sea-water the transported salt mediates fluid absorption from the urinary bladder. In toadfish it has been estimated that without utilizing the activity of the bladder they would need to drink, and desalinate, 10% more sea-water (Howe and Gutknecht 1978). In euryhaline teleosts it has been observed that such fluid absorption increases when they are in sea-water as compared to fresh water (Hirano et al. 1971; Johnson et al. 1974). Preinjection of marine gobies with cortisol has been found to result in an increase in Na^+ transport across the bladder (subsequently measured in vitro) (Doneen 1976). The reported effects of prolactin on such Na^+ transport are conflicting. Some have found it to increase such ion transport (Johnson et al. 1974; Hirano 1975), while others observed little or no change (Foster 1975; Owens et al. 1977). Differences between species and environmental and experimental conditions could be contributing to such variations in the response.

3.4 The Gut

The physiological role and functioning of the gut of marine fishes in relation to the desalination and absorption of imbibed sea-water has been described earlier in this chapter. In fresh water the intestine provides an extrabranchial avenue for the accumulation of salt from food and the little water that they imbibe. In fasting eels, less than 0.5% of the total sodium that is accumulated by fish kept in fresh water is absorbed through the gut (Maetz and Skadhauge 1968). However, fish that are feeding actively will be expected to gain more salt, due to its absorption from the food. The magnitude of this varies with the nature of the diet. In fresh water a herbivorous or detrital diet contains less sodium chloride than a carnivorous one. Jørgensen and Rosenkilde (1956a) calculated that the expected accumulation of chloride by a goldfish from its normal diet would be less than 25% the quantity that is normally absorbed by the gills. Carnivorous freshwater fish, like the pike, probably gain adequate salt from their diet.

4 Nitrogen Metabolism and Osmoregulation

Most teleost fish utilize their voluminous aqueous environment to excrete ammonia (NH_3 and NH_4^+) as the end product of their amino acid catabolism (see Chap. 1; Fig. 1.5). They are said to be ammonotelic. The elasmobranchs and coelacanth convert this ammonia to urea, which is less toxic. They utilize most such urea as an osmolyte to maintain the concentrations of their body fluids at a slightly hyperosmotic concentration to the sea-water in which most of them live. They are said to be ureosmotic. The elasmobranchs, in addition, use urea as a nitrogenous excretory product so they are also ureotelic. Lungfish are usually ammonotelic. However, when African lungfish aestivate, in mud cocoons, during periods of drought, they convert ammonia to urea, which acts as a low-toxicity osmolyte that is stored in their body fluids until water again becomes plentiful. Most teleosts form small amounts of urea as a result of nucleotide catabolism via the formation of purines and uric acid. Uricolysis then results in the urea formation. About 20 species of teleosts have been found to form larger amounts of urea from amino acid catabolism. These fish include some amphibious species, the gulf toadfish (Walsh 1997) and Lake Magadi tilapia (Randall et al. 1989). The latter are the sole teleosts in this east African lake, which is extremely alkaline (pH 10). Under these conditions, the loss of NH_3 by diffusion across the gills is inhibited. These fish are completely ureotelic.

Ammonia may be present in the body fluids of fish as NH_3 or NH_4^+. The former gas is highly soluble and can be sequestered in lipids. It is highly toxic, though apparently less so in fishes than in mammals. Ammonium (NH_4^+) at a physiological pH of 7 to 8 is predominant, and it is also the primary form of ammonia that is produced by cells. It is very water-soluble, but not lipid-soluble, and its toxicity is less than that of NH_3. Ammonia and NH_4^+ are mainly excreted across the gills though some appears in the urine (Wilkie 1997). In fresh water, NH_3 is the principal excretory product but in sea-water substantial amounts of NH_4^+ are excreted. As it is a cation, NH_4^+ can interact with the transport of other ions. These include the uptake of Na^+ across the apical plasma membrane of branchial cells which utilize the Na^+/H^+ cotransporter (NHE). The H^+ apparently protonates NH_3 to form NH_4^+, which can diffuse through paracellular pathways to the external medium. Such protonation, to form NH_4^+, may also occur in fresh water as a result of the activity of H^+-ATPase, which is present in the apical plasma membrane of branchial epithelial cells (Lin and Randall 1993). The NH_4^+ may also be transported across the basolateral membrane of branchial epithelial cells by substituting for K^+ in Na-K-ATPase and the cotransporter Na^+-K^+-$2Cl^-$ (Claiborne et al. 1982; Mallery 1983; Wilkie 1997).

In the elasmobranchs, lungfishes and probably the coelacanth, the utilization of ammonia to synthesize urea takes place in the ornithine-urea cycle. This biochemical mechanism is also present in mammals and the amphibians. Carbamoyl phosphate, formed by the activity of the mitochondrial enzyme carbamoyl phosphate synthetase (CPSase), is the primary metabolite to enter the ornithine-urea cycle. In tetrapods, CPSase utilizes NH_4^+ to form carbamoyl phosphate. The enzyme is called CPSase I. In teleosts, elasmobranchs and the coelacanth there is a variant of this enzyme called CPSase III. It utilizes NH_3 formed from glutamine (Mommsen and Walsh 1989; Anderson 1995). The glutamine is produced in the liver as a result of the activity of glutamine synthase. It is interesting that lungfish have CPSase I, like tetrapods. While most adult teleosts do not produce urea as a major end product of nitrogen catabolism, it appears that this process may be functioning for a short time in their embryos (Depeche et al. 1979; Wright and Land 1998). The principal enzymes of the ornithine-urea cycle have been identified in teleost embryos. The genes for these enzymes, including CPSase III, appear to have been retained by teleosts. However, they are usually not expressed.

The excretion of urea in fish occurs principally across the gills, but also in the urine. These processes can be restricted and controlled by the presence of specific urea transport proteins. They are important for the conservation of urea in ureosmotic elasmobranchs (see later). Gulf toadfish, *Opsanus beta*, are normally ammonotelic but under conditions of stress, such as crowding and polluted water, they become ureotelic (Walsh 1997; Wood et al. 1997). Under such conditions, the cortisol concentration in the plasma increases and appears to contribute to the activation of glutamine synthetase (Hopkins et al. 1995). This hepatic enzyme is rate-limiting in the process of ureogenesis. Cortisol could also be contributing to urea synthesis by promoting the production of catabolic nitrogen as a result of a stimulation of gluconeogenesis.

Toadfish excrete urea in discrete pulses, each lasting for up to 3 h (Wood et al. 1995). This secretion, which appears to occur across the gills, takes place once or twice a day. It appears to be initiated by a discrete stimulus (Wood et al. 1997; Pärt et al. 1999). The urea permeability of the gills increases about 30-fold, but the permeability to water is unchanged. The mechanism appears to utilize a specific urea transport protein, which has been compared to that present in the mammalian kidney (Wood et al. 1998; Pärt et al. 1999; Walsh et al. 2000). A similar urea transport protein is involved in the excretion of urea in trout embryos (Pilley and Wright 2000). The mammalian transport protein contributes to absorption of urea across the renal tubule and the maintenance of the urea concentration gradient in the renal medulla. Its activity is increased by vasopressin. Vasotocin, as described earlier, can also increase the permeability of the skin and urinary bladder of amphibians to urea. The pulse of urea excretion across the toadfish gills is preceded by a decline in plasma cortisol concentration (Wood et al. 1997). The administration of vasotocin has been shown to result in large increases in such urea excretion (Perry et al. 1998). It is possible that the pulsatile excretion of urea is regulated by several hormones, which could be acting synergistically, and include cortisol, vasotocin and adrenaline (Pärt et al. 1999). The effects of the other neurohypophysial peptide that is present in bony fish, isotocin, does not appear to have been investigated.

5 Osmoregulation and the Endocrines of Sharks and Rays

The osmoregulation of sharks, and skates and rays has excited special interest since it was observed that, unlike marine teleosts, they maintain their body fluids at a concentration that is similar to that of sea-water (Smith 1936). They are ure-osmotic and, as described in the last section, are also ureotelic. This process is complemented by the retention of other solutes, especially trimethylamine oxide (TMAO) (Table 1.2). The latter acts as a protective counteracting solute to antagonize possible toxic effects of the urea (see Chap. 1). Other aspects of their osmoregulation are also quite different to that of bony fishes. Some of the general information has been summarized in earlier sections of this book. The present account is mainly a synthesis of this information.

The cartilaginous fish (Chondrichthyes) apparently evolved and diverged from the main line of vertebrates about 300 million years ago. It was once considered likely that their ancestors lived in fresh water, but their origins are still controversial. Most species are marine, but a few venture into fresh water and are considered to be euryhaline. At least one small family, Potamotrygonidae, the freshwater stingrays, live permanently in fresh water and are apparently stenohaline. Marine elasmobranchs, unlike teleosts, do not normally drink sea-water in order to gain osmotically free water (Smith 1931). However, as they are slightly hyperosmotic to sea-water, they apparently gain small amounts of water by osmosis across their integument. As the concentrations of Na^+ and Cl^- in their body fluids are less than those in sea-water, they gain these ions by diffusion. However, the rate of such exchanges is much less than in marine teleosts (Table 7.4). The slower uptake of salt partly reflects the absence of significant drinking and, as the concentration of sodium chloride in their body fluids is higher than that in teleosts, a lower gradient for the diffusion of these ions. The permeability of the gills to salt is also much less than it is in teleosts, though permeability to water is greater. The gills of elasmobranchs have a very low permeability to urea, which aids its retention in the body. Water and salts are excreted by the kidneys. The urine is isosmotic or slightly hyposmotic to the plasma. The concentrations of Na^+ and Cl^- that are attainable in the urine are similar to those in the plasma, so that the kidneys do not appear to provide an osmotically economical avenue for salt excretion. Little or no urea is lost in the urine. The kidney remains the principal avenue for water excretion and probably, as in teleosts, divalent ions. In contrast to marine teleosts, the gills do not appear to be the site of significant salt excretion. It was once generally believed that the elasmobranch gills lacked chloride cells. However, these cells are now known to be present, though in small numbers. The elasmobranchs possess rectal salt glands which secrete a fluid containing much higher concentrations of sodium chloride than present in the plasma and even somewhat greater than those in sea-water (Table 7.5). The rectal gland provides an avenue for salt excretion. However, surgical removal of this gland does not seem to compromise the fish's osmotic balance (Burger 1962, 1965). Possibly other mechanisms for salt excretion exist, such as the gills, which can assume a salt excretory role in some circumstances.

The control and integration of the unique physiological mechanisms involved in urea retention and salt excretion in elasmobranchs is of considerable interest. These fish have a full complement of the vertebrate endocrine glands and the homologous hormones that they secrete. However, although several such hormones are known to contribute to the osmoregulation in other vertebrates, such roles have not been established in the elasmobranchs. A plethora of different neu-

Table 7.5. Comparison of the composition of the rectal gland secretion of *Squalus acanthius* with that of its plasma, urine and sea-water. (Burger and Hess 1960)

	Osmolarity mosmol l^{-1}	Sodium	Potassium mEq l^{-1}	Chloride
Sea-water	1000–1100	470	12	550
Rectal gland secretion	1018	540	7	533
Plasma	1018	289	5	246
Urine	780	339	2	203

rohypophysial peptides, including low concentrations of vasotocin, have been identified in their neurohypophyses (Acher et al. 1999). Their diversity and evolutionary persistence has been a source of wonder, but their physiological roles are still unknown. Elasmobranchs possess distinct adrenocortical interrenal glands which secrete adrenocorticosteroids (Chester Jones and Mosley 1980; Henderson and Garland 1980). Corticosterone has been identified in the plasma, but the predominant such steroid is 1α-hydroxycorticosterone, which appears to be unique to these fish. When it is administered to tetrapod vertebrates, the latter have "mineralocorticoid" activity (Grimm et al. 1969). However, such an effect has not been conclusively demonstrated in cartilaginous fish. Interrenalectomy in the skate, *Dasyatis sabina*, has been observed to result in a decrease in the concentration of urea and Na$^+$ in the plasma (DeVlaming et al. 1975). The adenohypophysis contains ACTH, which can stimulate adrenocorticosteroid synthesis (Hazon and Henderson 1985), prolactin and growth hormone. Hypophysectomy in dogfish has been observed to reduce the turnover of body water, though this effect was absent in the little skate (Payan and Maetz 1971; Payan et al. 1973). A renin-angiotensin system is present and a mammalian-like juxtaglomerular apparatus has been identified (Hazon et al. 1999). Elasmobranch angiotensin has a unique structure (Fig. 2.6). Angiotensin II has activities in elasmobranchs similar to those in other vertebrates. It can promote the secretion of 1α-hydroxycorticosterone (Hazon and Henderson 1985). It may also be contributing to cardiovascular homeostasis. Although elasmobranchs do not normally drink, or seem to need to, the administration of angiotensin II has been observed, in some circumstances, to promote drinking in the dogfish, *Scyliorhinus canicula* (Hazon et al. 1989). However, this response has not been observed in all elasmobranchs (Kobayashi and Takei 1997). The C-type natriuretic peptides (CNP) have been identified in the heart and brain of elasmobranchs and they are also present in the blood (Schofield et al. 1991; Evans 1995; Evans et al. 1989; Suzuki et al. 1994). The CNP may contribute to cardiovascular homeostasis and, as described below, appears to regulate secretion by the rectal salt gland. Other osmoregulatory roles may have evolved for the secretions of endocrine glands in cartilaginous fish, but it is possible that they may be mediating other types of responses.

Osmoregulation involving the retention of urea, accumulation of water and the excretion of salts appears to be an integrated process in elasmobranchs involving the gills, kidneys and rectal salt gland. The liver may also contribute to this process by changes in urea synthesis.

The Gills. The role and functioning of the gills in the osmoregulation of elasmobranch fishes differ considerably from marine teleosts. However, it is consistent with their ureotelic and ureosmotic habitude. The permeability of the gills to urea is very low, though that to water is high (Pärt et al. 1998). Mitochondria-rich cells, which appear to function like chloride cells, are present (Laurent and Dunel 1980). Na-K-ATPase is also present in the gills but at much lower levels of activity than in marine teleosts (Jampol and Epstein 1970). The exchange rate of Na^+, which mainly occurs across the gills, is much less in elasmobranchs than marine teleosts, but it is similar to that of freshwater teleosts (Table 7.4). Changes in blood flow may influence branchial exchanges of solutes and water. Small doses of noradrenaline have been observed to decrease gill blood flow in the blacktip reef shark (Chopin and Bennett 1995). This effect appears to be an α-adrenergic one. Higher doses of noradrenaline elicited a β-adrenergic response and an increase in the blood flow. Possible changes in the distribution of blood within the gills were not studied. Adrenaline has been observed to increase the diffusion permeability to water in a perfused preparation of dogfish gills (Pärt et al. 1998). Receptors for the C-type natriuretic peptide have been identified in the gills of the Japanese dogfish, *Triakis scyllia* (Sakaguchi and Takei 1998). Some of these receptors were linked to guanylate cyclase and the formation of cGMP. As suggested by Evans and his collaborators (Evans et al. 1989), this peptide could be contributing to the control of branchial blood flow.

The presence of specific Cl^- and/or Na^+ uptake or secretion mechanisms in the elasmobranch gill is uncertain. An active Cl^- secretion mechanism has been tentatively identified in dogfish gills (Maetz and Lahlou 1966; Bentley et al. 1976). The observed electrochemical gradient across the gills could not completely account for a passive efflux of the Cl^-. A saturable electroneutral Na^+ influx was also observed, which was decreased by reduction in external pH. Such processes appear to involve exchanges with H^+ and HCO_3^- and may primarily contribute to acid-base balance rather than osmoregulation (Evans 1982). An H^+-ATPase, which could be involved in such exchanges, has been identified in the mitochondria-rich cells in the gills of the spiny dogfish (Wilson et al. 1997).

The low permeability of the gills of elasmobranchs to urea is a major factor in maintaining their ureosmotic way of life. It has been proposed that the apical plasma membrane of branchial epithelial cells provides the barrier to such diffusion (Pärt et al. 1998). The accumulation of urea in these cells appears to be prevented by the activity of a urea-transporter protein, which is associated with the basolateral plasma membrane of these cells. This transporter passes urea that is accumulated in the cells back into the blood. Such a urea-transporter protein, named ShUT, has been identified and cloned in the spiny dogfish (Smith and Wright 1999). Its amino acid sequence has a 66% identity to a similar protein, UT-A2, which is present in the rat kidney. The ShUT is expressed in the kidney of the dogfish, but related proteins, as their mRNAs, were also identified in the gills. Roger Acher (Acher et al. 1999) has suggested that in elasmobranchs vasotocin, or related oxytocin-like peptides, could be involved in regulating the activity of such urea-transport proteins.

Specific binding proteins and receptors for 1α-hydroxycorticosterone and angiotensin II have been identified in the gills of elasmobranchs (Burton and

Idler 1986; Tierney et al. 1997). While such observations suggest that the receptors may mediate physiological activity and influence the functioning of the gills, the nature of such effects is unknown. Elasmobranch angiotensin II has been observed to constrict the branchial artery of Japanese dogfish (Hamano et al. 1998). Thus, it could be contributing to the regulation of branchial blood flow. By analogy with teleosts, one may suspect that 1α-hydroxycorticosterone could be influencing ion exchanges, but there is no direct evidence for such an effect.

The Kidneys. Elasmobranchs have kidneys with well-developed glomeruli and, in contrast to marine teleosts, a nephron that includes a distal tubule segment. The nephrons are long and contain five distinct segments that are arranged to form two complete hairpin loops (Boylan 1972; Lacy et al. 1985). This morphological arrangement appears to be a countercurrent mechanism. It may be contributing to the retention of urea and the ability of the tubular system to actively reabsorb this metabolite. This process may be utilizing the activity of the ShUT urea transport protein (Smith and Wright 1999). Such urea conservation is a major function of the kidneys in elasmobranchs.

The GFR in marine elasmobranchs is high compared to that of marine teleosts (Table 7.3). Urine flow is quite low, due to an efficient reabsorption of more than 90% of the glomerular filtrate. The urine is usually slightly hyposmotic and the Na^+ concentration is similar to that of the plasma. Thus, the kidney cannot effect significant changes in the plasma Na^+ concentration, but it could be mediating isosmotic adjustments of extracellular fluid volume.

The control of urine formation and its composition probably involves changes in the GFR and the reabsorption and secretion of fluid and solutes by the renal tubule. Such processes could involve neural and hormonal regulatory mechanisms. However, there is little direct information about such processes. A possible action of 1α-hydroxycorticosterone on the renal tubule of elasmobranchs is an attractive extension of the role of such corticosteroids in tetrapod vertebrates. It has been suggested (Armour et al. 1993). The administration of mammalian atrial natriuretic peptide into the spiny dogfish was found to decrease the blood pressure and the GFR, resulting in an antidiuresis and salt retention (Benyajati and Yokota 1990). This action contrasts with its natriuretic and diuretic effects in other vertebrates and may reflect the size of the dose of peptide given. Possible actions of neurohypophysial peptides on kidney function do not appear to have been reported in elasmobranchs.

The Rectal Gland. Elasmobranchs, as described in Chapter 1, can excrete concentrated solutions of sodium chloride in the secretions of the rectal gland. The fluid is nearly isosmotic to the plasma and contains almost pure sodium chloride at a concentration which is similar to that of sea-water (Table 7.5). It provides a mechanism for adjusting the sodium chloride concentration and the volume of the extracellular fluids. The principal stimulus for its secretion is an increase in the volume of these fluids rather than their concentration (Burger 1962, 1965). In contrast to the nasal salt glands of birds and reptiles, a direct neural control mechanism is not involved. Burger suggested that "volume receptors" may control secretion through the intervention of some "undefined hormone". Over 30 years later, this hormone has been shown to be the C-type natriuretic peptide (CNP). In elasmobranchs it is secreted by the heart in response to volume expansion of the extracellular fluid (Solomon et al. 1992).

The primary ion that is secreted by the rectal gland is Cl$^-$ with Na$^+$ following down its electrochemical gradient (see Chap. 1). The process of formation of the secretion appears to be the same as occurs in branchial chloride cells. High concentrations of Na-K-ATPase maintain low intracellular concentrations of Na$^+$ in the tubular cells of the rectal gland. This gradient allows an inward diffusion, across the basolateral surface, of the Na$^+$-K$^+$-2Cl$^-$ cotransport protein (Silva et al. 1999a). The accumulated Cl$^-$ diffuses out of the cell across its apical plasma membrane. This process occurs through specific CFTR-like chloride channels (see Chap. 1). The precise mechanism of action of CNP is quite complex (Silva et al. 1999b) and involves at least three processes. (Fig. 7.5). The CNP interacts with guanylate cyclase-linked receptors on the basal side of the secretory cells, resulting in the formation of cGMP. The latter alone cannot stimulate secretion but is aided, synergistically, by a parallel effect of CNP, which results in the activation of protein kinase C. Earlier studies had shown that vasoactive intestinal peptide (VIP) could, via activation of adenylate cyclase, stimulate secretion (Greger et al. 1988). However, this action was only observed in vitro (Solomon et al. 1985). Vasoactive intestinal peptide is associated with nerve cells in a local neural network in the rectal gland. Its release from these nerves is promoted by CNP. Thus, at least three interacting mechanisms are involved in the stimulation of rectal salt gland secretion by CNP.

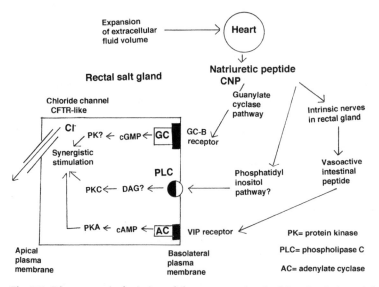

Fig. 7.5. Diagrammatic depiction of the processes involved in stimulation of the secretion of the rectal salt gland of elasmobranch fish by the C-type natriuretic peptide (CNP). The response appears to involve the combined parallel effects of activation of guanylate cyclase, adenylate cyclase and, possibly, phospholipase C. For more details see the text. (After Silva et al. 1999b)

Adaptation of Sharks and Rays to Fresh Water. Most elasmobranchs live exclusively in the sea, but a few species venture into brackish and even fresh water. Some apparently make only brief excursions into the mouths of rivers so that little adaptation is necessary (Price 1967). However, other species stay for longer periods of time in estuarine conditions and are considered to be euryhaline. The latter include some skates (genus *Raja*), which live along the east coast of the USA (Payan et al. 1973), and the sawfish, *Pristis microdon* (Smith 1936). The shark *Carcharhinus leucas* lives for extended periods in Lake Nicaragua, but, contrary to earlier impressions, it returns periodically to the sea. Stingrays belonging to the family Potamotrygonidae live permanently in freshwater rivers, far from the sea. These elasmobranchs include the Amazon stingray (*Potamotrygon* spp.). Other members of this family have been identified in African rivers, but at least some of these have now been shown to belong to the family Dasyatidae (Thorson and Watson 1975). The latter family are mainly marine, but a few species live in fresh water and their fossils have been identified in freshwater sediments.

With their well-developed glomeruli and a distal renal tubular segment for the reabsorption of Na^+, elasmobranchs appear to be well equipped to deal with the vicissitudes of fresh water. Under such conditions, the GFR and urine volume increases and the reabsorption of water in the renal tubule declines. Sodium reabsorption from the glomerular filtrate increases. The concentrations of urea in the plasma decline in fresh water but they are usually still maintained at levels of 100 to 200 mM. In skates, this change does not reflect an increase branchial permeability but an increase in the excretion of urea in the urine (Payan et al. 1973). Whether it is due to a regulated change in renal tubular reabsorption or an unavoidable consequence of an increased GFR is unknown. Urea synthesis in little skates (*Raja*) declines under such conditions (Goldstein and Forster 1971). In another euryhaline skate, *Dasyatis sabina*, transfer from sea-water to dilute (33%) sea-water was found to be associated with the accumulation of a prolactin-like activity in the pituitary glands (de Vlaming et al. 1975). This observation suggested that prolactin could be acting to reduce the permeability of the gills, as observed in teleosts. The Amazon stingray, *Potamotrygon*, lives exclusively in fresh water. The concentrations of urea in its plasma are even lower (about 1 mM) than in most mammals (Griffith et al. 1973; Gerst and Thorson 1977; Bittner and Lang 1980). Attempts to adapt these fish to 50% sea-water did not result in any significant accumulation of urea. They appear to lack the facility for its accumulation as an osmolyte. The Amazon stingray cannot survive in sea-water and lacks the renal tubular countercurrent system that apparently contributes to the conservation of urea by marine elasmobranchs. The rectal gland regresses and becomes non-functional in elasmobranchs living in fresh water (Oguri 1964). The Na^+ concentration in the plasma of elasmobranchs living in fresh water declines. This change could reflect unavoidable losses by diffusion across the gills and more voluminous urine. The manner by which they maintain adequate salts in their body fluids in such circumstances is unknown. Possibly salt in the diet is sufficient for their needs. However, teleosts living in fresh water utilize an active salt-uptake mechanism in their gills, and such a process would also appear to be appropriate for such elasmobranchs. Possibly the Na/H^+ exchange process, which appears to be primarily dedicated to acid-base balance (see above) is adequate for such salt uptake. Pang and his collaborators (1977) described "preliminary" experiments on freshwater stingrays in which an influx of Na^+ was observed. This process was saturable, suggesting that an active transport was occurring. The adaptations of such elasmobranch fish to life in fresh water have excited interest and curiosity since Homer Smith drew attention to their plight over 70 years ago. Almost nothing is known about the possible roles of endocrines in such adaptations.

6 The Hagfish

Hagfish (class Myxini) are jawless fishes (superclass Agnatha). They are exclusively marine with a worldwide distribution, often living at great depths (see Hardisty 1979). The two most common genera are *Myxine* and *Eptatretus*. Like sharks and rays, they maintain the concentrations of their body fluids at concentrations that are isosmotic or slightly hyperosmotic to sea-water (Table 7.6). However, in contrast to the cartilaginous fish, the Na^+ and Cl^- concentrations in their plasma are remarkably similar to those in sea-water. They do not utilize organic osmolytes to balance the osmotic concentrations of their extracellular fluids with that of their external bathing solution. This observation indicates that vertebrate life can exist when the cells are bathed by a concentration of salt similar to that of sea-water, just as occurs in marine invertebrates. The composition of their plasma, their lowly phyletic status and their relationship to fossil ostracoderm ancestors of the vertebrates suggests that they are the closest living relatives of early vertebrates. The lampreys (class Cephalaspidomorpha) are also agnathans and many are euryhaline. In sea-water, like teleosts, they maintain their body fluids at a concentration which is hyposmotic. The Agnatha appear to be diphyletic with the two classes sharing a common pre-Cambrian ancestor (Hardisty 1979). The lampreys may be closer to the main line of vertebrate evo-

Table 7.6. Concentration ($mM\,kg^{-1}$ water) of solute in the body fluids of Myxinoidea

	Sea-water[a]	*Myxine glutinosa* Serum[a]	*Myxine glutinosa* Muscle[b]	Urine[c]	*Eptatretus stouti* Mucous secretion[d]
Sodium	470	549	33	481	95
Potassium	12	11	142	13	207
Chloride	550	563	41	536	–
Calcium	8	5	2	10	11
Magnesium	50	20	13	28	39
Trimethylamine oxide	–	nil	87	–	–
Amino acids			291		

The concentrations of sea-water in all these observations was not as in reference a but the differences in composition are consistent.
References: [a] Bellamy and Chester Jones (1961). [b] Robertson (1976). [c] Morris (1965). [d] Munz and McFarland (1964).

lution, but the hagfish appear be more similar in their habitude to primaeval vertebrates.

It has been said that hagfish have no osmoregulation. Certainly, few osmotic exchanges are expected to occur between their extracellular fluids and sea-water. However, some solute gradients do exist, especially for Ca^{2+} and Mg^{2+}, which necessitate some regulation of the composition of the extracellular fluids. The composition of the intracellular fluid differs considerably from that of the extracellular one. The ratio for the ion concentrations cell water/serum water are $13:1$ for K^+ (Table 7.6). The pattern of a relatively high intracellular concentration of K^+ and a low one of Na^+ and Cl^- are similar to those in other animals. Intracellular osmotic balance is achieved by the presence of amino acids and TMAO. Thus, osmoregulation continues to be essential at the barrier of the extracellular and intracellular fluid but, due to the high extracellular ion concentrations, it may require special adaptations in hagfish.

Attempts have been made to perturb the osmotic balance of hagfish by placing them in hyposmotic and hyperosmotic solutions in order to see if any physiological adjustments may then occur (Munz and McFarland 1964). The fish rapidly succumb when they are transferred directly from sea-water to 25% sea-water. (This solution is equivalent to about 0.8% sodium chloride.) However, they can survive for at least 7 days when transferred directly from sea-water to 80 or 120% sea-water. When these hagfish were replaced in sea-water, the latter survived, but those from the dilute sea-water subsequently died. The acute responses were a shrinkage and loss of body water in the hyperosmotic solution and a swelling in the hyposmotic one. They appeared to behave like osmometers. The swelling that occurred in 80% sea-water was followed by a slow return, over several days, to their original weight. Nevertheless, as described above, these fish succumbed when replaced in sea-water. In 120% sea-water the hagfish lost 25% of their body weight and maintained this condition. When they were returned to sea-water they regained their original weight and survived. Changes in the external osmotic concentration have also been shown to result in compensatory increases in the concentrations of intracellular amino acids (Cholette et al. 1970). The amino acid levels in the extracellular fluids remained low. Regulation of the volume of the extracellular fluids under conditions of external osmotic change may depend on urinary excretion. Compensation may be "automatic" and result from adjustments of the GFR due to changes in the concentrations of plasma proteins and, possibly, the blood pressure (McFarland and Munz 1965; Alt et al. 1981).

Most vertebrate hormones appear to exist in hagfish, including those that are involved in osmoregulation in other vertebrates. Vasotocin was identified in the pituitaries of *Myxine glutinosa* by Follett and Heller (1964b). The structure of its precursor, provasotocin, has a homology in its amino acid sequence of only 47% with lamprey provasotocin (Suzuki et al. 1995). It was suggested that the hagfish provasotocin may have a greater homology to the precursor of a homologous peptide, conopressin, which is present in molluscs. Corticosterone and cortisol have been identified in the plasma of *Myxine glutinosa* (Phillips et al. 1962b; Idler and Truscott 1972). ACTH, which could be controlling such adenocorticosteroid secretion, appears to be present in the pituitary (Strahan 1959). Catecholamines have also been identified in hagfish (Randall and Perry 1992). However, the renin-angiotensin system has not been found in the Agnatha. Immunoreactive assays have observed natriuretic peptides in the plasma of *Myxine*, though its precise type is unknown (Evans et al. 1989). It appears that neither growth hormone nor prolactin is present in the Agnatha.

The integument (skin and gills) of hagfishes is very permeable to water (McFarland and Munz 1965; Rudy and Wagner 1970). However, exchanges of Na^+ are highly restricted Table 7.4. The skin is liberally supplied with mucous glands, which can secrete copious amounts of fluid containing K^+, Mg^{2+} and Ca^{2+} in higher concentrations than those in the plasma (Table 7.6). It has been suggested that these glands may provide a mechanism for the extrarenal excretion of the ions (Munz and McFarland 1964). The functional kidneys are mesonephric and have large, well-developed glomeruli that open directly into the archinephric duct. Such anephric glomeruli are unique among the vertebrates. Little fluid or sodium chloride reabsorption appears to occur in the archinephric duct (Munz and McFarland 1964; Alt et al. 1981). Thus, the kidney does not appear to contribute to the control of the overall osmotic concentration of the body fluids. However, it could be regulating the volume. The concentrations of other solutes in the urine suggest that K^+ and Mg^{2+} are secreted into the archinephric duct and that glucose is reabsorbed there. The kidney may also contribute to acid-base balance. The hydrostatic pressure in the glomeruli of *Myxine* is, at most times, not great enough to promote ultrafiltration against the opposing osmotic forces of the plasma proteins (Riegel 1998, 1999). A fluid secretory process may be involved in urine formation. Adrenaline and noradrenaline enhanced this process, possibly by increasing the blood flow through capillaries involved in the secretory process. Whether such an effect of catecholamines could be normally contributing to this urine formation is unknown. Receptors for natriuretic peptides have been identified in the kidney of *Myxine glutinosa* (Kloas et al. 1988; Toop et al. 1995a). They have been found in the glomeruli as well as the archinephric duct. These receptors are linked to guanylate cyclase, suggesting that they could be functioning just as in other vertebrates. Such natriuretic hormone receptors could be mediating changes in renal function that contribute to the regulation of the volume of the extracellular fluids. This action would be consistent with the previously suggested primaeval function of these hormones.

The gills of hagfish, despite earlier doubts, contain mitochondria-rich cells which appear to function like chloride cells (Elger and Hentschell 1983; Bartels 1985). They are the site of carbonic anhydrase activity (Mallatt et al. 1987) and Na-K-ATPase (Choe et al. 1999). The concentrations of the latter enzyme appear to be relatively low and like those in the gills of elasmobranchs. The chloride cells may be the site of the Na^+/H^+ and Cl^-/HCO_3^- exchanges which have been observed to take place across the integument of hagfish (Evans 1984). Such mechanisms exist in freshwater teleosts, where they contribute to acid-base balance. Natriuretic peptide dilates the ventral aorta of *Myxine*, and this response may lead to an increase in gill perfusion (Evans 1991). Receptors for this hormone have been identified in the aorta of this fish as well as in the gills (Toop et al. 1995b). These receptors in the gills were of two types; guanylate cyclase-linked ones and "clearance" (C-type) receptors. The latter are present in other vertebrates but their function is unknown. Natriuretic peptide receptors in the gills, via a vascular or even direct effect, could be mediating salt exchanges across the gills, as observed in teleosts. However, at this time, such an effect has not been demonstrated in hagfish. Few attempts appear to have been made, or reported, to study the effects of the administration of hormone preparations to hagfish. In 1962, Ian Chester Jones and his colleagues administered aldosterone and deoxycorticosterone, as well as ACTH, to hagfish that were placed in dilute (60%) sea-water. This strategy was used in an attempt to elicit conditions where an osmoregulatory response may be physiologically appropriate. No effects on plasma Na^+ were observed, but its concentration in muscle declined. Whether the effect was due to a decrease in intracellular Na^+ or a decline in the extracellular space of the muscle was not determined. The primary site of the effect is obscure, but it could be due to a decrease in plasma volume following increased Na^+ excretion. However, this is idle speculation. Currently, many assays are available for detecting receptors for hormones, which, apart from natriuretic peptides, include vasotocin and adrenocorticosteroids. A scan for such hormone receptors in various tissues in hagfish could be highly illuminating.

References

Aceves J, Erlij D, Edwards C (1968) Na$^+$ transport across the isolated skin of *Ambystoma mexicanum*. Biochim Biophys Acta 150:744–746

Acher R (1996) Molecular evolution of fish neurohypophysial hormones: neutral and selective evolutionary mechanisms. Gen Comp Endocrinol 102:157–172

Acher R, Chauvet J, Lenci MT, Morel F, Maetz J (1960) Présence d'une vasotocine dans le neurohypophysaire de la grenouille (*Rana esculenta* L.). Biochim Biophys Acta 42:379–380

Acher R, Chauvet J, Chauvet M-T, Crepy D (1964) Phylogénie des peptides neurohypophysaire: isolement de la mésotocine (Ileu^8-ocytocine) de la grenouille, intermédiare entre le Ser^4-Ileu^8-ocytocine des poissons osseux et l'ocytocine des mammiféres. Biochim Biophys Acta 90:613–614

Acher R, Chauvet J, Chauvet M-T (1969a) The neurohypophysial hormones of reptiles: comparison of the viper, cobra, and elaph active principles. Gen Comp Endocrinol 13:357–360

Acher R, Chauvet J, Chauvet M-T (1969b) Les hormones neurohypophysaire des reptiles: Isolement de la mésotocine et le vasotocine de la vipère, *Vivipera aspis*. Biochim Biophys Acta 154:255–257

Acher R, Chauvet J, Chauvet M-T (1970) Phylogeny of the neurohypophysial hormones. Eur J Biochem 17:509–513

Acher R, Chauvet J, Chauvet MT (1972) Reptilian neurohypophysial hormones: the active peptides of a saurian, *Iguana iguana*. Gen Comp Endocrinol 19:345–348

Acher R, Chauvet J, Rouillé Y (1997) Adaptive evolution of water homeostasis regulation in amphibians: vasotocin and hydrins. Biol Cell 89:283–291

Acher R, Chauvet J, Chauvet M-T, Rouillé Y (1999) Unique evolution of neurohypophysial hormones in cartilaginous fishes: possible implications for urea-based osmoregulation. J Exp Zool 284:475–484

Adolph EF (1933) Exchanges of water in the frog. Biol Rev 8:224–240

Adolph EF (1947) Physiology of man in the desert. Interscience, New York

Akester AR, Anderson RS, Hill KJ, Osbaldiston GW (1967) A radiographic study of urine flow in the domestic fowl. Br Poult Sci 8:209–212

Al-Baldawi NF, Stockand JD, Al-Khalili OK, Yue G, Eaton DC (2000) Aldosterone-induced Ras methylation in A6 epithelia. Am J Physiol 279:C429–C439

Alper SL, Kopito RR, Libresco SM, Lodish HF (1988) Cloning and characterization of a murine band 3-related cDNA from kidney and from a lymphoid cell line. J Biol Chem 263:17092–17099

Alt JM, Stolt H, Eisenbach GM, Walvig F (1981) Renal electrolyte and fluid excretion in the Atlantic hagfish, *Myxine glutinosa*. J Exp Biol 91:323–330

Alvarado RH (1972) The effects of dehydration of water and electrolytes in *Ambystoma tigrinum*. Physiol Zool 45:43–53

Alvarado RH, Johnson SR (1965) The effects of arginine vasotocin and oxytocin on sodium and water balance in *Ambystoma*. Comp Biochem Physiol 16:531–546

Alvarado RH, Johnson SR (1966) The effects of neurohypophysial hormones on water and sodium balance in larval and adult bullfrogs (*Rana catesbeiana*). Comp Biochem Physiol 18:549–561

Alvarado RH, Kirschner LB (1963) Effect of aldosterone on sodium fluxes in *Ambystoma tigrinum*. Nature (Lond) 202:922–923

Alvarado RH, Moody A (1970) Sodium and chloride transport in tadpoles of the bullfrog *Rana catesbeiana*. Am J Physiol 218:1510–1516

Ambrose SJ, Bradshaw SD (1988) The water and electrolyte metabolism of free-ranging and captive white-browed scrubwrens, *Sericornis frontalis* (Acanthizidae), from arid, semi-arid and mesic environements. Aust J Zool 36:29–51

Ameniya Y, Sogabe Y, Nozaki M, Takahashi A, Kawauchi H (1999) Somatolactin in the white sturgeon and African lungfish and its evolutionary significance. Gen Comp Endocrinol 114:181–190

Amer S, Brown JA (1995) Glomerular actions of arginine vasotocin in the in situ perfused trout kidney. Am J Physiol 269:R775-R780

Ames RG, Van Dyke HB (1950) Antidiuretic hormone in the urine and pituitary of the kangaroo rat. Proc Soc Exp Biol Med (NY) 75:417–420

Ames RG, Van Dyke HB (1952) Antidiuretic activity in the serum or plasma of rats. Endocrinology 50:350–360

Amour KJ, O'Toole LB, Hazon N (1993)The effects of dietary protein restriction on the secretory dynamics of 1α-hydroxycorticosterone and urea in the dogfish, *Scyliorhinus canicula*. A possible role for 1α-hydroxycorticosterone in sodium retention. J Endocrinol 138: 275–282

Andersen B, Ussing HH (1957) Solvent drag on non-electrolytes during osmotic flow through isolated toad skin and its response to antidiuretic hormone. Acta Physiol Scand 39:229–239

Anderson PM (1995) Urea cycle in fish: molecular and mitochondrial studies. In: Wood C, Shuttleworth TJ (eds) Cellular and molecular approaches to fish ionic regulation. Academic Press, San Diego, pp 57–83

Ando M, Kondo K, Takei Y (1992) Effects of eel natriuretic peptide on NaCl and water transport across intestine of the seawater eel. J Comp Physiol B 162:436–439

Andre R, Crabbé J (1966) Stimulation by insulin of active sodium transport by toad skin: influence of aldosterone and vasopressin. Arch Int Physiol Biochim 73:538–540

Andrews RM, Rose BR (1994) Evolution of viviparity: constraints on egg retention. Physiol Zool 67:1006–1024

Aramburg C, Carranza M, Martinez H, Luna M (1997) A comparative overview of growth hormone proteins and genes. In: Kawashima S, Kikuyama S (eds) XIIIth International Congress of Comparative Endocrinology, vol 2. Yokohama (Japan), Monduzzi Editore, Bologna, pp 893–898

Árnason SS (1997) Aldosterone and the control of lower intestinal Na$^+$ absorption and Cl$^-$ secretion in chickens. Comp Biochem Physiol 118A:257–259

Árnason SS, Skadhauge E (1991) Steady-state sodium absorption and chloride secretion of colon and coprodeum, and plasma levels of osmoregulatory hormones in hens in relation to sodium intake. J Comp Physiol B 161:1–14

Árnason SS, Rice GE, Chadwick A, Skadhauge E (1986) Plasma levels of arginine vasotocin, prolactin, aldosterone and corticosterone during prolonged dehydration in the domestic fowl: effect of dietary NaCl. J Comp Physiol B 156:383–397

Arnold J, Shield J (1970) Oxygen consumption and body temperature of the chuditch (*Dasyurus geoffroii*). J Zool (Lond) 160:391–404

Arnold-Reed DE, Balment RJ (1989) Steroidogenic role of the caudal neurosecretory system in the flounder, *Platichthys flesus*. Gen Comp Endocrinol 76:267–273

Arnold-Reed DE, Balment RJ (1994) Peptide hormones influence in vitro interrenal secretion of cortisol in the trout, *Oncorhynchus mykiss*. Gen Comp Endocrinol 96:85–91

Arnold-Reed DE, Hazon N, Balment RJ (1991) Biological action of atrial natriuretic factor in flatfish. Fish Physiol Biochem 9:271–277

Arsaky A (1813) De piscium cerebro et medulla spinali. Dissertatio inaugeralis, Halae

Asher C, Wald HW, Rosen OM, Garty H (1996) Aldosterone-induced increase in abundance of Na$^+$ channel subunits. Am J Physiol 271:C605–C611

Atlas SA, Maack T (1992) Atrial natriuretic factor. In: Handbook of physiology. Sect 8, Renal physiology. Oxford University Press, New York, pp 1577–1672

Augee ML, McDonald IR (1969) The effect of cold stress on adrenalectomized echidnas (*Tachglossus aculeatus*). Aust J Exp Biol Med Sci 47:P1

234

Auperin B, Rentier-Delrue F, Martial JA, Prunet P (1994a) Evidence that two tilapia (*Oreochromis niloticus*) prolactins have different osmoregulatory functions during adaptation to a hyper-osmoric environment. J Mol Endocrinol 12:13–24

Auperin B, Rentier-Delrue F, Martial JA, Prunet P (1994b) Characterization of a single prolactin (PRL) in tilapia (*Oreochromis niloticus*) which binds both PRL_I and PRL_{II}. J Mol Endocrinol 13:241–251

Babiker MM, Rankin JC (1973) Effects of neurohypophysial hormones on renal function in the freshwater- and sea-water-adapted eel *Anguilla anguilla* L. J Endocrinol 57:xi–xii

Babiker MM, Rankin JC (1978) Neurohypophysial hormonal control of kidney function in the European eel (*Anguilla anguilla* L.) adapted to sea-water or fresh water. J Endocrinol 76:347–358

Baddouri K, Butlen D, Imbert-Teboul M, Le Bouffant K, Marchetti J, Chabardes D, Morel F (1984) Plasma antidiuretic hormone levels and kidney responsiveness to vasopressin in the jerboa, *Jaculus orientalis*. Gen Comp Endocrinol 54:203–215

Baker CA, Hillyard SD (1992) Capacitance, short-circuit current and osmotic water flow across different regions of the isolated toad skin. J Comp Physiol B 162:707–713

Bakker HR, Bradshaw SD (1977) Effect of hypothalamic lesions on water metabolism of the toad *Bufo marinus*. J Endocrinol 75:161–172

Bakker HR, Waring H (1976) Experimental diabetes insipidus in a marsupial, *Macropus eugenii* (DEMAREST). J Endocrinol 69:149–157

Balinsky JB, Chortiz EL, Coe CGL, Schans VD (1967) Amino acid metabolism and urea synthesis in naturally aestivating *Xenopus laevis*. Comp Biochem Physiol 22:59–68

Ball JN, Ensor DM (1965) Effects of prolactin on plasma sodium in the teleost, *Poecilia latipinna*. J Endocrinol 32:269–270

Ball JN, Ensor DM (1969) Specific action of prolactin on plasma sodium levels in hypophysectomized *Poecilia latipinna* (Teleostei). Gen Comp Endocrinol 8:432–440

Ballantyne JS, Moyes CD, Moon TW (1987) Compatible and counteracting solutes and the evolution of ion and osmoregulation in fishes. Can J Zool 65:1883–1888

Balment RJ, Loveridge JP (1989) Endocrine and osmoregulatory mechanisms in the nile croocodile, *Crocodylus niloticus*. Gen Comp Endocrinol 73:361–367

Bankir L, Trinh-Trang-Tan M-M (2000) Renal urea transporters. Direct and indirect regulation by vasopressin. Exp Physiol 85S:243S–252S

Barbry P, Hofman P (1997) Molecular biology of Na^+ absorption. Am J Physiol 273:G571–G585

Bardack D, Zangerl R (1968) First fossil lamprey: a record from the Pennsylvania of Illinois. Science 162:1265–1267

Bartels H (1985) Assemblies of linear arrays of particles in the apical plasma membrane of mitochondria-rich cells in the gill epithelium of the Atlantic hagfish (*Myxine glutinosa*). Anat Rec 211:229–238

Bartholomew GA (1956) Temperature regulation in the macropod marsupial *Setonix brachyurus*. Physiol Zool 29:26–40

Bartholomew GA, Cade TJ (1963) The water economy of land birds. Auk 80:504–539

Bastide F, Jard S (1968) Actions de la noradrénaline et de l'ocytocine sur le transport actif de sodium et la perméabilité a l'eau de la peau de grenouille. Rôle du 3,5-AMP cyclique. Biochim Biophys Acta 150:113–123

Bathgate RAD, Sernia C (1995) Characterization of vasopressin and oxytocin receptors in an Australian marsupial. J Endocrinol 144:19–29

Bathgate RAD, Sernia C, Gemmel RT (1992) Brain content and plasma concentrations of arginine vasopressin in an Australian marsupial, the brushtail possum *Trichosurus vulpecula*. Gen Comp Endocrinol 88:217–223

Bazan JF (1990) Structural design and molecular evolution of a cytokine receptor superfamily. Proc Natl Acad Sci USA 87:6934–6938

Beesley AH, Hornby D, White SJ (1998) Regulation of distal nephron K^+ channels (ROMK) mRNA expression by aldosterone in rat kidney. J Physiol (Lond) 509:629–634

Belfry CS, Cowan FBM (1995) Peptidergic and adrenergic innervation of the lachrymal gland in the euryhaline turtle, *Malaclemys terrapin*. J Exp Zool 273:363–375

Bellamy D (1961) Movements of potassium, sodium and chloride in incubated gills from the silver eel. Comp Biochem Physiol 3:125–135

Bellamy D, Chester Jones I (1961) Studies on *Myxine glutinosa*. 1. The chemical composition of the tissues. Comp Biochem Physiol 3:175–183

Benedict FG (1932) The physiology of large reptiles. Carnegie Institution of Washington, Washington

Bennett DC, Hughes MR, De Sobrina CN, Gray DA (1997) Interaction of osmotic and volemic components in initiating salt-gland secretion in Pekin ducks. Auk 114:242–248

Bennett MB, Rankin JC (1986) The effects of neurohypophysial hormones on the vascular resistance of the isolated perfused gill of the European eel, *Anguilla anguilla* L. Gen Comp Endocrinol 64:60–66

Bentley PJ (1955) Some aspects of the water metabolism of an Australian marsupial *Setonyx brachyurus*. J Physiol (Lond) 127:1–10

Bentley PJ (1958) The effects of neurohypophysial extracts on water transfer across the wall of the isolated bladder of the toad *Bufo marinus*. J Endocrinol 17:201–209

Bentley PJ (1959) Studies on the water and electrolyte metabolism of the lizard *Trachysaurus rugosus* (Gray). J Physiol (Lond) 145:37–47

Bentley PJ (1960) Evaporative water loss and temperature regulation in the marsupial *Setonyx brachyurus*. Aust J Exp Biol Med 38:301–306

Bentley PJ (1961) Directional differences in the permeability to water of the isolated urinary bladder of the toad, *Bufo marinus*. J Endocrinol 22:95–100

Bentley PJ (1962a) Permeability of the skin of the cyclostome *Lampetra fluviatilis* to water and electrolytes. Comp Biochem Physiol 6:95–97

Bentley PJ (1962b) Studies on the permeability of the large intestine and urinary bladder of the tortoise (*Testudo graeca*). Gen Comp Endocrinol 2:323–328

Bentley PJ (1963) Composition of the urine of the fasting humback whale (*Megaptera nodosa*). Comp Biochem Physiol 10:257–259

Bentley PJ (1964) Physiological properties of the isolated frog bladder in hyperosmotic solutions. Comp Biochem Physiol 12:233–239

Bentley PJ (1965) Rectification of water flow across the bladder of the toad: effect of vasopressin. Life Sci 4:133–140

Bentley PJ (1966) The physiology of the urinary bladder of Amphibia. Biol Rev 41:275–316

Bentley PJ (1968) Amiloride: a potent inhibitor of sodium transport across the toad bladder. J Physiol (Lond) 195:317–330

Bentley PJ (1969a) Neurohypophysial hormones in Amphibia: a comparison of their actions and storage. Gen Comp Endocrinol 13:39–44

Bentley PJ (1969b) Neurohypophysial function in Amphibia: hormone activity in the plasma. J Endocrinol 43:359–369

Bentley PJ (1971a) Endocrines and osmoregulation. Springer, Berlin Heidelberg New York

Bentley PJ (1971b) The permeability of the urinary bladder of a urodele amphibian (the mudpuppy *Necturus maculosus*): studies with vasotocin and aldosterone. Gen Comp Endocrinol 16:356–362

Bentley PJ (1973a) Osmoregulation in the aquatic urodeles *Amphiuma* means (the congo eel) and *Siren lacertina* (the mud eel). Effects of vasotocin. Gen Comp Endocrinol 20:386–391

Bentley PJ (1973b) Role of the skin in amphibian sodium metabolism. Science 181:686–687

Bentley PJ (1975) The electrical P.D. across the integument of some neotenous urodele amphibians. Comp Biochem Physiol 50A:639–643

Bentley PJ (1998) Comparative vertebrate endocrinology. Cambridge University Press, Cambridge

Bentley PJ, Baldwin GF (1980) Comparison of transcutaneous permeability in skins of larval and adult salamanders (*Ambystoma tigrinum*). Am J Physiol 230:R505–R508

Bentley PJ, Blumer WFC (1962) Uptake of water by the lizard *Moloch horridus*. Nature (Lond) 194:699–700

Bentley PJ, Bradshaw SD (1972) Electrical potential difference across the cloaca and colon of the Australian lizards *Amphibolurus ornatus* and *A. inermis*. Comp Biochem Physiol 42A: 465–471

Bentley PJ, Ferguson DR (1967) The role of the toad urinary bladder in the amphibian "water balance effect" of neurohypophysial hormones. J Endocrinol 37:349–350

Bentley PJ, Follett BK (1963) Kidney function in a primitive vertebrate, the cyclostome *Lampetra fluviatilis*. J Physiol (Lond) 169:902–918

Bentley PJ, Greenwald L (1970) Neurohypophysial function in bullfrog (*Rana catesbeiana*) tadpoles. Gen Comp Endocrinol 14:412–415

Bentley PJ, Heller H (1964) The action of neurohypophysial hormones on the water and sodium metabolism of urodele amphibians. J Physiol (Lond) 171:434–453

Bentley PJ, Heller H (1965) The water-retaining action of vasotocin on the fire salamander (*Salamandra maculosa*): the role of the urinary bladder. J Physiol (Lond) 181:124–129

Bentley PJ, Main AR (1972) Zonal differences in permeability of the skin of some anuran Amphibia. Am J Physiol 223:361–363

Bentley PJ, Schmidt-Nielsen K (1965) Permeability to water and sodium of the crocodilian *Caiman sclerops*. J Cell Comp Physiol 66:303–309

Bentley PJ, Schmidt-Nielsen K (1966) Cutaneous water loss in reptiles. Science 151:1547–1549

Bentley PJ, Schmidt-Nielsen K (1967) The role of the kidney in water balance of the echidna. Comp Biochem Physiol 20:285–290

Bentley PJ, Schmidt-Nielsen K (1970) Comparison of water exchange in two aquatic turtles, *Trionyx spinifer* and *Pseudemys scripta*. Comp Biochem Physiol 32:363–365

Bentley PJ, Schmidt-Nielsen K (1971) Acute effects of sea water on frogs (*Rana pipiens*). Comp Biochem Physiol 40A:547–548

Bentley PJ, Shield J (1962) Metabolism and kidney function in the pouch young of the macropod marsupial *Setonix brachyurus*. J Physiol (Lond) 164:127–137

Bentley PJ, Yorio T (1977) The permeability of the skin of the neotenous urodele amphibian, the mudpuppy *Necturus maculosus*. J Physiol (Lond) 265:537–547

Bentley PJ, Yorio T (1979) Do frogs drink? J Exp Biol 79:41–46

Bentley PJ, Lee AK, Main AR (1958) Comparison of dehydration and hydration two genera of frogs (*Heleioporus* and *Neobatrachus*) that live in areas of varying aridity. J Exp Biol 35:677–684

Bentley PJ, Bretz WL, Schmidt-Nielsen K (1967a) Osmoregulation in the diamondback terrapin *Malaclemys terrapin centrata*. J Exp Biol 46:161–167

Bentley PJ, Herreid CF, Schmidt-Nielsen K (1967b) Respiration of a monotreme, the echidna *Tachyglossus aculeatus*. Am J Physiol 212:957–961

Bentley PJ, Maetz J, Payan P (1976) A study of the unidirectional fluxes of Na and Cl across the gills of the dogfish *Scyliorhinus canicula* (Chondrichthyes). J Exp Biol 64:629–637

Benyajati S, Yokota SD (1990) Renal effects of atrial natriuretic peptide in a marine elasmobranch. Am J Physiol 258:R1204–R1206

Bereiter-Hahn J, Matoltsy AG, Richards KS (1984) Biology of the integument. Vol 2. Vertebrates. Springer, Berlin Heidelberg New York

Bernstein MH (1971) Cutaneous water loss in small birds. Condor 73:468–469

Beuchat CA (1996) Structure and concentrating ability of the mammalian kidney; correlations with habitat. Am J Physiol 271:R157–R179

Beyenbach KW (1984) Water-permeable and -impermeable barriers of snake distal tubules. Am J Physiol 246:F290–F299

Biebach H (1990) Strategies of Trans-Sahara migrants. In: Gwinner E (ed) Bird migration. Springer, Berlin Heidelberg New York, pp 352–367

Binder HJ, Sandle GI (1994) Electrolyte transport in the mammalian colon. In: Johnson LR (ed) Physiology of the gastrointestinal tract, vol 2. Raven Press, New York, pp 2133–2171

Bindslev N, Skadhauge E (1971) Sodium chloride absorption and solute-linked water flow across the epithelium of the coprodeum and large intestine in normal and dehydrated fowl (*Gallus domesticus*). In vivo perfusion studies. J Physiol (Lond) 216:753–768

Birnbaumer M, Seibold A, Gilbert S, Ishido M, Barberis C, Antaramian A, Braber P, Rosenthal W (1992) Molecular cloning for the receptor for human antidiuretic hormone. Nature (Lond) 357:333–339

Bittner A, Lang S (1980) Some aspects of the osmoregulation of Amazonian freshwater stingrays (*Potamotrygon hystrix*)-I. Serum osmolality, sodium and chloride content, water content, hematocrit and urea level. Comp Biochem Physiol 67A:9–13

Björnsson BT, Yamauchi K, Nishioka RS, Deftos LJ, Bern HA (1987) Effects of hypophysectomy and subsequent hormonal replacement therapy on hormonal and osmoregulatory status of coho salmon, *Oncorhynchus kisutch*. Gen Comp Endocrinol 68:421–430

Blair-West JR, Gibson A (1980) The renin-angiotensin system in marsupials. In: Schmidt-Nielsen K, Bolis L, Taylor CR (eds) Comparative physiology: primitive mammals. Cambridge University Press, Cambridge, pp 297–306

Blair-West JR, Coghlan JP, Denton DA, Wright RD (1967) Effects of endocrines on salivary glands. Handbook of physiology Sect 6. Alimentary canal. American Physiological Society, Bethesda, pp 633–664

Blair-West JR, Coghlan JP, Denton DA, Nelson JF, Orchard E, Scoggins BA, Wright RD, Meyers K, Junqueira LCU (1968) Physiological, morphological and behavioural adaptation to a sodium-deficient diet by wild native Australian and introduced species of animals. Nature (Lond) 217:922–928

Blair-West JR, Coghlan JP, Denton DA, Funder J, Scoggins BA, Wright RD (1971) The effect of heptapeptide (2–8) and the hexapeptide (3–8) fragments of angiotensin II on aldosterone secretion. J Clin Endocrinol Metab 32:574–578

Blair-West JR, Gibson A, McKinley MJ, Nelson JF (1983) Water drinking and the effect of angiotensin and renin in a dasyurid marsupial (*Antechinus stuartii*). Comp Endocrinol 52:388–394

Blanco G, Mercer RW (1998) Isozymes of the Na-K-ATPase: heterogeneity in structure, diversity in function. Am J Physiol 275:F633–F650

Blaylock LA, Ruibal R, Platt-Aloia K (1976) Skin structure and wiping behavior of phylomedusine frogs. Copeia 1976:283–295

Blazer-Yost BL, Cox M, Furlanetto R (1989) Insulin and IGF I receptor-mediated Na$^+$ transport in toad urinary bladders. Am J Physiol 257:C612–C620

Bogert CM, Cowles RB (1947) Moisture loss in relation to habitat selection in some Floridian reptiles. Am Mus Novit 1358:1–34

Boim MA, Ho K, Shuck ME, Bienkowski MJ, Block JH, Slightom JL, Yang Y, Brenner BM, Hebert SC (1995) ROMK inwardly rectifying ATP-sensitive K$^+$ channel. II. Cloning and distribution of alternative forms. Am J Physiol 268:F1132–F1140

Bøkenes L, Mercer JB (1995) Salt gland function in the common eider duck (*Somateria mollissima*). J Comp Physiol 165:255–267

Bøkenes L, Mercer JB (1998) A morphometric study of the salt gland in freshwater- and saltwater-adapted ducks (*Somateria mollissima*). J Exp Zool 280:295–402

Bolton JP, Collie NL, Kawauchi H, Hirano T (1987) Osmoregulatory actions of growth hormone in rainbow trout (*Salmo gairdneri*). J Endocrinol 112:63–68

Borgnia M, Nielsen S, Engel A, Agre P (1999) Cellular and molecular biology of the aquaporin water channels. Annu Rev Biochem 68:425–458

Bourguet J, Maetz J (1961) Arguments en faveur de l'indépendence des mécanismes d'action de divers peptides neurohypophysaires sur le flux osmotique d'eau et sur le transport actif de sodium au sein d'un même récepteur: études sur la vessie et sur la peau de *Rana esclulenta*. Biochim Biophys Acta 52:552–565

Bourguet J, Lahlou B, Maetz J (1964) Modifications experimentale de l'équilibre hydrominérale et osmorégulation chez *Carassius auratus*. Gen Comp Endocrinol 4:563–576

Bourguet J, Chevalier J, Hugon JS (1976) Alterations in membrane-associated particle distribution during antidiuretic challenge in frog urinary bladder epithelium. Biophys J 16:627–639

Boylan JW (1972) A model for passive urea absorption in the elasmobranch kidney. Comp Biochem Physiol 42:27–30

Bradley SE, Mudge GH, Blake WD (1954) The renal excretion of sodium, potassium, and water by the harbor seal (*Phoca vitulina* L.): effects of apnea, sodium, potassium, and water loading; pitressin; and mercurial diuresis. J Cell Comp Physiol 43:1–22

Bradshaw FJ, Bradshaw SD (1996) Arginine vasotocin: locus of action along the nephron of the ornate dragon lizard, *Ctenophorus ornatus*. Gen Comp Endocrinol 103:281–300

Bradshaw SD (1970) Seasonal changes in the water and electrolyte metabolism of *Amphibolurus* lizards in the field. Comp Biochem Physiolol 36:689–719

Bradshaw SD (1975) Osmoregulation and pituitary-adrenal function in desert reptiles. Gen Comp Endocrinol 25:230–248

Bradshaw SD (1976) Effect of hypothalamic lesions on kidney and adrenal function in the lizard *Amphibolurus ornatus*. Gen Comp Endocrinol 29:285

Bradshaw SD (1997) Homeostasis in desert reptiles. Springer, Berlin Heidelberg New York

Bradshaw SD (1999) Ecophysiological studies on desert mammals: insights from stress physiology. Aust Mammal 21:55-65

Bradshaw SD, Grenot CJ (1976) Plasma aldosterone levels in two reptilian species, *Uromastix acanthinurus* and *Tiliqua rugosa*, and the effect of several experimental treatments. J Comp Physiol 111:71-76

Bradshaw SD, Rice GE (1981) The effects off pituitary and adrenal hormones on renal and-postrenal absorption of water in the lizard *Varanus gouldii* (Gray). Gen Comp Endocrinol 44:82-93

Bradshaw SD, Shoemaker VH (1967) Aspects of water and electrolyte changesi n a field population of *Amphibolurus* lizards. Comp Biochem Physiol 20:855-865

Bradshaw SD, Shoemaker VH, Nagy KA (1972) The role of adrenal corticosteroids in the regulation of kidney function in the desert lizard *Dipsosaurus dorsalis*. Comp Biochem Physiol 43A:621-635

Bradshaw SD, Lemire M, Vernet R, Grenot CJ (1984a) Aldosterone and the control of the secretion of the nasal salt gland of the North African desert lizard, *Uromastix acanthinurus*. Gen Comp Endocrinol 54:314-323

Bradshaw SD, Tom JA, Bunn SE (1984b) Corticosteroids and control of nasal salt gland function in the lizard *Tilqua rugosa*. Gen Comp Endocrinol 54:308-313

Bradshaw SD, Morris KD, Bradshaw FJ (2001) Water and electrolyte homeostasis and kidney function of desert-dwelling marsupial wallabies in Western Australia. J Comp Physiol B 171:23-32

Braun EJ (1976) Intrarenal blood flow distribution in the desert quail following salt loading. Am J Physiol 231:1111-1118

Braun EJ (1999a) Integration of organ systems in avian osmoregulation. J Exp Zool 283:702-707

Braun EJ (1999b) Integration of renal and gastrointestinal function. J Exp Zool 283:495-499

Braun EJ, Dantzler WH (1972) Function of mammalian-type and reptilian-type nephrons in kidney of desert quail. Am J Physiol 222:617-629

Braun EJ, Dantzler WH (1997) Vertebrate renal system. In: Handbook of physiology. Sect 13 Comparative physiology. Vol I. Oxford University Press, New York for the American Physiological Society, pp 481-576

Bray AA (1985) The evolution of the terrestrial vertebrates: environmental and physiological considerations. Philos Trans R Soc Lond B 309:289-322

Braysher M (1971) The structure and function of the nasal salt gland from the Australian sleepy lizard *Trachydosaurus* (formerly *Tiliqua*) *rugosus*: family Scincidae. Physiol Zool 44:129-136

Braysher M, Green B (1970) Absorption of water and electrolytes from the cloaca of an Australian lizard, *Varanus gouldii* (Gray). Comp Biochem Physiol 35:606-613

Brekke DR, Hillyard SD, Winokur RM (1991) Behavior associated with the water absorption response by the toad, *Bufo punctatus*. Copeia 1991:393-401

Brenner BM, Ballermann BJ, Gunning ME, Zeidel ML (1990) Diverse biological actions of atrial natriuretic peptide. Physiol Rev 70:665-699

Bres O, Eales JG (1988) High-affinity, limited capacity triiodothyronine-binding sites in nuclei from various tissues of the rainbow trout (*Salmo gairdneri*). Gen Comp Endocrinol 69:71-79

Brewer KJ, Ensor DM (1980) Hormonal control of osmoregulation in the Chelonia. 1. The effects of prolactin and interrenal steroids in freshwater chelonians. Gen Comp Endocrinol 42:304-309

Broadhead CL, O'Sullivan UT, Deacon CF, Henderson IW (1992) Atrial natriuretic peptide in the eel *Anguilla anguilla* L.; its cardiac distribution, receptors and actions on isolated branchial cells. J Mol Endocrinol 9:103-114

Brodsky WA, Schilb TP (1960) Electrical and osmotic characteristics of the isolated turtle bladder. J Clin Invest 39:974

Brodsky WA, Schilb TP (1965) Osmotic properties of the isolated turtle bladder. Am J Physiol 208:46-57

Brown D, Katsura T, Gustafson CE (1998) Cellular mechanisms of aquaporin trafficking. Am J Physiol 275:F328-F331

Brown GD, Dawson TJ (1977) Seasonal variations in the body temperatures of unrestrained kangaroos (Macropodidae: Marsupialia). Comp Biochem Physiol 56A:59–67

Brown JA, Rankin JC (1999) Lack of glomerular intermittency in the river lamprey *Lampetra fluviatilis* acclimated to sea water and following acute transfer to iso-osmotic brackish water. J Exp Biol 202:939–946

Brown JA, Oliver JA, Henderson IW, Jackson BA (1980) Angiotensin and single nephron glomerular function in the trout *Salmo gairdneri*. Am J Physiol 239:R509–R514

Brown PS, Brown SC (1973) Prolactin and thyroid hormone interactions in salt and water balance in the newt, *Notophthalmus viridescens*. Gen Comp Endocrinol 20:456–466

Brown PS, Brown SC (1987) Osmoregulatory actions of prolactin and other adenohypophysial hormones. In: Pang PKT, Schreibman MP (eds) Vertebrate endocrinology: fundamentals and biomedical implications. Vol 2, Regulation of water and electrolytes. Academic Press, San Diego, pp 45–84

Brown SC, Brown PS (1980) Water balance in the California newt, *Taricha torosa*. Am J Physiol 238:R113–R118

Brunn F (1921) Beitrag zur Kenntnis der Wirkung von Hypophysenextrakt auf den Wasserhaushalt des Frosches. Z Gesamte Exp Med 25:170–175

Budtz PE (1977) Aspects of moulting in anurans and its control. In: Spearman RIC (ed) Comparative biology of skin. Academic Press, London, pp 317–334

Burden CE (1956) The failure of the hypophysectomized *Fundulus heteroclitus* to survive in fresh water. Biol Bull (Woods Hole) 110:8–28

Burg M (1995) Molecular basis of osmotic regulation. Am J Physiol 268: F983–F996

Burger JW (1962) Further studies on the rectal gland in the spiny dogfish. Physiol Zool 35: 205–217

Burger JW (1965) Roles of the rectal gland and kidneys in salt and water excretion in the spiny dogfish. Physiol Zool 38:191–196

Burger JW, Hess WN (1960) Function of the rectal gland in the spiny dogfish. Science 131: 670–671

Burger JW, Tosteson DC (1966) Sodium influx and efflux in the spiny dogfish, *Squalus acanthias*. Comp Biochem Physiol 19:649–653

Burgess WW, Harvey AM, Marshall EK (1933) The site of the antidiuretic action of pituitary extract. J Pharmacol Exp Ther 49:237–248

Burns J, Copeland DE (1950) Chloride excretion in the head region of *Fundulus heteroclitus*. Biol Bull (Woods Hole) 99:381–385

Burns TW (1956) Endocrine factors in the water metabolism of the desert mammal, *G. gerbilus*. Endocrinology 58:773–781

Burton M, Idler DR (1986) The cellular location of 1α-hydroxycorticosterone binding protein in skate. Gen Comp Endocrinol 64:260–266

Bush IE (1953) Species differences in adrenocortical secretion. J Endocrinol 9:95–100

Butler DG (1972) Antidiuretic effect of arginine vasotocin in the western painted turtle (*Chrysemys picta belli*). Gen Comp Endocrinol 18:121–125

Butler DG (1999) Mecamylamine blocks the [Asp[1],Val[5]]-ANG II-induced attenuation of salt gland activity in Pekin ducks. Am J Physiol 277: R836–R842

Butler DG, Knox WH (1970) Adrenalectomy of the painted turtle (*Chrysemys picta belli*): effect on ion regulation and tissue glycogen. Gen Comp Endocrinol 14:551–566

Butler DG, Youson JH (1988) Kidney function in the bowfin (*Amia calva* L.). Comp Biochem Physiol 89A:343–345

Butler DG, Siwanowicz H, Puskas D (1989) A re-evaluation of experimental evidence for the hormonal control of avian nasal salt glands. In: Hughes MR, Chadwick A (eds) Progress in avian osmoregulation. Leeds Philosophical and Literary Society, Leeds (UK), pp 127–141

Buttemer WA, Dawson TJ (1989) Body temperature, water flux and estimated energy expenditure of incubating emus (*Dromaius novaehollandiae*). Comp Biochem Physiol 94A:21–24

Buttle JM, Kirk RL, Waring H (1952) The effects of complete adrenalectomy on the wallaby (*Setonyx brachyurus*). J Endocrinol 8:281–290

Buus O, Larsen LO (1975) Absence of known corticosteroids in blood of river lampreys (*Lampetra fluviatilis*) after treatment with mammalian corticotropin. Gen Comp Endocrinol 26:96–99

Cade T, Greenwald L (1966) Nasal salt gland secretion in falconiform birds. Condor 68:338–350

Cade TJ, Dybas JA (1962) Water economy of the budgerygah. Auk 79:345–364

Cade TJ, Maclean GL (1967) Transport of water by adult sandgrouse to their young. Condor 69:323–342

Cain JR, Lien RJ (1985) A model for drought inhibition of bobwhite quail (*Colinus virginianus*) reproductive systems. Comp Biochem Physiol 82A:925–930

Calder WA (1964) Gaseous metabolism and water relations of the zebra finch. Physiol Zool 243:400–413

Campbell JW, Vorhaben JE, Smith DD (1987) Uricotely: its nature and origin during the evolution of tetrapod vertebrates. J Exp Zool 243:349–363

Canessa CM (1996) What is new about the structure of the epithelial Na^+ channel? News Physiol Sci 11:195–201

Capelli JP, Wesson LG, Aponte GE (1970) A phylogenetic study of the renin-angiotensin system. Am J Physiol 218:1171–1178

Carballeira A, Brown JW, Fishman LM, Trujillo D, Odell DK (1987) The adrenal gland of stranded whales (*Kogia breviceps* and *Mesoplodon europaeus*): morphology, hormonal contents, and biosynthesis of corticosteroids. Gen Comp Endocrinol 68:293–303

Carlisky NJ, Botbol V, Barrio A, Sadnik LI (1968) Renal handling of urea in three preferentially terrestrial species of amphibian anura. Comp Biochem Physiol 26:573–578

Carpenter RE (1968) Salt and water metabolism in the marine fish-eating bat, *Pizonyx vivesi*. Comp Biochem Physiol 24:951–964

Carstensen H, Burgers ACJ, Li CH (1961) Demonstration of aldosterone and corticosterone as the principal steroids formed in incubates of adrenal glands of the American bullfrog (*Rana catesbeiana*) and stimulation of their production by mammalian adrenocorticotrophin. Gen Comp Endocrinol 1:37–50

Carter-Su C, Schwartz J, Smit LS (1996) Molecular mechanism of growth hormone action. Annu Rev Physiol 58:187–207

Castel M, Abraham M (1969) Effects of a dry diet on the hypothalamic neurohypophyseal neurosecretory system in spiny mice as compared to the albino rat and mouse. Gen Comp Endocrinol 12:231–241

Castel M, Borut A, Haines H (1974) Blood titres of vasopressin in various murids (Mammalian: Rodentia). Isr J Zool 23:208–209

Castillo GA, Orci GG (1997) Response of frog and toad skin to norepinephrine. Comp Biochem Physiol 118A:1145–1150

Chadwick CS (1940) Identity of prolactin with water drive factor in *Titurus viridescens*. Proc Soc Exp Biol Med (NY) 45:335–337

Chadwick CS (1941) Further observations of the water drive in *Triturus viridescens* II. Induction of the water drive with the lactogenic factor. J Exp Zool 86:175–187

Chakraborti PK, Weisbart M, Chakraborti A (1987) The presence of corticosteroid receptor activity in the gills of the brook trout, *Salvelinus fontinalis*. Gen Comp Endocrinol 66:323–332

Challis JRG, Lye SJ (1994) Parturition. In: Knobil E, Neill JD (eds) The physiology of reproduction. Raven Press, New York, pp 985–1032

Chan DKO, Chester Jones I, Henderson IW, Rankin JC (1967a) Studies on the experimental alteration of water and electrolyte composition of the eel (*Anguilla anguilla* L.). J Endocrinol 37:297–317

Chan DKO, Phillips JG, Chester Jones I (1967b) Studies on electrolyte changes in the lip shark, *Hemiscyllium plagiosum* (Bennett), with special reference to the hormonal influence on the rectal gland. Comp Biochem Physiol 23:185–198

Chan DKO, Chester Jones I, Mosley W (1968) Pituitary and adrenocortical factors in the control of the water and electrolyte composition of the freshwater European eel (*Anguilla anguilla* L.). J Endocrinol 42:91–98

Chan SJ, Nagamatus S, Cao Q-P, Steiner DF (1992) Structure and evolution of insulin and insulin-like growth factors in chordates. Prog Brain Res 92:15–24

Chan STH, Edwards BR (1970) Kinetic studies on the biosynthesis of corticosteroids in vitro from exogenous precursor by the interrenal glands of the normal, corticotropin-treated and adenohypopophysectomized *Xenopus laevis* Daudin. J Endocrinol 47:183–195

Chang EB, Rao MC (1994) Intestinal water and electrolyte transport. Mechanisms of physiological and adaptive response. In: Johnson LR (ed) Physiology of the gastrointestinal tract, vol 2. Raven Press, New York, pp 2027–2082

Chauvet J, Lenci M-T, Acher R (1960) Présence de deux vasopressines dans la neurohypophyses du poulet. Biochim Biophys Acta 38:571–573

Chauvet J, Chauvet M-T, Acher R (1963) Les hormone neurohypophysaires des mammiferes: isolement et caracterisation de l'ocytocine et del la vasopressine de la baleine (*Balaenoptera physalus* L.) Bull Soc Chim Biol 45:1369–1378

Chauvet MT, Colne T, Hurpet D, Chauvet J, Acher J (1983) Marsupial neurohypophysial hormones: identification of mesotocin, lysine vasopressin, and phenypressin in the quokka wallaby (*Setonix brachyurus*). Gen Comp Endocrinol 52:309–315

Chauvet J, Michel G, Ouedraogo Y, Chou J, Chait BT, Acher R (1995) A new neurohypophyseal peptide, seritocin ([Ser^5,Ile^8]-oxytocin) identified in a dryness-resistant African toad *Bufo regularis*. International J Pept Protein Res 45:482–487

Chen J-Y, Dzik J, Edgecombe GD, Ramskold L, Zhou G-Q (1995) A possible early Cambrian chordate. Nature (Lond) 377:720–722

Chen S, Bhargave A, Mastroberardino L, Meijer OC, Wang J, Buse P, Firestone GL, Verrey F, Pearce D (1999) Epithelial sodium channels regulated by aldosterone-induced protein sgk. Proc Natl Acad Sci USA 96:2514–2519

Chen TT, Marsh A, Shamblott M, Chan K-M, Tang Y-L, Cheng CM, Yang B-Y (1994) Structure and evolution of fish growth hormone and insulin-like growth factor genes. In: Sherwood NM, Hew CY (eds) Fish physiology, vol XIII, molecular endocrinology. Academic Press, San Diego, pp 179–209

Chester Jones I, Mosley W (1980) The interrenal gland of Pisces. Part 1. Structure. In: Chester Jones I, Henderson IW (eds) General, comparative and clinical endocrinology of the adrenal cortex. Academic Press, London, pp 395–472

Chester Jones I, Phillips JG, Holmes WN (1959) Comparative physiology of the adrenal Cortex. In: Gorbman A (ed) Comparative endocrinology. Wiley, New York, pp 582–612

Chester Jones I, Phillips JG, Bellamy D (1962) Studies on water and electrolytes in cyclostomes and teleosts with special reference to *Myxine glutinosa* L. (the hagfish) and *Anguilla anguilla* L. (the European eel). Gen Comp Endocrinol Suppl 1:36–47

Chester Jones I, Robertson JIS, Tree M (1966) Pressor activity in extracts of the corpuscles of Stannius from the European eel (*Anguilla anguilla* L.). J Endocrinol 34:393–408

Chester Jones I, Chan DKO, Rankin JC (1969) Renal function in the European eel (*Anguilla anguilla* L) II. Effects of the caudal neurosecretory system, corpuscles of Stannius, neurohypophysial peptides and vasoactive substances. J Endocrinol 43:21–31

Cheung L, Leung DW (1997) Elevated plasma levels of human adrenomedullin in cardiovascular, respiratory, hepatic and renal disorders. Clin Sci 92: 59–62

Chevalier J, Bourguet J, Hugon JS (1974) Membrane associated particles: distribution in frog urinary bladder epithelium at rest and after oxytocin treatment. Cell Tissue Res 152:129–140

Chevalier J, Adragna N, Bourguet J, Gobin R (1981) Fine structure of intramememranous particle aggregates in ADH-treated frog urinary bladder and skin: influence of glutaraldehyde and N-ethyl maleimide. Cell Tissue Res 152:129–140

Ching CAT,Hughes MR, Poon AMS, Pang SF (1999) Melatonin receptors and melatonin inhibition of duck salt gland secretion. Gen Comp Endocrinol 116:229–240

Chinkers M, Garbers DM, Chang M-S, Lowe DG, Chin H, Goeddel DV, Schulz S (1989) A membrane form of guanylate cyclase is an atrial natriuretic peptide receptor. Nature (Lond) 338: 78–83

Choe KP, Edwards S, Morrison-Shetlar AI, Toop T, Claiborne JB (1999) Immunolocalization of Na$^+$/K$^+$-ATPase in mitochondria-rich cells of the Atlantic hagfish (*Myxine glutinosa*) gill. Comp Biochem Physiol 124A:161–168

Cholette C, Gagnon A, Germain P (1970) Isosmotic adaptation in *Myxine glutinosa* L. I. Variations of some parameters and role of the amino acid pool in the muscle cells. Comp Biochem Physiol 33:333–336

Chopin LK, Bennett MB (1995) The regulation of branchial blood flow in the blacktip reef shark, *Carcharhinus melanopterus* (Carcharhinidae: Elasmobranchii). Comp Biochem Physiol 112A:35–41

Chosniak I, Munck BG, Skadhauge E (1977) Sodium chloride transport across the chicken copre-odeum. Basic characteristics and dependence on sodium chloride intake. J Physiol (Lond) 271:489–504

Christensen BM, Zelenina M, Aperia A, Nielsen S (2000) Localization and regulation of PKA-phosphorylated AQP2 in response to V_2-receptor agonist/antagonist treatment. Am J Physiol 278:F29–F42

Christensen CU (1974a) Adaptations in the water economy of some anuran amphibia. Comp Biochem Physiol 47A:1035–1049

Christensen CU (1974b) Effect of arterial perfusion on net water flux and active sodium trans-port across the isolated skin of *Bufo bufo* (L.). J Comp Physiol B 93:93–104

Christian JHB, Waltho JA (1962) Solute concentrations within cells of halophilic and non-halophilic bacteria. Biochim Biophys Acta 65:506–508

Churchill TA, Storey KB (1994) Effects of dehydration on organ metabolism in the frog *Pseudacris crucifer*: hyperglycemic responses to dehydration mimic freezing-induced cry-oprotectant production. J Comp Physiol B 164:492–498

Claiborne JB (1998) Acid-base regulation. In: Evans DH (ed) The physiology of fishes, 2nd edn. CRC Press, Boca Raton, pp 177–198

Claiborne JB, Evans DH, Goldstein L (1982) Fish branchial Na^+/NH_4^+ exchange is via basolateral Na^+-K^+-activated ATPase. J Exp Biol 96:431–434

Clarke WC, Farmer SW, Hartwell KM (1977) Effect of teleost pituitary growth hormone on growth of *Tilapia mossambica* and on growth and seawater adaptation of sockeye salmon (*Oncorhynchus nerka*). Gen Comp Endocrinol 33:174–178

Claussen DL (1967) Studies on water loss in two species of lizards. Comp Biochem Physiol 20: 115–130

Claussen DL (1969) Studies on water loss and rehydration in anurans. Physiol Zool 42:1–14

Cobb CS, Brown JA (1992) Angiotensin II binding to tissues of the rainbow trout, *Oncorhynchus mykiss*, studied by autoradiography. J Comp Physiol B 162:197–202

Cofré G, Crabbé J (1965) Stimulation by aldosterone of active sodium transport by the colon of the toad, *Bufo marinus*. Nature (Lond) 207:1299–1300

Cofré G, Crabbé J (1967) Active sodium transport by the colon of *Bufo marinus*: stimulation by aldosterone and antidiuretic hormone. J Physiol (Lond)188:177–190

Coghlan JP, Scoggins BA (1967) The measurement of aldosterone, cortisol and corticosterone in the blood of the wombat (*Vombatus hirsutus* Perry). J Endocrinol 39:445–448

Cole PM, Chester Jones I, Bellamy D (1963) Observations on the excretion of water and elec-trolytes in the desert rat (*Dipodomys spectabilis* Perry) and in the laboratory rat. J Endocrinol 25:515–532

Conlon JM, O'Hare F, Smith DD, Tonon M-C, Vaudry H (1992) Isolation and primary structure of urotensin II from the brain of a tetrapod, the frog *Rana ridibunda*. Biochem Biophys Res Commun 188:578–583

Conlon JM, Yano Y, Waugh D, Hazon N (1996) Distribution and molecular forms of urotensin II and its role in cardiovascular regulation in vertebrates. J Exp Zool 275:226–238

Conn JW, Louis LH, Johnston MW, Johnson BJ (1947) The electrolyte content of thermal sweat as an index of adrenal function. J Clin Invest 27:529–530

Conte FP, Lin DHY (1967) Kinetics of cellular morphogenesis in gill epithelium during sea water adaptation of *Oncorhynchus* (Walbaum). Comp Biochem Physiol 23:945–957

Conte FP, Wagner HH (1965) Development of osmotic and ionic regulation in juvenile steelhead trout *Salmo gairdneri*. Comp Biochem Physiol 14:603–620

Conte FP, Wagner HH, Fessler J, Gnose C (1966) Development of osmotic and ionic regulation in juvenile coho salmon *Oncorhynchus kisutch*. Comp Biochem Physiol 18:1–15

Cooperstein IL, Hogben CAM (1959) Ionic transfer across the isolated frog large intestine. J Gen Physiol 42:461–473

Costa DP (1982) Energy, nitrogen, and electrolyte flux and sea water drinking in the sea otter *Enhydra lutris*. Physiol Zool 55:35–44

Coulouarn Y, Lihrmann I, Jegou S, Anouar Y, Tostivini H, Beauvillain JC, Conlon JM, Bern HA, Vaudry H (1998) Cloning of the cDNA encoding the urotensin II precursor in frog and human reveal intense expression of the urotensin II gene in motoneurons of the spinal cord. Proc Natl Acad Sci USA 95:15803–15808

Cowley AW (2000) Control of renal medullary circulation by vasopressin V_1 and V_2 receptors in the rat. Exp Physiol 85S:223S–231S

Cox TH, Alvarado RH (1983) Nystatin studies of the skin of larval *Rana catesbeiana*. Am J Physiol 244:R58–R65

Crabbé J (1961a) Stimulation of active sodium transport by the isolated toad bladder with aldosterone in vitro. J Clin Invest 40:2103–2110

Crabbé J (1961b) Stimulation of active sodium transport after injection of aldosterone to the animal. Endocrinology 69:673–682

Crabbé J (1963) The sodium retaining action of aldosterone. Academique Européennes, Bruxelles

Crabbé J (1964) Stimulation by aldosterone of active sodium transport across the isolated ventral skin of Amphibia. Endocrinology 75:809–811

Crabbé J, de Weer P (1964) Action of aldosterone on the bladder and skin of the toad. Nature (Lond) 202:278–279

Crabbé J, Francois B (1967) Stimulation par l'insuline du transport actif de sodium à travers les membranes épithéliales du crapaud, *Bufo marinus*. Ann Endocrinoe (Paris) 28:713–715

Crane RK (1965) Na^+-dependent transport in the intestine and other animal tissues. Fed Proc 24:1000–1006

Curran PF (1965) Ion transport in the intestine and its coupling to other transport processes. Fed Proc 24:993–999

Curtis BJ, Wood CM (1991) The function of the urinary bladder in vivo in the freshwater rainbow trout. J Exp Biol 155:567–583

Cutler CP, Brezillon S, Bekir S, Sanders IL, Hazon N, Cramb G (2000) Expression of a duplicate Na, K-ATPase β1-isoform in the European eel (*Anguilla anguilla*). Am J Physiol 279: R222–R229

Dang Z, Balm PHM, Flik G, Wendelaar Bonga SE, Lock RAC (2000) Cortisol increases Na^+/K^+-ATPase density in plasma membranes of gill chloride cells in freshwater tilapia *Oreochromis mossambica*. J Exp Biol 203:2349–2355

Dangé AD (1986) Branchial Na^+-K^+-ATPase activity in freshwater or salt-water acclimated tilapia, *Oreochromis* (*Saratheradon*) *mossambicus*: effects of cortisol and thyroxine. Gen Comp Endocrinol 62:341–343

Dantzler WH (1967) Glomerular and tubular effects of arginine-vasotocin in water snakes (*Natrix sipedon*). Am Physiol 212:83–91

Dantzler WH (1982) Reptilian glomerular and tubular functions and their control. Fed Proc 41:2371–2376

Dantzler WH (1989) Comparative physiology of the vertebrate kidney. Springer, Berlin Heidelberg New York

Dantzler WH (1992) Comparative physiology of the kidney. In: Handbook of physiology Sect 8: Renal physiology, vol I. Oxford University Press, New York, for the American Phyiological Society, pp 415–474

Dantzler WH (1997) Vertebrate renal system. In: Handbook of physiology Sect 13 Comparative physiology. Oxford University Press, New York, for the American Physiological Society, pp 481–575

Dantzler WH, Schmidt-Nielsen B (1966) Excretion in fresh-water turtle (*Pseudemys scripta*) and desert tortoise (*Gopherus agassizii*). Am J Physiol 210:198–210

Darlington PJ (1957) Zoogeography: the geographical distribution of animals. Wiley, New York

Darwin C (1839) Journal of researches into the natural history and geology of the countries visited during the voyage of H.M.S. Beagle round the world. George Routledge, London

Davies SJJF (1982) Behavioural adaptations of birds to environments where evaporation is high and water is in short supply. Comp Biochem Physiol 71A:557–566

Dawson AB (1951) Functyional and egenrate or rudimentary glomeruli in the kidney of two species of Australian frog, *Cyclorana* (*Chiroleptes*) and *Alboguttatus* (*Gunther*). Anat Rec 109:417–424

Dawson TJ (1973) Thermoregulatory responses in the arid zone kangaroos, *Megaleia rufa* and *Macropus robustus*. Comp Biochem Physiol 46A:153–169

Dawson TJ, Hulbert AJ (1969) Standard energy metabolism of marsupials. Nature (Lond) 221:383

Dawson TJ, Robertshaw D, Taylor CR (1974) Sweating in the kangaroo: a cooling mechanism during exercise, but not in heat. Am J Physiol 227:494–498

Dawson TJ, Denny MJS, Russell EM, Ellis B (1975) Water usage and diet preferences of free ranging kangaroos, sheep and feral goats in the Australian arid zone during summer. J Zool (Lond) 177:1–23

Dawson WR (1984) Physiological studies of desert birds: present and future considerations. J Arid Environ 7:133–15

Dawson WR, Shoemaker VH (1965) Observations on the metabolism of sodium chloride in the red crossbill. Auk 82:606–623

de Bold AJ (1985) Atrial natriuretic factor: a hormone produced by the heart. Science 230: 767–770

Deen PMT, van Os CH (1998) Epithelial aquaporins. Curr Opin Cell Biol 10:435–442

Degani G, Silanikove N, Shkolnik A (1984) Adaptation of green toad (*Bufo viridis*) to terrestrial life by urea accumulation. Comp Biochem Physiol 77A:585–587

Degen AA (1997) Ecophysiology of small desert mammals. Springer, Berlin Heidelberg New York

Degen AA, Duke GE, Reynhout JK (1994) Gastroduodenal motility and glandular stomach function in young ostriches. Auk 111:750–755

Delson J, Whitford WG (1973) Adaptation of the tiger salamander, *Ambystoma tigrinum*, to arid habitats. Comp Biochem Physiol 46A:631–638

Denault DL, Féjes-Toth G, Naray-Féjes-Toth A (1996) Aldosterone regulation of sodium channel γ-subunit mRNA in cortical collecting duct cells. Am J Physiol 271:C423–C428

Denny MJS, Dawson TJ (1975) Comparative metabolism of tritiated water by macropodid marsupials. Am J Physiol 228:1794–1799

Denton DA (1965) Evolutionary aspects of the emergence of aldosterone secretion and salt appetite. Physiol Rev 45:245–295

Denton DA (1982) The hunger for salt. Springer, Berlin Heidelberg New York

Denton DA, Nelson JF, Tarjan E (1985) Waterand salt intake of wild rabbits (*Oryctolagus cuniculus* (L)) following dipsogenic stimuli. J Physiol (Lond) 362:285–301

Denton DA, McKinley MJ, Weisinger RS (1996) Hypothalamic integration of body fluid regulation. Proc Natl Acad Sci USA 93:7397–7404

Denton DA, Blair-West JR, McBurnie MI, Miller JAP, Weisinger RS, Williams RM (1999) Effects of adrenocorticotrophic hormone on sodium appetite in mice. Am J Physiol 277:R1033–R1040

Depeche J, Gilles R, Daufresne S, Chiapello H (1979) Urea content and urea production via the ornithine-urea cycle pathway during ontogenic development of two teleost fishes. Comp Biochem Physiol 63A:51–56

DeRoos CC, Bern HA (1961) The corticoids of the adrenal of California sea lion (*Zalophus californianus*). Gen Comp Endocrinol 1:275–285

Derst C, Karschin A (1998) Evolutionary link between prokaryotic and eukaryotic K$^+$ channels. J Exp Biol 201:2791–2799

De Ruyter ML, Stiffler DF (1986) Interrenal function in larval *Ambystoma tigrinum*. II. Control of aldosterone secretion and electrolyte balance by ACTH. Gen Comp Endocrinol 62:298–305

de Vlaming VL, Sage M, Beitz B (1975) Pituitary, adrenal and thyroid influences on osmoregulation in the euryhaline elasmobranch, *Dasyatis sabina*. Comp Biochem Physiol 52A:505–513

Dharmamba M, Maetz J (1976) Branchial sodium exchange in seawater-adapted *Tilapia mossambica*: effects of prolactin and hypophysectomy. J Endocrinol 70:293–299

Dharmamba M, Mayer-Gostan N, Maetz J, Bern HA (1973) Effect of prolactin on sodium movement in *Tilapia mossambica* adapted to sea water. Gen Comp Endocrinol 21:179–187

Diaz M, Lorenzo A (1992) Aldosterone regulation of active sodium transport in the lizard colon (*Gallotia galloti*). J Comp Physiol B 162:189–196

DiBona GF (2000) Neural control of the kidney: functionally specific renal sympathetic nerves. Am J Physiol 279:R1517–R1524

DiBona GF, Kopp UC (1997) Neural control of renal function. Physiol Rev 77:75–197

Dicker SE, Eggleton MG (1964) Renal function in the primitive mammal *Aplodontia rufa*, with some observations on squirrels. J Physiol (Lond) 170:186–194

Dicker SE, Elliott AB (1966) Ultrastructure of the urinary bladder of *Rana cancrivora*. J Physiol (Lond) 209:23P–24P

Dicker SE, Elliott AB (1969) Effects of neurohypophysial hormones on the skin permeability of *Rana cancrivora* in vivo. J Physiol (Lond) 203:74P–75P

Dicker SE, Elliott AB (1970) Water uptake by the crab-eating frog *Rana cancrivora*, as affected by osmotic gradients and neurohypophysial hormones. J Physiol (Lond) 207:119–132

Dicker SE, Elliott AB (1973) Neurohypophysial hormones and homeostasis in the crab-eating frog, *Rana cancrivora*. Horm Res 4:224–260

Dicker SE, Haslam J (1966) Water diuresis in the domestic fowl. J Physiol (Lond)183:225–235

Dicker SE, Nunn J (1957) The role of antidiuretic hormone during water deprivation in rats. J Physiol (Lond) 136:235–248

Dicker SE, Tyler C (1953) Vasopressor and oxytocic activities of the pituitary glands of rats, guinea pigs and cats and of human foetuses. J Physiol (Lond) 121:205–214

Dickoff WW (1993) Hormones metamorphosis and smolting. In: Schreibman MP, Scanes CG, Pang PKT (eds) Endocrinology of growth, development and metabolism. Academic Press, San Diego, pp 519–540

Dmi'el R, Zilber B (1971) Water balance in a desert snake. Copeia 1971:754–755

Donald JA, Vomachka AJ, Evans DH (1992) Immunohistochemical localization of natriuretic peptides in the brains and hearts of spiny dogfish *Squalus acanthias* and the Atlantic hagfish *Myxine glutinosa*. Cell Tissue Res 270:535–545

Donald JA, Toop T, Evans DH (1994) Localization and analysis of natriuretic peptide receptors in the gills of the toadfish, *Opsanus beta* (teleostei). Am J Physiol 267:R1437–R1444

Doneen BA (1976) Water and ion movement in the urinary bladder of the gobbid teleost *Gillichthys mirabilis* in response to prolactin and cortisol. Gen Comp Endocrinol 28:33–41

Dores RM, Hoffman NE, Chilcutt-Ruth T, Lancha A, Brown C, Marra L, Youson JH (1996) A comparative analysis of somatolactin-related immunoreactivity in the pituitaries of four neopterygian fishes and one chondrostean fish: an immunohistochemical study. Gen Comp Endocrinol 102:79–87

Dove H, Cork SJ, Christian KR (1989) Recyling of water during lactation and its effects on the estimation of milk intake in a developing macropodid, the tammar wallaby (*Macropus eugenii*). In: Grigg G, Jarman P, Hume I (eds) Kangaroos, wallabies and rat-kangaroos. Surrey Beatty, New South Wales (Australia), pp 231–237

Drewes RC, Hillman SS, Putnam RW, Sokol OM (1977) Water, nitrogen and ion balance in the African treefrog *Chiromantis petersi* Boulenger (Anura: Rhacophoridae), with comments on the structure of the integument. J Comp Physiol B 116:257–267

Duan C (1997) The insulin-like growth factor system and its biological actions in fish. Am Zool 37:491–503

Ducouret B, Tujague M, Ashraf J, Mouchel N, Servel N, Valotaire Y, Thompson EB (1995) Cloning of fish glucocorticoid receptor shows that it contains a different deoxyribonucleic acid-binding domain from that of mammals. Endocrinology 136:3774–3783

Duff DW, Olson KR (1986) Trout vascular and renal responses to atrial natriuretic factor and heart extracts. Am J Physiol 251:R639–R642

Duggan RT, Lofts B (1979) The pituitary-adrenal axis in the sea snake *Hydrophis cyanocinctus* Daudin. Gen Comp Endocrinol 38:374–383

Dunson WA (1967) Sodium fluxes in freshwater turtles. J Exp Zool 165:171–182

Dunson WA (1968) Salt gland secretion in the pelagic sea snake *Pelamis*. Am J Physiol 216: 1512–1517

Dunson WA (1969) Electrolyte excretion by the salt gland of the Galapagos marine iguana. Am J Physiol 215:995–1002

Dunson WA (1978) Role of the skin in sodium and water exchange of aquatic snakes placed in seawater. Am J Physiol 235:R151–R159

Dunson WA, Dunson MK (1974) Interspecific differences in fluid concentration and secretion rate of sea snake salt glands. Am J Physiol 227:430–438

Dunson WA, Taub AM (1967) Extrarenal salt excretion in sea snakes (*Laticauda*). Am Physiol 213:975–982

Dunson WA, Weymouth RD (1965) Active uptake of sodium by softshell turtles (*Trionyx spinifer*). Science 149:67–69

Dupont W, Leboulenger F, Vaudry H, Vaillant R (1976) Regulation of aldosterone secretion in the frog *Rana esculenta* L. Gen Comp Endocrinol 29:51–60

Ealey EHM, Bentley PJ, Main AR (1965) Studies on water metabolism of the hill kangaroo, *Macropus robustus* (Gould) in Northwest Australia. Ecology 46:473–479

Edmonds CJ, Marriott J (1970) Sodium transport and short-circuit current in the rat colon in vivo. J Physiol (Lond) 210:1021–1039

Eggena P (1986) Disk method for measuring effects of neurohypophysial hormones on urea permeability of toad bladder. Am J Physiol 250:E31–E34

Ehrenfeld J, Garcia-Romeu F, Harvey BJ (1985) Electrogenic active proton pump in *Rana esculenta* skin and its role in sodium ion transport. J Physiol (Lond) 359:331–355

Elger M, Hentschel H (1983) Morphological evidence for ionocytes in the gill epithelium of the hagfish, *Myxine glutinosa* L. Bull Mt Desert Isl Biol Lab 23:4–8

El-Husseini M, Haggag G (1974) Antidiuretic hormone and water conservation in desert rodents. Comp Biochem Physiol 47A:347–350

Elizondo RS, LeBrie SJ (1969) Adrenal-renal function in water snakes, *Natrix cyclopion*. Am J Physiol 217:419–425

Elliott AB (1968) Effect of adrenaline on water uptake in *Bufo*. J Physiol (Lond) 197:87P–88P

Enami M (1955) Caudal neurosecretory system in the eel (*Anguilla japonica*). Gunma J Med Sci 4:23–26

Enemar A, Hanstrom B (1956) The relative sizes of neural and glandular lobes some rodents. Lund University, Lund

Ensor DM, Ball JN (1972) Prolactin and osmoregulation in fishes. Fed Proc 31:1615–1623

Epple HJ, Amasheh S, Mankertz J, Goltz M, Schulzke JD, Fromm M (2000) Early aldosterone effect in distal colon by transcriptional regulation of ENaC subunits. Am J Physiol 278: G718–G724

Epstein F (1999) The sea within us. J Exp Zool 284:50–54

Epstein FH, Katz AI, Pickford GE (1967) Sodium and potassium-activated adenosine triphosphatase of gills; role in adaptation of teleosts to salt water. Science 156:1245–1247

Epstein FH, Cynamon M, McKay W (1971) Endocrine control of Na-K-ATPase and seawater adaptation in *Anguilla rostrata*. Gen Comp Endocrinol 16:323–328

Etheridge K (1990) Water balance in estivating sirenid salamanders (*Siren lacertina*). Herpetologica 46:400–406

Evans DH (1982) Mechanisms of acid extrusion by two marine fishes; the teleost, *Opsanus beta*, and the elasmobranch, *Squalus acanthias*. J Exp Biol 97:289–299

Evans DH (1984) Gill Na$^+$/H$^+$ and Cl$^-$/HCO$_3^-$ exchange systems evolved before the vertebrates entered fresh water. J Exp Biol 113:465–469

Evans DH (1991) Rat atriopeptin dilates vascular smooth muscle of the ventral aorta from the shark (*Squalus acanthias*) and hagfish (*Myxine glutinosa*). J Exp Biol 157:551–555

Evans DH (1995) The roles of natriuretic peptide hormones in fish osmoregulation and hemodynamics. In: Heisler N (ed) Advances in comparative and environmental physiology. Springer, Berlin Heidelberg New York, pp 119–152

Evans DH, Hooks C (1983) Sodium fluxes across the hagfish, *Myxine glutinosa*. Bull Mt Desert Isl Biol Lab 23:61–62

Evans DH, Forster MA, Rudy PP, Howells GP (1967) Sodium and water balance in the cichlid teleost, *Tilapia mossambica*. J Exp Biol 47:461–470

Evans DH, Chipouras E, Payne JA (1989) Immunoreactive atriopeptin in plasma fishes: its potential role in gill hemodynamics. Am J Physiol 257:R939–R945

Evans DH, Piermarini PM, Potts WTM (1999) Ionic transport in the fish gill epithelium. J Exp Zool 283:641–652

Ewart HS, Klip A (1995) Hormonal regulation of the Na$^+$-K$^+$-ATPase: mechanisms underlying rapid and sustained changes in pump activity. Am J Physiol 269:C295–C311

Ewer RF (1952a) The effect of pituitrin on fluid distribution in *Bufo regularis* Reuss. J Exp Biol 29:173–177

Ewer RF (1952b) The effects of posterior pituitary extracts on water balance in *Bufo carens* and *Xenopus laevis*. J Exp Biol 29:429–439

Fagerlund UHM, Higgs DA, McBride JR, Plotnikoff MD, Dosanjh BS (1980) The potential for using the anabolic hormones 17α-methyltestosterone and (or) 3′,5,3-triiodo-L-thyronine in the fresh water rearing of coho salmon (*Oncorhynchus kisutch*) and the effects on subsequent seawater performance. Can J Zool 58:1424–1432

Fanestil DD, Porter GA, Edelman IS (1967) Aldosterone stimulation of sodium transport. Biochim Biophys Acta 135:74–88

Fänge R, Fugelli K (1962) Osmoregulation in chimaeroid fish. Nature (Lond) 196:689

Fänge R, Schmidt-Nielsen K, Robinson M (1958) Control of secretion from the avian salt gland. Am J Physiol 195:321–326

Fergusson B, Bradshaw SD (1991) Plasma arginine vasotocin, progesterone and luteal development during pregnancy in viviparous lizards, *Tiliqua rugosa*. Gen Comp Endocrinol 82: 140–151

Field M, Graf LH, Laird WJ, Smith PL (1978) Heat-stable endotoxin in *Escherichia coli*: in vitro effects on guanylate cyclase activity, cyclic GMP concentration, and ion transport in small intestine. Proc Natl Acad Sci USA 75:2800–2804

Figler RA, MacKenzie DS, Owens DW, Licht P, Amoss MS (1989) Increased levels arginine vasotocin and neurophysin during nesting in sea turtles. Gen Comp Endocrinol 73:223–232

Findlay JD, Robertshaw D (1965) The role of the sympatho-adrenal system in the control of sweating in the ox (*Bos taurus*). J Physiol (Lond) 179:285–297

Finidori J, Kelly PA (1995) Cytokine receptor signalling through two novel families of transducer molecules: janus kinases, and signal transducers and activators of transcription. J Endocrinol 147:11–23

Finn JT, Grunwald ME, Yau K-W (1996) Cyclic nucleotide-gated ion channels: an extended family with diverse functions. Annu Rev Physiol 58:395–426

Fischbarg J, Kuang K, Vera JC, Arant S, Silverstein SC, Loike J, Rosen OM (1990) Glucose transporters serve as water channels. Proc Natl Acad Sci USA 87:3244–3247

Fisher CD, Lindgren E, Dawson WR (1972) Drinking patterns and behavior of Australian birds in relation to their ecology and abundance. Condor 74:111–136

Fitzsimons JT (1998) Angiotensin, thirst and sodium appetite. Physiol Rev 78:583–686

Follett BK (1963) Mole ratios of the neurohypophysial hormones in the vertebrate neural lobe. Nature (Lond) 198:693–694

Follett BK (1967) Neurohypophysial hormones of marine turtles and of the grass snake. J Endocrinol 39:293–294

Follett BK, Farner DS (1966) The effects of daily photoperiod on gonadal growth, neurohypophysial hormone content, and neurosecretion in the hypothalamo-hypophysial system of the Japanese quail (*Coturnix coturnix japonica*). Gen Comp Endocrinol 7:111–124

Follett BK, Heller H (1964a) The neurohypophysial hormones of lungfishes and amphibians. J Physiol (Lond) 172:92–106

Follett BK, Heller H (1964b) The neurohypophysial hormones of bony fishes and cyclostomes. J Physiol (Lond) 172:72–91

Fontaine M (1956) The hormonal control of water and salt-electrolyte metabolism in fish. Mem Soc Endocrinol 5:69–82

Fontaine M, Koch H (1950) Les variations d'euryhalinité et d'osmorégulation chez les poissons. J Physiol (Paris) 42:287–318

Forrest JN, Cohen JD, Schon DA, Epstein FH (1973a) Na transport and Na-K-ATPase in gills during adaptation to seawater: effects of cortisol. Am J Physiol 224:709–713

Forrest JN, MacKay WC, Gallagher B, Epstein FH (1973b) Plasma cortisol response to saltwater adaptation in the American eel *Anguilla rostrata*. Am J Physiol 224:714–717

Forster RY (1954) Active cellular transport of urea by frog renal tubules. Am J Physiol 179: 372–377

Forte LR, Hamra FK (1996) Guanylin and uroguanylin: intestinal peptide hormones that regulate epithelial transport. News Physiol Sci 11:17–24

Forte LR, Thorne PK, Eber SL, Krause WJ, Freeman RH, Francis SH, Corbin JD (1992) Stimulation of intestinal Cl⁻ transport by heat-stable enterotoxin: activation of cAMP-dependent protein kinase by cGMP. Am J Physiol 263:C607–C615

Forte LR, Eber SL, Fan X, London RM, Wang Y, Rowland LM, Chin DT, Freeman RH, Krause WJ (1999) Lymphoguanylin: cloning and characterization of a unique member of the guanylin peptide family. Endocrinology 140:1800–1806

Forte LR, London RM, Freeman RH, Krause WJ (2000) Guanylin peptides: renal actions mediated by cyclic GMP. Am J Physiol 278:F180–F191

Foskett JK (1987) The chloride cell. In: Kirsch R, Lahlou B (eds) Comparative physiology of environmental adaptation. Karger, Basel, pp 83–91

Foskett JK, Scheffey C (1982) The chloride cell: definitive identification as the salt-secretory cell in teleosts. Science 215:164–166

Foskett JK, Bern HA, Machen TE, Connor M (1983) Chloride cells and the hormonal control of teleost fish osmoregulation. J Exp Biol 106:255–281

Fossat B, Lahlou B (1979) The mechanism of coupled transport of sodium and chloride in isolated urinary bladder of the trout. J Physiol (Lond) 294:211–222

Foster RC (1975) Changes in urinary bladder and kidney function in the starry flounder (*Platichthys stellatus*) in response to prolactin and freshwater transfer. Gen Comp Endocrinol 27:153–161

Fowler MA, Chester Jones I (1955) The adrenal cortex in the frog *Rana temporaria* and its relation to water and salt electrolyte metabolism. J Endocrinol 13:vi–vii

Franklin CE, Holmgren S, Taylor GC (1996) A preliminary investigation of the effects of vasoactive intestinal peptide on secretion from the lingual glands of *Crocodylus porosus*. Gen Comp Endocrinol 102:74–78

Freeman JD, Bernard RA (1990) Atrial natriuretic peptide and salt adaptation in the sea lamprey *Petromyzon marinus*. Physiologist 33:A38

Fridberg G, Bern HA (1968) The urophysis and the caudal neurosecretory system of fishes. Biol Rev 43:175–199

Friedman M, Kaplan A, Williams E (1942) Studies concerning the site of renin formation in the kidney. II. Absence of renin from the kidneys of marine fish. Proc Soc Exp Biol Med (NY) 50:199–202

Friedman PA, Hebert SC (1990) Diluting segment in kidney of dogfish shark. I. Localization and characterization of chloride absorption. Am J Physiol 258:R398–R408

Fromm PO (1963) Studies on renal and extra-renal excretion in a fresh-water teleost, *Salmo gairdneri*. Comp Biochem Physiol 10:121–128

Fryer JN, Lederis K, Rivier J (1983) CRF-like neuropeptide stimulates ACTH release from the teleost pituitary. Endocrinology 113:2308–2310

Fuhrman FA, Ussing HH (1951) A characteristic response of the frog skin potential to neurohypophysial principles and its relation to the transport of sodium and water. J Cell Comp Physiol 38:109–130

Fukuzawa A, Watanabe TX, Itahara Y, Nakajima K, Yoshizawa-Kumagaye K, Takei Y (1996) B-type natriuretic peptide isolated from frog cardiac ventricles. Biochem Biophys Res Commun 222:323–329

Funder J, Myles K (1996) Exclusion of corticosterone from epithelial mineralocorticoid receptors is insufficient for selectivity of aldosterone action: in vivo binding studies. Endocrinology 137:5264–5268

Fushimi K, Uchida S, Hara Y, Hirata Y, Marumo F, Sasaki S (1993) Cloning and expression of apical membrane water channel of rat kidney collecting tubule. Nature (Lond) 361:549–552

Gallardo R, Pang PKT, Sawyer WH (1980) Neural influences on bullfrog renal function. Proc Soc Exp Biol Med (NY) 165:233–240

Gallis J-L, Lasserre P, Belloc F (1979) Freshwater adaptation in the euryhaline teleost, *Chelon labrosus*. I. Effects of adaptation, prolactin, cortisol and actinomycin D on plasma osmotic balance and (Na$^+$-K$^+$)ATPase in gill and kidney. Gen Comp Endocrinol 38:1–10

Galton VA (1990) Mechanisms underlying the acceleration of thyroid-induced tadpole metamorphosis by corticosterone. Endocrinology 127:2997–3002

Gamba G, Saltzberg SN, Lombardi M, Miyanoshita A, Lytton J, Hediger MA, Brenner BM, Hebert SC (1993) Primary structure and functional expression of a cDNA encoding the thiazide-sensitive, electroneutral sodium-chloride cotransporter. Proc Natl Acad Sci USA 90: 2749–2753

Garcia Romeu F, Masoni A (1970) Sur la mise en évidence des cellules à chlorure de la branchie des poissons. Arch Anat Microsc 59:289–294

Garland HO, Henderson IW (1975) Influence of environmental salinity on renal and adrenocortical function in the toad, *Bufo marinus*. Gen Comp Endocrinol 27:136–143

Garty H, Palmer LG (1997) Epithelial sodium channels: function, structure and regulation. Physiol Rev 77:359–396

Geck P, Pietrzyk C, Burckhardt B-C, Pfeiffer B, Heinz E (1980) Electrically silent cotransport of Na, K and Cl in Ehrlich cells. Biochim Biophys Acta 600:432–447

Geering K, Girardet M, Bron C, Kraehenbuhl J-P, Rossier BC (1982) Hormonal regulation of (Na^+K^+)-ATPase biosynthesis in the toad bladder. J Biol Chem 257:10338–10343

Geiser F (1994) Hibernation and daily torpor in marsupials: a review. Aust J Zool 42:1–16

Gentry RL (1981) Seawater drinking in eared seals. Comp Biochem Physiol 68A:81–86

Gerst JW, Thorson TB (1977) Effects of saline acclimation on plasma electrolytes, urea excretion, and hepatic urea synthesis in a freshwater stingray, *Potamotrygon* sp. Garman, 1877. Comp Biochem Physiol 56A:87–93

Gerstberger R, Simon E, Gray DA (1984) Salt gland and kidney responses to intracerebral osmotic stimulation in salt- and water-loaded ducks. Am J Physiol 247:R1022–R1028

Giebisch G (1998) Renal potassium transport: mechanisms and regulation. Am J Physiol 274: F817–F833

Giladi I, Goldstein DL, Pinshow B, Gerstberger R (1997) Renal function and plasma levels of arginine vasotocin during free flight in pigeons. J Exp Biol 200:3203–3211

Giorgio M, Giacoma C, Vellano C, Mazzi V (1982) Prolactin and sexual behaviour in the crested newt (*Triturus cristatus carnifex* Laur.). Gen Comp Endocrinol 47:139–147

Gislen T, Kauri H (1959) Zoogeography of Swedish amphibians and reptiles with notes on their growth and ecology. Acta Vertebr 1:197–397

Gist DH, DeRoos R (1966) Corticoids of the alligator adrenal gland and the effects of ACTH and progesterone on their production in vitro. Gen Comp Endocrinol 7:304–313

Goecke CS, Goldstein DL (1997) Renal glomerular and tubular effects of antidiuretic hormone and two antidiuretic hormone analogues in house sparrows (*Passer domesticus*). Physiol Zool 70:283–291

Goetz KL (1991) Renal natriuretic peptide (urodilatin?) and atriopeptin: evolving concepts. Am J Physiol 261:F921–F932

Goldstein DL (1993) Influence of dietary sodium and other factors on plasma aldosterone concentrations and in vitro properties of the lower intestine in house sparrows (*Passer domesticus*). J Exp Biol 176:159–174

Goldstein DL (1995) Effects of water restriction during growth and adulthood on renal function of bobwhite quail, *Colinus virginianus*. J Comp Physiol B 164:663–670

Goldstein DL (1997) Osmoregulation by free-living birds: the need for field osmoregulatory physiology. In: Harvey S, Etches RJ (eds) Perspectives in avian endocrinology. Journal of Endocrinology Ltd., Bristol, pp 310–314

Goldstein DL, Bradshaw SD (1998) Regulation of water and sodium balance in the field by Australian honeyeaters (Aves: Meliphagidae). Physiol Zool 71:215–225

Goldstein DL, Braun EJ (1988) Contributions of the kidneys and intestines to water conservation, and plasma levels of antidiuretic hormone, during dehydration in house sparrows (*Passer domesticus*). J Comp Physiol B 158:353–361

Goldstein DL, Hughes MR, Braun EJ (1986) Role of the lower intestine in the adaptation of gulls (*Larus glaucescens*) to sea water. J Exp Biol 123:345–357

Goldstein L, Forster RP (1971) Osmoregulation and urea metabolism in the little skate, *Raja erinacea*. Am J Physiol 220:742–746

Goldstein O, Asher C, Garty H (1997) Cloning and induction of low NaCl intake of avian intestine Na^+ channel subunits. Am J Physiol 272:C270–C277

Gordon MS (1962) Osmotic regulation in the green toad (*Bufo viridis*). J Exp Biol 39:261–270

Gordon MS, Tucker VA (1965) Osmotic regulation in the tadpoles of the crab-eating frog (*Rana cancrivora*). J Exp Biol 42:437–445

Gordon MS, Schmidt-Nielsen K, Kelly HM (1961) Osmotic regulation of the crab-eating frog (*Rana cancrivora*). J Exp Biol 38:659–678

Gorin MB, Yancey SB, Cline J, Revel JP, Horwitz J (1984) The major intrinsic protein (MIP) of the bovine lens fiber membrane: characterization and structure based on cDNA cloning. Cell 39:49–59

Goss GG, Perry SF, Fryer JN, Laurent P (1998) Gill morphology and acid-base regulation in freshwater fishes. Comp Biochem Physiol 119A:107–115

250

Grau EG (1987) Thyroid hormones. In: Pang PKT, Schreibman MP, Sawyer WH (eds) Vertebrate endocrinology: fundamentals and biomedical implications. Vol 2, Regulation of water and electrolytes. Academic Press, Orlando, pp 85–102

Grau EG, Richman NH, Borski IRJ (1994) Osmoreception and a simple endocrine reflex of the prolactin cell of the tilapia *Oreochromis mossambica*. In: Davy KG, Peter RE, Tobe SS (eds) Perspectives in comparative endocrinology. National Research Council of Canada, Ottawa, pp 251–256

Gray CJ, Brown JA (1985) Renal and cardiovascular effects of angiotensin II in the rainbow trout, *Salmo gairdneri*. Gen Comp Endocrinol 59:375–381

Gray DA (1993) Plasma atrial natriuretic factor concentrations and renal actions in the domestic fowl. J Comp Physiol B 163:519–523

Gray DA (1994) Role of endogenous atrial natriuretic peptide in volume expansion diuresis and natriuresis of the Pekin duck. J Endocrinol 140:85–90

Gray DA, Erasmus T (1988a) Glomerular filtration changes during vasotocin-induced antidiuresis in kelp gulls. Am J Physiol 255:R936–R939

Gray DA, Erasmus T (1988b) Plasma arginine vasotocin and angiotensin II in the water deprived kelp gull (*Larus dominicus*), cape gannet (*Sula capensis*) and jackass penguin (*Spheniscus demersus*). Comp Biochem Physiol 91A:727–732

Gray DA, Erasmus T (1989a) Control of renal and extrarenal salt excretion by plasma angiotensin II in the kelp gull (*Larus dominicanus*). J Comp Physiol B 158:651–660

Gray DA, Erasmus T (1989b) Control of arginine vasotocin in kelp gulls (*Larus dominicanus*): roles of osmolality, volume and plasma angiotensin II. Gen Comp Endocrinol 74:110–119

Gray DA, Naudé RJ, Erasmus T (1988) Plasma arginine vasotocin and angiotensin II in the water deprived ostrich (*Struthio camelus*). Comp Biochem Physiol 89A:251–256

Gray DA, Simon E (1983) Mammalian and avian antidiuretic hormone: studies related to possible species variation in osmoregulatory systems. J Comp Physiol B 151:241–246

Gray DA, Simon E (1987) Dehydration and arginine vasotocin and angiotensin II in CSF and plasma in Pekin ducks. Am J Physiol 253:R285–R291

Gray DA, Schütz H, Gerstberger R (1991) Interaction of atrial natriuretic factor and osmoregulatory hormones in the Pekin duck. Gen Comp Endocrinol 81:246–255

Greenwald L (1972) Sodium balance in amphibians from different habitats. Physiol Zool 45:229–237

Greenwald L (1989) The significance of renal medullary relative thickness. Physiol Zool 62:1005–1014

Greger R, Gögelein H, Schlatter E (1988) Stimulation of NaCl secretion by the rectal gland of the dogfish *Squalus acanthias*. Comp Biochem Physiol 90A:733–737

Gregory PT (1982) Reptilian hibernation. In: Gans C, Pough FH (eds) Biology of the Reptilia, vol 13. Academic Press, New York, pp 53–154

Grenot C (1967) Observations physio-écologique sur la régulation thermique chez le lézard agamide *Uromastyx acanthinurus*, Bell. CR Acad Sci (Paris) 266:1871–1874

Griffith RW (1987) Freshwater or marine origin of the vertebrates. Comp Biochem Physiol 87A:523–531

Griffith RW, Pang PKT, Srivastava AK, Pickford GE (1973) Serum composition of freshwater stingrays (Potamotrygonidae) adapted to fresh and dilute sea water. Biol Bull (Woods Hole) 144:304–320

Griffiths M (1978) The biology of the Monotremes. Academic Press, New York

Grigg GC, K J, Harlow P, Taplin LE (1986) Facultative aestivation in a tropical freshwater turtle *Chelodina rugosa*. Comp Biochem Physiol 83A:321–323

Grimm AS, O'Halloran MJ, Idler DR (1969) Stimulation of sodium transport across the isolated toad bladder by 1α-hydroxycorticosterone from an elasmobranch. J Fish Res Board Can 26:1823–1835

Grubb BR, Bentley PJ (1987) Aldosterone-induced amiloride inhibitable short-circuit current in the avian ileum. Am J Physiol 253:G211–216

Grubb BR, Bentley PJ (1989) Avian colonic ion transport: effects of corticosterone and dexamethasone. J Comp Physiol B 159:131–138

Grubb BR, Bentley PJ (1992) Effects of corticosteroids on short-circuit current across the cecum of the domestic fowl, *Gallus domesticus*. J Comp Physiol B 162:690–695

Guibbolini MB, Henderson IW, Mosley W, Lahlou B (1988) Arginine vasotocin binding to isolated branchial cells of the eel: effect of salinity. J Mol Endocrinol 1:125–130

Guibbolini ME, Lahlou B (1987) Neurohypophyseal peptide inhibition of adenylate cyclase activity in fish gills. FEBS Lett 220:98–102

Guppy M, Withers P (1999) Metabolic depression in animals: physiological perspectives and biochemical generalizations. Biol Rev 74:1–40

Haggag G, El-Husseini M (1974) The adrenal cortex and water conservation in desert rodents. Comp Biochem Physiol 47A:351–359

Haines H (1964) Salt tolerance and water requirements in the salt-water harvest mouse. Physiol Zool 37:266–272

Halley A, Loveridge JP (1997) Metabolic depression during dormancy in the African tortoise *Kinixys spekii*. Can J Zool 75:1328–1335

Halstead LB (1985) The vertebrate invasion of fresh water. Philos Trans R Soc Lond B 309: 243–258

Hamano K, Tierney ML, Ashida K, Hazon N (1998) Direct vasoconstricter action of homologous angiotensin II on isolated arterial ring preparations in an elasmobranch fish. J Endocrinol 158:419–423

Hammel HT, Maggert JE (1983) Nasal salt gland secretion inhibited by angiotensin II. Physiologist 26:A58

Hammel HT, Simon-Oppermann C, Simon E (1980) Properties of body fluids influencing salt gland secretion in Pekin ducks. Am J Physiol 239:R489–R496

Handa RK, Ferrario CM, Strandhoy JW (1996) Renal actions of angiotensin (1–7): in vivo and in vitro studies. Am J Physiol 270:F141–F147

Handler JS, Preston AS, Orloff J (1969) Effect of adrenal steroid hormones on the toad's urinary bladder response to vasopressin. J Clin Invest 48:823–833

Hanke W (1997) The adrenal gland of fish. Morphology and regulation of the gland, function of corticosteroids. In: Kawashima S, Kikuyama S (eds) XIIIth International Congress of Comparative Endocrinology, vol 2. Yokohama (Japan), Monduzzi Editore, Bologna, pp 1261–1266

Hardisty MW (1956) Some aspects of osmotic regulation in lampreys. J Exp Biol 33:431–447

Hardisty MW (1979) Biology of cyclostomes. Chapman and Hall, London

Hart WM, Essex HE (1942) Water metabolism of the chicken with special reference to the role of the cloaca. Am J Physiol 136:657–658

Haruta K, Yamashita T, Kawashima S (1991) Changes in arginine vasotocin content in the pituitary of the medaka (*Oryzias latipes*) during osmotic stress. Gen Comp Endocrinol 83:327–336

Harvey BJ (1992a) Energization of sodium absorption by the H^+-ATPase pump in mitochondria-rich cells of frog skin. J Exp Biol 172:289–309

Harvey BJ (1992b) Physiology of V-ATPases. J Exp Biol 172:1–17

Harvey S, Klandorf H, Radke WJ, Few JD (1984) Thyroid and adrenal responses of ducks (*Anas platyrhynchos*) during saline adaptation. Gen Comp Endocrinol 55:46–53

Harvey S, Hall TR, Chadwick A, Ensor DM (1989) Osmoregulation and the physiology of prolactin. In: Hughes MR, Chadwick A (eds) Progress in avian osmoregulation. Leeds Philosophical and Literary Society, Leeds (UK), pp 81–109

Hazon N, Henderson IW (1985) Factors affecting the secretory dynamics of 1α-hydroxycorticosterone in the dogfish, *Scyliorhinus canicula*. J Endocrinol 59:50–55

Hazon N, Balment RJ, Perrott MN, O'Toole LB (1989) The renin-angiotensin system and vascular and dipsogenic regulation in elasmobranchs. Gen Comp Endocrinol 74:230–236

Hazon N, Tierney ML, Takei Y (1999) Renin-angiotensin system in elasmobranch fish: a review. J Exp Zool 284:526–534

Hediger MA, Coady MJ, Ikeda TS, Wright EM (1987) Expression cloning and cDNA sequencing of the Na^+/glucose co-transporter. Nature (Lond) 330:379–381

Heller H (1941) Differentiation of an (amphibian) water balance principle from the antidiuretic principle of the pituitary gland. J Physiol (Lond) 100:125–141

Heller H (1972) The effect of neurohypophysial hormones on the female reproductive tract of lower vertebrates. Gen Comp Endocrinol Suppl 3:703–714

Heller H, Bentley PJ (1965) Phylogenetic distribution of the effects of neurohypophysial hormones on water and sodium metabolism. Gen Comp Endocrinol 5:96–108

Heller H, Pickering AD (1961) Neurohypophysial hormones in non-mammalian vertebrates. J Physiol (Lond) 155:98–114

Heller H, Pickering BT, Maetz J, Morel F (1961) Pharmacological characterization of the oxytocic peptides in the pituitary of a marine teleost (Pollachius virens). Nature (Lond) 191: 670–671

Henderson IW (1997) Endocrinology of the vertebrates. In: Handbook of physiology, Sect 13, Comparative physiology. Oxford University Press, New York, pp 623–749

Henderson IW, Chester Jones I (1967) Endocrine influences on the net extrarenal fluxes of sodium and potassium in the European eel (Anguilla anguilla L.). J Endocrinol 37:319–325

Henderson IW, Garland HO (1980) The interrenal gland in Pisces. Part 2. Physiology. In: Chester Jones I, Henderson IW (eds) General, comparative and clinical endocrinology of the adrenal cortex, vol 3. Academic Press, London, pp 473–523

Henderson IW, Kime DE (1987) The adrenal cortical hormones. In: Pang PKT, Schreibman MP, Sawyer WH (eds) Vertebrate endocrinology: fundamentals and biomedical implications, vol 2. Regulation of water and electrolytes. Academic Press, San Diego, pp 121–142

Henderson IW, Wales NAM (1974) Renal diuresis and antidiuresis after injections of arginine vasotocin in the freshwater eel (Anguilla anguilla L.). J Endocrinol 61:487–500

Henderson IW, Jotisankasa V, Mosley W, Oguri M (1976) Endocrine and environmental influences upon the plasma cortisol concentrations and plasma renin activity in the eel, Anguilla anguilla L. J Endocrinol 70:81–95

Henderson IW, Oliver JA, McKeever A, Hazon N (1981) Phylogenetic aspects of the renin-angiotensin system. In: Pethes G, Frenyo VL (eds) Advances in animal and comparative physiology. Pergamon Press, New York, pp 355–363

Heney HW, Stiffler DF (1983) The effects of aldosterone on sodium and potassium metabolism in larval Ambystoma tigrinum. Gen Comp Endocrinol 49:122–127

Herndon TM, McCormick SD, Bern HA (1991) Effects of prolactin on chloride cells in opercular membrane from seawater-adapted tilapia. Gen Comp Physiol 83:283–289

Herrera FC (1965) Effect of insulin on short-circuit current and sodium transport across toad urinary bladder. Am J Physiol 209:819–824

Heymann JB, Engel A (1999) Aquaporins: phylogeny, structure, and physiology of water channels. News Physiol Sci 14:187–193

Hickman CP (1965) Studies on renal function in freshwater teleost fish. Trans R Soc Can 3: 213–236

Hiebert SM, Salvante KG, Ramenofsky M (2000) Corticosterone and nocturnal torpor in the rufous humming bird (Selasphorus rufus). Gen Comp Endocrinol 120:220–234

Hillman SS (1974) The effect of arginine vasopressin on water and sodium balance in the urodele amphibian Aneides lugubris. Gen Comp Endocrinol 24:74–82

Hillman SS (1978) Some effects of dehyration on internal distributions of water and solutes in Xenopus laevis. Comp Biochem Physiol 61A:303–307

Hillyard SD, von Sechendorff Hoff K, Propper C (1998) The water absorption response: a behavioral assay for physiological processes in terrestrial amphibians. Physiol Zool 71: 127–138

Hirano T (1964) Further studies on the neurohypophysial hormones in the avian median eminence. Endocrinol Jpn 11:87–95

Hirano T (1966) Neurohypophysial hormones in the median eminence of the bullfrog, turtle and duck. Endocrinol Jpn 13:59–74

Hirano T (1967) Effect of hypophysectomy on water transport in isolated intestine of the eel, Anguilla japonica. Proc Jpn Acad 43:793–796

Hirano T (1974) Some factors regulating water intake by the eel, Anguilla japonica J Exp Biol 61:737–747

Hirano T (1975) Effects of prolactin on osmotic and diffusion permeability of the urinary bladder of the flounder, Platichthys flesus. Gen Comp Endocrinol 27:88–94

Hirano T (1980) Effects of cortisol and prolactin on ion permeability of the eel oesophagus. In: Lahlou B (ed) Epithelial transport in lower vertebrates. Cambridge University Press, Cambridge, pp 143–149

Hirano T, Mayer-Gostan N (1976) Eel esophagus as an osmoregulatory organ. Proc Natl Acad Sci USA 73:1348–1350

Hirano T, Utida S (1968) Effects of ACTH and cortisol on water movement in isolated intestine of the eel, *Anguilla japonica*. Gen Comp Endocrinol 11:373–380

Hirano T, Johnson DW, Bern HA (1971) Control of water movement in flounder urinary bladder by prolactin. Nature (Lond) 230:469–471

Hoar WS (1951) Hormones in fish. Publ Ont Fish Res Lab 71:1–51

Hoffman J, Katz U (1997) Salt and water balance in the toad *Bufo viridis* during recovery from two different osmotically stressful conditions. Comp Biochem Physiol 117A:147–154

Hoffman J, Katz U (1998) Glyconeogenesis and urea synthesis in the toad *Bufo viridis* during acclimation to water restriction. Physiol Zool 71:85–92

Hoffman J, Katz U (1999) Elevated plasma osmotic concentration stimulates water absorption response in a toad. J Exp Zool 284:168–173

Hoffman TCM, Walsberg GE (1999) Inhibiting ventilatory evaporation produces an adaptive increase in cutaneous evaporation in mourning doves *Zenaida macroura*. J Exp Biol 202: 30021–33028

Holmes WN (1959) Studies on the hormonal control of sodium metabolism in the rainbow trout (*Salmo gairdneri*). Acta Endocrinol 31:587–602

Holmes WN, Cronshaw J (1993) Some actions of angiotensin II in gallinaceous and anseriform birds. In: Sharp PJ (ed) Avian endocrinology. Journal of Endocrinology Ltd., Bristol, pp 201–216

Holmes WN, McBean RL (1963) Studies on the glomerular filtration rate of the rainbow trout (*Salmo gairdneri*). J Exp Biol 40:335–341

Holmes WN, McBean RL (1964) Some aspects of the electrolyte excretion in the green turtle, *Chelonia mydas mydas*. J Exp Biol 41:81–90

Holmes WN, Butler DG, Phillips JG (1961) Observations on the effects of maintaining glaucous-winged gulls (*Larus glaucescens*) on fresh water and sea water for long periods. J Endocrinol 22:53–61

Hopkins TE, Wood CM, Walsh PJ (1995) Interactions of cortisol and nitrogen metabolism in the ureogenic gulf toadfish *Opsanus beta*. J Exp Biol 198:2229–2235

Horisberger J-D (1998) Amiloride-sensitive Na channels. Curr Opin Cell Biol 10:443–449

Horowitz M, Adler JH (1983) Plasma volume regulation during heat stress: albumin synthesis vs. capillary permeability. A comparison between desert and non-desert species. Comp Biochem Physiol 74A:105–110

Horowitz M, Borut A (1970) Effect of acute dehydration on body fluid compartments in three rodent species, *Rattus norvegicus*, *Acomys cahirinus* and *Meriones crassus*. Comp Biochem Physiol 35:283–290

Horowitz M, Samueloff S (1988) Cardiac output distribution in thermally dehydrated rodents. Am J Physiol 254:R109–R116

Horseman N (1997) The biology of prolactin signalling at the cellular and organismal levels. In: Kawashima S, Kikuyama S (eds) XIIIth International Congress of Comparative Endocrinology, vol 2. Yokohama (Japan), Monduzzi Editore, Bologna, pp 969–978

House CR (1963) Osmotic regulation in the brackish water teleost, *Blennius pholis*. J Exp Biol 40:87–104

Houston AH (1959) Osmoregulatory adaptation of steelhead trout (*Salmo gairdneri, Richardson*) to sea water. Can J Zool 37:728–748

Howe A, Jewell PA (1959) Effects of water deprivation on the neurosecretory material of the desert rat (*Meriones meriones*) compared to the laboratory rat. J Endocrinol 19:118–124

Howe D, Gutknecht J (1978) Role of urinary bladder in osmoregulation in marine teleost, *Opsanus tau*. Am J Physiol 23:R48-R54

Howes NH (1940) The response of the water regulating mechanism of developmental stages of the common toad *Bufo bufo bufo* (L.) to treatment with extracts of the posterior lobe of the pituitary. J Exp Biol 17:128–138

Hughes MR, Chadwick A (eds) (1989) Progress in avian osmoregulation. Leeds Philosophical Society, Leeds (UK)

Hughes MR, Kojwang D, Zenteno-Savin T (1992) Effects of caecal ligation and saline acclimation on plasma concentration and organ mass in male and female Pekin ducks, *Anas platyrhynchos*. J Comp Physiol B 162:625–631

Hughes MR, Bennett DC, Sullivan TM, Hwang H (1999) Retrograde movement of urine into the gut of salt water acclimated Mallards (*Anas platyrhynchos*). Can J Zool 77:342–346

Hui CA (1981) Seawater consumption and water flux in the common dolphin *Delphinus delphis*. Physiol Zool 54:430–440

Hunter T (2000) Signaling – 2000 and beyond. Cell 100:113–127

Hyodo S, Urano A (1991) Changes in expression of provasotocin genes during adaptation to hyper- and hypo-osmotic environments in rainbow trout. J Comp Physiol B 161:549–556

Idler DR, Truscott B (1972) Corticosteroids in fish. In: Idler DR (ed) Steroids in nonmammalian vertebrates. Academic Press, New York, pp 341–389

Idler DR, Sangalang GB, Truscott B (1972) Corticosteroids in the South American lungfish. Gen Comp Endocrinol Suppl 3:238–244

Immelmann K (1971) Ecological aspects of periodic reproduction. In: Farner DS, King JR, Parkes KC (eds) Avian biology vol 1. Academic Press, New York, pp 341–389

Inoue T, Terris J, Ecelbarger CA, Chou C-L, Nielsen S, Knepper MA (1999) Vasopressin regulates apical targeting of aquaporin-2 but not UT1 urea transporter in renal collecting duct. Am J Physiol 276:F559–F566

Inui Y, Miwa S, Yamano K, Hirano T (1994) Hormonal control of flounder metamorphosis. In: Davey KG, Peter RE, Tobe SS (eds) Perspectives in comparative endocrinology. National Research Council of Canada, Ottawa, pp 408–411

Isaia J (1979) Non-electrolyte permeability of trout gills: effect of temperature and adrenaline. J Physiol (Lond) 286:361–373

Isaia J (1984) Water and nonelectrolye permeation. In: Hoar WS, Randall DJ (eds) Fish physiology, vol X, part B gills. Academic Press, Orlando, pp 1–38

Isenring P, Jacoby SC, Payne JA, Forbush B III (1998) Comparison of Na-K-Cl cotransporters. NKCC1, NKCC2, and HEK cell Na-K-Cl cotransporter. J Biol Chem 273:11295–11301

Ishii S, Hirano T, Kobayashi H (1962) Neurohypophysial hormones in avian median eminence and pars nervosa. Gen Comp Endocrinol 2:433–440

Ismail-Beigi F (1993) Thyroid hormone regulation of Na, K-ATPase expression. Trends Endocrinol Metab 4:152–155

Jackson DC, Schmidt-Nielsen K (1964) Countercurrent heat exchanges in the respiratory passages. Proc Natl Acad Sci USA 51:1192–1197

Jaisser F, Beggah AT (1999) The nongastric H^+-K^+-ATPases: molecular and functional properties. Am J Physiol 276:F812–F824

Jamison RL, Gehrig JJ (1992) Urinary concentration and dilution: physiology. In: Handbook of physiology sect 8, Renal physiology vol II. Oxford University Press, New York, for the American Physiological Society, pp 1219–1279

Jampol LM, Epstein FH (1970) Sodium-potassium-activated adenosinetriphosphatase and osmotic regulation by fishes. Am J Physiol 218:607–611

Jan LY, Jan YN (1994) Potassium channels and their evolving gates. Nature (Lond) 371:119–122

Jard S (1966) Etudes des effets de la vasotocine sur l'excretion de l'eau et des électrolytes par la rein de la grenouille *Rana esculenta* L.: analyse à l'aide d'analogues artificiels de l'hormone naturelle des caractères structuraux requis pour son activité biologique. J Physiol (Paris) Suppl 15:1–124

Jard S, Morel F (1963) Actions of vasotocin and some of its analogues on salt and water excretion by the frog. Am J Physiol 204:222–226

Jard S, Maetz J, Morel F (1960) Action de quelques analogues de l'ocytocine sur différents récepteurs intervenant dans l'osmorégulation de *Rana esculenta*. CR Acad Sci (Paris) 251:788–790

Jard S, Bastide F, Morel F (1968) Analyse de la relation "dose-effet biologique" par l'action de l'ocytocine et de la noradrénaline sur la peau et la vessie de la grenouille. Biochim Biophys Acta 150:124–130

Jared C, Navas CA, Toledo RC (1999) An appreciation of the physiology and morphology of the caecilians (Amphibia: Gymnophiona). Comp Biochem Physiol 123A:313–328

Jensen BL, Gambaryan S, Schmaus E, Kutz A (1998) Effects of dietary salt on adrenomedullin and its receptor mRNAs in rat kidney. Am J Physiol 275: F55–F61

Jentsch TJ, Friedrich T, Schriever A, Yamada H (1999) The CLC chloride channel family. Eur J Physiol 437:783–795

Johnson DW, Hirano T, Sage M, Foster RC, Bern HA (1974) Time course of response of starry flounder (*Platichthys stellatus*) to prolactin and to salinity transfer. Gen Comp Endocrinol 24:373–380

Johnston CI, Davis JO, Wright SF, Howards SS (1967) Effects of renin and ACTH on adrenal steroid production in the American bullfrog. Am J Physiol 213:393–399

Jolivet Jaudet GJ, Leloup Hatey J (1984) Variations in aldosterone and corticosterone plasma levels during metamorphosis in *Xenopus laevis* tadpoles. Gen Comp Endocrinol 56:59–65

Jones MEE, Bradshaw SD, Fergusson B, Watts R (1990) Effect of available surface water on levels of antidiuretic hormone (lysine vasopressin) and water and electrolyte metabolism in the Rottnest Island quokka (*Setonix brachyurus*). Gen Comp Endocrinol 77:75–87

Jones RM (1980) Metabolic consequences of accelerated urea synthesis during seasonal dormancy of spadefoot toads, *Scaphiopus couchi* and *Scaphopus multiplicatus*. J Zool 212: 255–267

Jørgensen CB (1993) Role of pars nervosa of the hypophyysis in amphibian water economy: a re-assessment. Comp Biochem Physiol 109A:311–324

Jørgensen CB (1994) Water economy in a terrestrial toad (*Bufo bufo*), with special reference to cutaneous drinking and urinary bladder function. Comp Biochem Physiol 109A:311–324

Jørgensen CB (1997a) 200 years of amphibian water economy; from Robert Townson to the present. Biol Rev 72:153–237

Jørgensen CB (1997b) Urea and amphibian water economy. Comp Biochem Physiol 117A: 161–170

Jørgensen CB (1998) Role of urinary and cloacal bladder in chelonian water economy: historical and comparative perspectives. Biol Rev 73:347–366

Jørgensen CB, Rosenkilde P (1956a) On regulation of concentration and content of chloride in goldfish. Biol Bull (Woods Hole) 110:300–305

Jørgensen CB, Rosenkilde P (1956b) Relative effectiveness of dehydration and neurohypophysial extracts in enhancing water absorption in toads and frogs. Biol Bull (Woods Hole) 110: 306–309

Jørgensen CB, Wingstrand KG, Rosenkilde P (1956) Neurohypophysis and water metabolism in the toad *Bufo bufo* (L.). Endocrinology 59:601–610

Jørgensen CB, Rosenkilde P, Wingstrand KG (1969) Role of the preoptico-neurohypophysial system in the water economy of the toad, *Bufo bufo* L. Gen Comp Endocrinol 12:91–98

Joss JMP, Arnold-Reed DE, Balment RJ (1994) The steroidogenic response to angiotensin II in the Australian lungfish *Neoceratodus forsteri*. J Comp Physiol B 164:378–382

Joss JMP, Itahara Y, Watanabe TX, Nakajima K, Takei Y (1999) Teleost-type angiotensin is present in Australian lungfish, *Neoceratodus forsteri*. Gen Comp Endocrinol 114:206–212

Jougasaki M, Wei C-M, Aarhus LL, Heublein DM, Sandberg SM, Burnett JC (1995) Renal localization and actions of adrenomedullin: a natriuretic peptide. Am J Physiol 268:F657–F663

Jungreis AM (1976) Partition of excretory nitrogen in amphibia. Comp Biochem Physiol 53A: 133–141

Junqueira LCU, Malnic G, Monge C (1966) Reabsorptive function of the ophidian cloaca and large intestine. Physiol Zool 29:151–159

Kahn CR, White MF, Shoelson SE, Backer JM, Arahi E, Cheatham B, Csermely P, Folli F, Goldstein BJ, Huertas P, Rothenberg PI, Saad MJA, Siddle K, Sun K-L, Wilden PA, Yamada, K, Kahn SA (1993) The insulin receptor and its substrate: molecular determinants of early events in insulin action. Recent Prog Horm Res 48:291–339

Kakizawa S, Kaneko T, Hasegawa S, Hirano T (1993) Activation of somatolactin cells in the pituitary of the rainbow trout, *Oncorhynchus mykiss* by low environmental calcium. Gen Comp Endocrinol 91:298–306

Kakizawa S, Kaneko T, Hirano T (1996) Elevation of plasma somatolactin concentrations during acidosis in rainbow trout (*Oncorhynchus mykiss*). J Exp Biol 199:1043–1051

Kamiya M (1972) Sodium-potassium activated adenosinetriphosphatase in isolated chloride cells from eel gills. Comp Biochem Physiol 43B:611–617

Kamiya M, Utida S (1969) Sodium-potassium-activated adenosinetriphosphatase activity in gills of freshwater, marine and euryhaline teleosts. Comp Biochem Physiol 31:671–674

Karnaky KJ, Kinter WB (1977) Killifish opercular skin: a flat epithelium with a high density of chloride cells. J Exp Zool 199:355–364

Katsura T, Ausiello DA, Brown D (1996) Direct demonstration of aquaporin-2 water channel recycling in stably transfected LLC-PK1 epithelial cells. Am J Physiol 270:F548–F553

Katz U (1973) Studies on the adaptation of the toad *Bufo viridis* to high salinities: oxygen consumption, plasma concentration and water content of the tissues. J Exp Biol 58:785–796

Kennett R, Christian KR (1994) Metabolic depression in estivating long-neck turtles (*Chelodina rugosa*). Physiol Zool 67:1087–1102

Kenyon CJ, McKeever A, Oliver JA, Henderson IW (1985) Control of renal and adrenocortical function by the renin-angiotensin system in two euryhaline teleost fishes. Gen Comp Endocrinol 58:93–100

Keys A (1931) Chloride and water secretion and absorption by the gills of the eel. Z Vergl Physiol 15:364–388

Keys A (1933) The mechanism of adaptation to varying salinity in the common eel and the general problem of osmotic regulation in fishes. Proc R Soc Lond B 112:184–199

Keys A, Bateman JB (1932) Branchial responses to adrenaline and pitressin in the eel. Biol Bull (Woods Hole) 63:327–336

Keys A, Willmer FN (1932) "Chloride-secreting cells" in the gills of fishes, with special reference to the common eel. J Physiol (Lond) 76:368–378

Kim G-H, Masilamani S, Turner R, Mitchell C, Wade JB (1998) The thiazide-sensitive Na-Cl cotransporter is an aldosterone-induced protein. Proc Natl Acad Sci USA 95:14552–14557

Kimura T, Tanizawa O, Mori K, Brownstein MJ, Okayama H (1992) Structure and expression of a human oxytocin receptor. Nature (Lond) 356:526–529

King JR, Farner DS (1964) Terrestrial animals in humid heat. In: Handbook of physiology. Sect 4, Adaptation to the environment. American Physiological Society, Washington pp 603–624

Kinnear JE, Purohit KG, Main AR (1968) The ability of the Tammar wallaby, *Macropus eugenii* (Marsupialia), to drink sea water. Comp Biochem Physiol 25:761–782

Kirsch R, Meister MF (1982) Progressive processing of ingested water in gut of sea-water teleosts. J Exp Biol 98:67–81

Kitamura K, Kangawa K, Kawamoto M, Ichiki Y, Nakamura S, Matsuo H, Eto T (1993a) Adenomedullin: a novel hypotensive peptide isolated from human pheochromocytoma. Biochem Biophys Res Commun 192:553–560

Kitamura K, Sakata J, Kangawa K, Kojima M, Matsuo H, Eto T (1993b) Cloning and characterization of cDNA encoding a precursor of human adrenomedullin. Biochem Biophys Res Commun 194:720–725

Kitamura K, Kangawa K, Matsuo H, Eto T (1995) Adrenomedullin. Implications for hypertension research. Drugs 49:485–495

Klingbeil CK (1985) Effects of chronic changes in dietary electrolytes and acute stress on plasma levels of corticosterone and aldosterone in the duck (*Anas platyrhynchos*). Gen Comp Endocrinol 58:10–19

Kloas W, Hanke W (1992) Localization and quantification of atrial natriuretic factor binding sites in the kidney of *Xenopus laevis*. Gen Comp Endocrinol 85:26–35

Kloas W, Flugge G, Fuchs E, Stolte H (1988) Binding sites for atrial natriuretic peptide in the kidney and aorta of the hagfish (*Myxine glutinosa*). Comp Biochem Physiol 91A:685–688

Kobayashi H (1981) Angiotensin-induced drinking in parrots. Gen Comp Endocrinol 43:399–401

Kobayashi H, Takei Y (1982) Mechanisms for induction of drinking with special reference to angiotensin II. Comp Biochem Physiol 71A:485–494

Kobayashi H, Takei Y (1997) The renin-angiotensin system. Comparative aspects. Springer, Berlin Heidelberg New York

Kobayashi H, Uemura H, Wada M, Takei Y (1979) Ecological adaptation of angiotensin-induced thirst mechanism in tetrapods. Gen Comp Endocrinol 38:93–104

Kobayashi H, Uemura H, Takei Y, Itatsu N, Ozawa M, Ichinohe K (1983) Drinking induced by angiotensin II in fishes. Gen Comp Endocrinol 49:295–306

Kobayashi H, Owada K, Yamada H, Okawara Y (1986) The caudal neurosecretory system in fishes. In: Pang PKT, Schreibman MP (eds) Vertebrate endocrinology: fundamentals and biomedical implications. Vol 1, Morphological considerations. Academic Press, Orlando, pp 147–174

Kobayashi T, Cohen P (1999) Activation of serum- and glucocorticoid-regulated protein kinase by agonists that activate phosphatidylinositide 3-kinase is mediated by 3-phosphoinositide-dependent protein kinase-1 (PDK1) and PDK2. Biochem J 339:319–328

Kobelt F, Linsenmair KE (1992) Adaptations of the reed frog, *Hyperolius viridiflavus* (Amphibia: Anura: Hyperoliidae) to its arid environment. VI. The iridophores in the skin as radiation reflectors. J Comp Physiol B 162:314–326

Koefoed-Johnsen V, Ussing HH (1953) The contribution of diffusion and flow. The effects of neurohypophysial hormone on isolated anuran skin to the passage of D_2O through living membranes. Acta Physiol Scand 28:60–76

Koefoed-Johnsen V, Ussing HH, Zerahn K (1952) The origin of the short-circuit current in the adrenaline stimulated frog skin. Acta Physiol Scand 27:38–48

Komourdjian MP, Saunders RL, Fenwick JC (1976) The effect of porcine somatotrophin on growth and survival in seawater of Atlantic salmon (*Salmo salar*) parr. Can J Zool 54:531–535

Kooistra TA, Evans DH (1976) Sodium balance in the green turtle, *Chelonia mydas*, in seawater and freshwater. J Comp Physiol 107:229–240

Kopito RR, Lodish HF (1985) Primary structure and transmembrane orientation of the murine anion exchange protein. Nature (Lond) 316:234–238

Krag B, Skadhauge E (1972) Renal salt and water excretion in the budgerygah (*Melopsittacus undulatus*). Comp Biochem Physiol 41A:667–683

Krakauer T, Gans C, Paganelli CV (1968) Ecological correlation of water loss in burrowing reptiles. Nature (Lond) 218:659–660

Krogh A (1937) Osmotic regulation in the frog (*R. esculenta*) by active absorption of chloride ions. Skand Arch Physiol 76:60–74

Krogh A (1939) Osmotic regulation in aquatic animals. Cambridge University Press, Cambridge

Krug EC, Honn KV, Battista J, Nicoll CS (1983) Corticosteroids in serum of *Rana catesbeiana* during development and metamorphosis. Gen Comp Endocrinol 52:232–241

Kuchel LJ, Franklin CE (1998) Kidney and cloaca function in the estuarine crocodile (*Crocodylus porosus*) at different salinities: evidence for solute-linked water uptake. Comp Biochem Physiol 119A:825–831

Kulczykowska E (1997) Response of circulating arginine vasotocin and isotocin to rapid osmotic challenge in rainbow trout. Comp Biochem Physiol 118A:773–778

Lacy ER (1991) Functional morphology of the large intestine. In: Field M, Frizzell RA (eds) Handbook of physiology Sect 6. The gastrointestinal system. American Physiological Society, Bethesda, pp 121–194

Lacy ER, Reale E (1990) The presence of a juxtaglomerular apparatus in elasmobranch fish. Anat Embryol 182:249–262

Lacy ER, Reale E, Schlusselberg DS, Smith WK, Woodward DJ (1985) A renal countercurrent system in marine elasmobranch fish: a computer-assisted reconstruction. Science 227:1351–1354

Lahlou B (1967) Excrétion rénale chez un poisson euryhalin, le flet (*Platichthys flesus* L.): caractéristiques de l'urine normale en eau douce et en eau de mer et effets des changements de milieu. Comp Biochem Physiol 20:925–938

Lahlou B, Fossat B (1971) Mécanisme du transport de l'eau et du sel à travers le vessie urinaire d'un poisson téléostéen en eau douce, la truite arc-en-ciel. CR Hebd Seances Acad Sci 273:2108–2110

Lahlou B, Sawyer WH (1969) Sodium exchanges in the toadfish, *Opsanus tau*, a euryhaline aglomerular teleost. Am J Physiol 216:1273–1278

Lahlou B, Henderson IW, Sawyer WH (1969a) Sodium exchanges in goldfish (*Carassius auratus* L.) adapted to hypertonic saline solution. Comp Biochem Physiol 28:1427–1433

Lahlou B, Henderson IW, Sawyer WH (1969b) Renal adaptations by *Opsanus tau*, a euryhaline aglomerular teleost, to dilute media. Am J Physiol 216:1266–1272

Lam TJ (1972) Prolactin and hydromineral metabolism in fishes. Gen Comp Endocrinol Suppl 3:328–338

Lam TJ, Hoar WS (1967) Seasonal effects of prolactin on freshwater osmoregulation of the marine form (*trachurus*) of the stickleback *Gasterosteus aculeatus*. Can J Zool 45:509–516

Lam TJ, Leatherland JF (1969) Effect of prolactin on freshwater survival of the marine form (*trachurus*) of the threespine stickleback, *Gasterosteus aculeatus*, in early winter. Gen Comp Endocrinol 12:385–394

Lange R, Fugelli K (1965) The osmotic adjustment of the euryhaline teleosts, the flounder *Pleuronectes flesus* L. and the three-spined stickleback *Gasterosteus aculeatus* L. Comp Biochem Physiol 15:283–292

LaPointe JA (1969) Effects of ovarian steroids and neurohypophysial hormones on the oviduct of the viviparus lizard, *Klauberina riversiana*. J Endocrinol 43:197–205

Laragh JH, Sealey JE (1992) Renin-angiotensin-aldosterone system and the renal regulation of sodium, potassium, and blood pressure homeostasis. In: Handbook of physiology. Sect 8: Renal physiology. Vol II. Oxford University Press, New York, for the American Physiological Society, pp 1409–1541

Larsen EH (1991) Chloride transport by high-resistance heterocellular epithelia. Physiol Rev 71:235–283

Larson B, Bern HA (1987) The urophysis and osmoregulation. In: Pang PKT, Schreibman MP, Sawyer WH (eds) Vertebrate endocrinology: fundamentals and biomedical implications. Vol 2, Regulation of water and electrolytes. Academic Press, San Diego, pp 143–156

Lasiewski RC, Bernstein MH, Ohmart RD (1971) Cutaneous water loss in the roadrunner and poor-will. Condor 73:470–472

Laurén DJ (1985) The effect of chronic saline exposure on the electrolyte balance, nitrogen metabolism, and corticosterone titer in the American alligator, *Alligator mississippiensis*. Comp Biochem Physiol 81A:217–223

Laurent P (1984) Gill internal morphology. In: Hoar WS, Randall DJ (eds) Fish physiology, X Gills Part A. Academic Press, Orlando, pp 73–183

Laurent P, Dunel S (1980) Morphology of gill epithelia in fish. Am J Physiol 23:R147–R159

Laverty G (1989) Renal tubular transport in avian kidney. In: Hughes MR, Chadwick A (eds) Progress in avian osmoregulation. Leeds Philosophical and Literary Society, Leeds (UK), pp 243–261

Laverty G, Skadhauge E (1999) Physiological roles and regulation of transport activities in the avian lower intestine. J Exp Zool 283:480–494

Leaf A, Hays RM (1962) Permeability of the isolated toad bladder to solutes and its modification by vasopressin. J Gen Physiol 45:921–932

Leaf A, Anderson J, Page LB (1958) Active sodium transport by the isolated toad bladder. J Gen Physiol 41:657–668

Lebel JM, Leloup J (1989) Triiodothyronine binding to putative solubilized nuclear thyroid hormone receptor in liver and gill of the brown trout (*Salmo trutta*) and the European eel (*Anguilla anguilla*). Gen Comp Endocrinol 75:301–309

Le Bras YM (1984) Circadian variations of catecholamine levels in brain, heart and plasma of the eel, *Anguilla anguilla* L., at three different times of the year. Gen Comp Endocrinol 55:472–479

LeBrie SJ, Elizondo RS (1969) Saline loading and aldosterone in water snakes, *Natrix cyclopion*. Am J Physiol 217:426–430

Lederis K, Fryer JN, Okawara Y, Schönrock C, Richter D (1994) Corticotropin-releasing factors acting on the fish pituitary: experimental and molecular analysis. In: Sherwood NM, Hew CL (eds) Fish physiology. XIII Molecular endocrinology of fish. Academic Press, San Diego, pp 67–100

Lee AK, Mercer JB (1967) Cocoon surounding desert-dwelling frogs. Science 157:87–88

Lee J, Malvin RL (1987) Natriuretic response to homologous heart extract in aglomerular toadfish. Am J Physiol 252:R1055–R1058

Lee P, Schmidt-Nielsen K (1971) Respiratory and cutaneous evaporation in the zebra finch: effect on water balance. Am J Physiol 220:1598–1605

Lee W-S, Hebert SC (1995) ROMK inwardly rectifying ATP-sensitive K^+ channel. I. Expression in rat distal nephron segments. Am J Physiol 268:F1124–F1131

LeFevre MD (1973) Effects of aldosterone on the isolated substrate-depleted turtle bladder. Am J Physiol 225:1252–1256

Leung DW, Loo DDF, Hirayama BA, Zeuthen T, Wright EM (2000) Urea transport by cotransporters. J Physiol (Lond) 528:251–257

Levinsky NG, Sawyer WH (1953) Significance of neurohypophysis in regulation of fluid balance in the frog. Proc Soc Exp Biol Med (NY) 82:272–274

Levitan R, Ingelfinger I (1965) Effect of D-aldosterone on salt and water absorption from the intact human colon. J Clin Invest 44:801–809

Licht P, Bradshaw SD (1969) A demonstration of corticotropic activity and its distribution in the pars distalis of the reptile. Gen Comp Endocrinol 13:226–235

Light DB, Schwiebert EM, Karlson KH, Stanton BA (1989) Atrial natriuretic peptide inhibits cation channels in renal inner medullary collecting duct cells. Science 243:383–385

Lihrmann I, Netchitailo P, Feuilloley M, Cantin M, Delarue C, Leboulenger F, De Lean A, Vaudry H (1988) Effect of atrial natriuretic factor on corticosteroid production by perifused frog interrenal slices. Gen Comp Endocrinol 71:55–62

Lillywhite HB, Maderson PFA (1982) Skin structure and permeability. In: Gans C (ed) Biology of the Reptilia. Vol 12, Physiology, C Physiological ecology. Academic Press, London, pp 397–481

Lillywhite HB, Mittal AK, Garg TK, Agrawal N (1997) Wiping behaviour and its ecophysiological significance in the Indian tree frog *Polypedates maculatus*. Copeia 1992:88–100

Lin H, Randall DJ (1993) H[+]-ATPase activity in crude homogenates of fish gill tissue: inhibitor sensitivity and environmental and hormonal regulation. J Exp Biol 180:163–174

Lin H, Randall D (1995) Proton pumps in fish gills. In: Wood CM, Shuttleworth TJ (eds) Cellular and molecular approaches to fish ionic regulation. Academic Press, San Diego, pp 229–255

Lodi G, Biciotti M, Viotto B (1982) Cutaneous osmoregulation in *Triturus cristatus carnifex* (Laur.) (Urodela). Gen Comp Endocrinol 46:452–457

Lofts B (1984) Amphibians. In: Lamming GE (ed) Marshall's physiology of reproduction vol 1. Churchill Livingstone, Edinburgh, pp 127–205

Logan AG, Moriarty AJ, Rankin JC (1980) A micropuncture study of kidney function in the river lamprey, *Lampetra fluviatilis*, adapted to fresh water. J Exp Biol 85:137–147

Lolait SJ, O'Carroll A-M, McBride OW, Konig M, Morel A, Brownstein MJ (1992) Cloning and characterization of the vasopressin V_2 receptor and possible link to nephrogenic diabetes insipidus. Nature (Lond) 357:336–339

Loo DDF, Hirayama BA, Meinild A-K, Chandy G, Zeuthen T, Wright EM (1999a) Passive water and ion transport by cotransporters. J Physiol (Lond) 518:195–202

Loo DDF, Wright EM, Meinild A-K, Klaerke DA, Zeuthen T (1999b) Commentary on "Epithelial fluid transport-a century of investigation". News Physiol Sci 14:98–100

Loretz CA (1995) Electrophysiology of ion transport in teleost intestinal cells. In: Wood CM, Shuttleworth TJ (eds) Cellular and molecular approaches to fish ionic regulation. Academic Press, San Diego, pp 25–56

Loretz CA (1996) Inhibition of goby posterior intestinal NaCl absorption by natriuretic peptides and by cardiac extracts. J Comp Physiol B 166:484–491

Loretz CA, Bern HA (1981) Stimulation of sodium transport across the teleost bladder by urotensin II. Gen Comp Endocrinol 43:325–330

Loretz CA, Howard ME, Seigel AJ (1985) Ion transport by goby intestine: cellular mechanism of urotensin II stimulation. Am J Physiol 249:G284–G293

Lovatt Evans C, Smith DFG, Weil-Malherbe H (1956) The relationship between sweating and the catecholamine content of the blood in the horse. J Physiol (Lond) 132:542–552

Lovejoy DA, Balment RJ (1999) Evolution and physiology of the corticotropin-releasing factor (CRF) family of neuropeptides in vertebrates. Gen Comp Endocrinol 115:1–22

Loveridge JP (1970) Observations on nitrogenous excretion and water relations in *Chiromantis xerampelina* (Amphibia, Anura). Arnoldia 5:1–6

Lowy RJ, Schreiber JH, Ernst SA (1987) Vasoactive intestinal peptide stimulates ion transport in avian salt gland. Am J Physiol 253:R801–R808

Lutz PL, Robertson JD (1971) Osmotic constituents of the coelacanth *Latimeria chalumnae* Smith. Biol Bull (Woods Hole) 141:553–560

MacAulay N, Gether U, Klaerke DA, Zeuthen T (2001) Water transport by the human Na[+]-coupled glutamate cotransporter expressed in *Xenopus oocytes*. J Physiol (Lond) 530:367–378

Macchi IA, Phillips JG (1966) In vitro effect of adrenocorticotropin on corticoid secretion in the turtle, snake, and bullfrog. Gen Comp Endocrinol 6:170–182

MacFarlane WV (1964) Terrestrial animals in dry heat: ungulates. In: Dill DB, Adolph EF, Wilbur CG (eds) Handbook of physiology. Sect 5. Adaptation to the environment. American Physiological Society, Washington, pp 509–539

Maclean GL (1996) Ecophysiology of desert birds. Springer, Berlin Heidelberg New York

MacMillen RE, Lee AK (1969) Water metabolism of Australian hopping mice. Comp Biochem Physiol 28:493–514

MacMillen RE, Lee AK (1970) Energy metabolism and pulmocutaneous water of Australian hopping mice. Comp Biochem Physiol 35A:335–369

Madara JL (1991) Functional morphology of epithelium of the small intestine. In: Field M, Frizzell RA (eds) Handbook of physiology. Sect 6. The gastrointestinal tract. American Physiological Society, Bethesda, pp 83–120

Madsen SS (1990a) Effects of repetitive cortisol and thyroxine injections on chloride cell number and Na$^+$/K$^+$-ATPase activity in gills of freshwater adapted rainbow trout, *Salmo gairdneri*. Comp Biochem Physiol 95A:171–175

Madsen SS (1990b) The role of cortisol and growth hormone in seawater adaptation and development of hypoosmoregulatory mechanisms in sea trout parr (*Salmo trutta trutta*). Gen Comp Endocrinol 79:1–11

Madsen SS, Bern HA (1993) In-vitro effects of insulin-like growth factor-I on gill Na$^+$,K$^+$-ATPase in coho salmon, *Oncorhynchus kisutch*. J Endocrinol 138:23–30

Madsen SS, Jensen MK, Nohr J, Kristiansen K (1995) Expression of Na$^+$,K$^+$-ATPase in the brown trout, *Salmo trutta*: *In vivo* modulation by hormones and seawater. Am J Physiol 269: R1339–R1345

Maetz J (1959) Le contrôle endocrinien du transport actif de sodium à travers la peau de grenouille. In: Coursaget J (ed) La méthode des indicateurs nucléaires dans l'etude de transport actif d'ions. Pergamon Press, London, pp 185–196

Maetz J (1963) Physiological aspects of neurohypophysial function in fishes, with some reference to the Amphibia. Symp Zool Soc Lond 9:107–140

Maetz J (1969) Sea water teleosts: evidence for sodium-potassium exchange in the branchial sodium excreting pump. Science 166:613–615

Maetz J, Bourguet J, Lahlou B, Hourdry J (1964) Peptides neurohypophysaires et osmorégulation chez *Carassius auratus*. Gen Comp Endocrinol 4:508–522

Maetz J, Lahlou B (1966) Les échanges de sodium et de chlore chez un elasmobranche, *Scyliorhinus*, mesurés à l'aide des isotopes ^{24}Na et ^{36}Cl. J Physiol (Paris) 58:249

Maetz J, Rankin JC (1969) Quelques aspects du rôle biologiques des hormones neurohypophysaires chez les poissons. Colloq CNRS Paris 177:45–55

Maetz J, Skadhauge E (1968) Drinking rates and gill ionic turnover in relation to external salinities in the eel. Nature (Lond) 217:371–373

Maetz J, Mayer N, Chartier-Baraduc MM (1967a) La balance minérale du sodium chez *Anguilla anguilla* en eau de mer, en eau douce et au cours de transfert d'un milieu à l'autre: effets de l'hypophysectomie et de la prolactine. Gen Comp Endocrinol 8:177–188

Maetz J, Sawyer WH, Pickford GE, Mayer N (1967b) Evolution de la balance minérale du sodium chez *Fundulus heteroclitus* au cours du transfert d'eau de mer en eau douce: effets de hypophysectomie et de la prolactine. Gen Comp Endocrinol 8:163–176

Maetz J, Nibelle J, Bornancin M, Motais R (1969) Action sur l'osmorégulation de l'anguille de divers antibiotiques inhibiteurs de la synthèse de proteines ou du renouvellement cellulaire. Comp Biochem Physiol 30:1125–1151

Mahlmann S, Meyerhof W, Hausmann H, Heierhorst J, Schönrock C, Zwiers H, Lederis K, Richter D (1994) Structure, function, and phylogeny of [Arg8]vasotocin receptors from teleost fish and toad. Proc Natl Acad Sci USA 91:1342–1345

Main AR, Bentley PJ (1964) Comparison of dehydration and hydration of burrowing desert frogs and tree frogs of the genus *Hyla*. Ecology 45:379–382

Main AR, Littlejohn MJ, Lee AK (1959) Ecology of Australian frogs. Monogr Biol 8:396–411

Mallatt J, Conley DM, Ridgeway RL (1987) Why do hagfish have gill "chloride cells" when they need not regulate plasma NaCl levels? Can J Zool 65:1956–1965

Mallery CH (1983) A carrier enzyme basis for ammonium excretion in teleost fish. NH$_4^+$-stimulated Na-dependent ATPase activity in *Opsanus beta*. Comp Biochem Physiol 74A:889–897

Malvin GM, Hood L, Sanchez M (1992) Regulation of blood flow through ventral pelvic skin by environmental water and NaCl in the toad *Bufo woodhousei*. Physiol Zool 65:540–553

Malvin RL, Rayner M (1968) Renal function and blood chemistry in Cetacea. Am J Physiol 214:187–191

Malvin RL, Bonjour-J-P, Ridgeway SM (1971) Antidiuretic hormone levels in some cetaceans. Proc Soc Exp Biol Med (NY) 136:1203–1205

Malvin RL, Ridgeway SM, Cornell L (1978) Renin and aldosterone levels in dolphins and sea lions. Proc Soc Exp Biol Med (NY) 157:665–668

Mancera JM, McCormick SD (2000) Rapid activation of gill Na$^+$, K$^+$-ATPase on the euryhaline teleost *Fundulus heroclitus*. J Exp Zool 287:263–274

Marchand S, Safi R, Escriva H, Van Rompaey E, Prunet P, Laudet V (2001) Molecular cloning and characterization of thyroid hormone receptors in teleost fish. J Mol Endocrinol 26:51–65

Marenzi AD, Fustinoni O (1938) El potasio de la sangre y de los tejidos de los sapos suprareno-privos. Rev Soc Argent Biol 14:118–122

Marples D, Schroer TA, Ahrens N, Taylor A, Knepper MA, Nielsen S (1998) Dynein and dynactin colocalize with AQP2 water channels in intracellular vesicles from kidney collecting duct Am J Physiol 274:F384–F394

Marples D, Frokiaer J, Nielsen S (1999) Long-term regulation of aquaporins in the kidney. Am J Physiol 276:F331–F339

Marshall AJ (1961) Breeding seasons and migration. In: Marshall AJ (ed) The biology and comparative physiology of birds vol 2. Academic Press, New York, pp 307–339

Marshall EK, Smith H (1930) The glomerular development of the vertebrate kidney in relation to habitat. Biol Bull (Woods Hole) 59:135–153

Marshall WS (1977) Transepithelial potential and short-circuit current across the isolated skin of *Gillichthys mirabilis* (Teleostei: Gobiidae), acclimated to 5% and 100% seawater. J Comp Physiol B 114:157–165

Marshall WS (1986) Independent Na$^+$ and Cl$^-$ active transport by urinary bladder epithelium of brook trout. Am J Physiol 250:R227–R234

Marshall WS (1995) Transport processes in isolated teleost epithelia: opercular epithelium and urinary bladder. In: Wood CM, Shuttleworth TJ (eds) Cellular and molecular approaches to fish ionic regulation. Academic Press, San Diego, pp 1–23

Marshall WS, Bern HA (1980) Ion transport across the isolated skin of the teleost *Gillichthys mirabilis*. In: Lahlou B (ed) Epithelial transport in lower vertebrates. Cambridge University Press, Cambridge, pp 337–350

Marshall WS, Bern HA (1981) Active chloride transport by the skin of a marine teleost is stimulated by urotensin I and inhibited by urotensin II. Gen Comp Endocrinol 43:484–491

Marshall WS, Bryson SE (1998) Transport mechanisms of seawater teleost chloride cells: an inclusive model of a multifunctional cell. Comp Biochem Physiol 119A:97–106

Marshall WS, Bryson SE, Darling P, Whitten C, Patrick M, Wilkie M, Wood CM, Buckland-Nicks J (1997) NaCl transport and ultrastructure of opercular epithelium from a freshwater-adapted euryhaline teleost, *Fundulus heteroclitus*. J Exp Zool 277:23–37

Marsigliante S, Acierno R, Maffia M, Muscella A, Vinson CP (1997) Immunolocalization of angiotensin II receptors in icefish (*Chionodraco hamatus*) tissues. J Endocrinol 154:193–200

Marsigliante S, Muscella A, Barker S, Storelli C (2000) Angiotensin II modulates the activity of Na$^+$/K$^+$ ATPase in eel kidney. J Endocrinol 165:147–156

Martin CJ (1902) Temperature and metabolism in monotremes and marsupials. Philos Trans R Soc B (Lond) 195:1–37

Martinez A, Unsworth EJ, Cuttitta F (1996) Adrenomedullin-like immunoreactivity in the nervous system of the starfish, *Marthasterias glacialis*. Cell Tissue Res 283:169–172

Marver D (1992) Corticosteroids and the kidney. In: Handbook of physiology. Sect 8. Renal physiology. Vol II. Oxford University Press, New York, for the American Physiological Society, pp 1545–1576

Massi M, Epstein AN (1990) Angiotensin/aldosterone synergy governs the salt appetite of the pigeon. Appetite 14:181–192

Mastroberardino L, Spindler B, Forster I, Loffing J, Assandri R, May A, Verrey F (1998) Ras pathway activates epithelial Na$^+$ channels and decreases its surface expression in *Xenopus* oocytes. Mol Cell Biol 9:3417–3427

Mathies T, Andrews RM (1996) Extended egg retention and its influence on embryonic development and egg water balance: implications for the evolution of viviparity. Physiol Zool 69: 1021–1035

Mayer N (1963) Nouvelle recherches sur l'adaptation des grenouilles vertes *Rana esculenta* à des milieux de salinité variée. Etude spéciale de l'excrétion rénale de l'eaux et des électrolytes. Diplômes d'Etudes Supérieures, Paris

Mayer N (1969) Adaptation de *Rana esculenta* à des milieux variés. Etude spéciale de l'excretion rénale de l'eau et des électrolytes au cours des changements de milieux. Comp Biochem Physiol 29:27–50

Mayer N, Maetz J, Chan DKO, Forster MF, Chester Jones I (1967) Cortisol, a sodium-excreting factor in the eel (*Anguilla anguilla* L.) adapted to sea water. Nature (Lond) 214:1118–1120

Mayer-Gostan N, Wendelaar Bonga SE, Balm PHM (1987) Mechanisms of hormone actions on gill transport. In: Pang PKT, Schreibman MP, Sawyer WH (eds) Vertebrate endocrinology: fundamentals and biomedical implications. Vol 2, Regulation of water and electrolytes. Academic Press, San Diego, pp 211–238

McAfee RD (1972) Survival of *Rana pipiens* in deionized water. Science 178:183–185

McAfee RD, Locke W (1961) Effect of certain steroids on bioelectric current in isolated frog skin. Am J Physiol 200:787–800

McCarron HCK, Dawson TJ (1989) Thermal relations of Macropodoidea in hot environments. In: Grigg G, Jarman P, Hume I (eds) Kangaroos, wallabies and rat-kangaroos. Surrey Beatty, New South Wales (Australia), pp 256–263

McClanahan L (1967) Adaptations of the spadefoot toad, *Scaphiopus couchi*, to desert conditions. Comp Biochem Physiol 20:73–99

McClanahan L, Baldwin R (1969) Rate of water uptake through the integument of the desert toad, *Bufo cognatus*. Comp Biochem Physiol 28:381–389

McClanahan LL (1972) Change in body fluid of burrowed spadefoot toads as a function of soil water potential. Copeia 1972:209–216

McCormick SD (1995) Hormonal control of gill Na$^+$,K$^+$-ATPase and chloride cell function. In: Wood CM, Shuttleworth TJ (eds) Cellular and molecular approaches to fish ionic regulation. Academic Press, San Diego, pp 285–315

McCormick SD (1996) Effects of growth hormone and insulin-like growth factor I on salinity tolerance and gill Na$^+$,K$^+$-ATPase in Atlantic salmon (*Salmo salar*): interaction with cortisol. Gen Comp Endocrinol 101:3–11

McCormick SD, Sakamoto T, Hasegawa S, Hirano T (1991) Osmoregulatory actions of insulin-like growth factor-I in rainbow trout (*Oncorhynchus mykiss*). J Endocrinol 130:87–90

McDevitt V, Kenedy A, Parsons RH (1995) Empty bladder and dehydrated pelvic patch water uptake in *Bufo marinus*: inhibition by captopril. Comp Biochem Physiol 111A:47–50

McDonald IR (1974) Adrenal insufficiency in the red kangaroo (*Megaleia rufa Desm*). J Endocrinol 62:689–690

McDonald IR, Augee ML (1968) Effects of bilateral adrenalectomy in the monotreme *Tachyglossus aculeatus*. Comp Biochem Physiol 27:669–678

McDonald IR, Bradshaw SD (1993) Adrenalectomy and steroid replacement in a small macropodid marsupial, the quokka (*Setonix brachyurus*): metabolic and renal effects. Gen Comp Endocrinol 90:64–77

McDonald IR, Martin IK (1989) Adrenocortical functions in macropodid marsupials. In: Grigg G, Jarman P, Hume I (eds) Kangaroos, wallabies and rat-kangaroos. Surrey Beatty, New South Wales (Australia), pp 265–275

McDonald IR, Than KA, Evans B (1988) Glucocorticoids in the blood of the platypus *Ornithorynchus anatinus*. J Endocrinol 118:407–415

McFarland WN, Munz FW (1965) Regulation of body weight and serum composition by hagfish in various media. Comp Biochem Physiol 14:383–398

McKie AT, Goecke IA, Naftalin RJ (1991) Comparison of fluid absorption by bovine and ovine descending colon in vitro. Am J Physiol 261:G433–G442

McLeese JM, Johnsson J, Huntley FM, Clarke WC, Weisbart M (1994) Seasonal changes in osmoreulation, cortisol, and cortisol receptor activity in the gills of parr/smolt of steelhead trout and steelhead-rainbow trout hybrids, *Oncorhynchus mykiss*. Gen Comp Endocrinol 93:103–113

Medica PA, Bury RB, Luckenbach RA (1980) Drinking and construction of water catchments by the desert tortoise, *Gopherus agassizii*, in the Mojave desert. Herpetologica 36:301–304

Meier S, Donald J (1997) Natriuretic peptide regulation of the kidney and bladder of the amphibian, *Bufo marinus*. In: Kawashima S, Kikuyama S (eds) XIIIth International congress of comparative endocrinology vol 2. Yokohama (Japan), Monduzzi Editore, Bologna, pp 1231–1235

Middler SA, Kleeman CR, Edwards E (1967) Neurohypophysial function in the toad *Bufo marinus*. Gen Comp Endocrinol 14:38–48

Middler SA, Kleeman CR, Edwards E, Brody D (1969) Effect of adenohypophysectomy on salt and water metabolism of the toad *Bufo marinus*: studies in hormonal replacment. Gen Comp Endocrinol 12:290–230

Miller RA, Riddle O (1942) The cytology of the adrenal cortex of normal pigeons and in experimentally induced atrophy and hypertrophy. Am J Physiol 71:311–341

Miller T, Bradshaw SD (1979) Adrenocortical function in a field population of a macropodid marsupial (*Setonix brachyurus*, Quoy and Gaimard). J Endocrinol 82:159–170

Minnich JE (1970) Water and elecrolyte balance of the desert iguana, *Dipsosaurus dorsalis*, in its natural habitat. Comp Biochem Physiol 35:921–923

Minnich JE (1972) Excretion of urate salts by reptiles. Comp Biochem Physiol 41A:535–549

Mitic LL, Anderson JM (1998) Molecular architecture of tight junctions. Annu Rev Physiol 60:112–142

Miwa S, Inui Y (1985) Effects of L-thyroxine and ovine growth hormone on smoltification of amago salmon (*Oncorhychus rhodurus*). Gen Comp Endocrinol 58:436–442

Miwa S, Tagawa M, Inui Y, Hirano T (1988) Thyroxine surge in metamorphosing flounder larvae. Gen Comp Endocrinol 70:158–163

Miyata A, Minamino N, Kangawa K, Matsuo H (1988) Identification of a 29-amino acid natriuretic peptide in chicken heart. Biochem Biophys Res Commun 155:1330–1337

Mommsen TP, Walsh PJ (1989) Evolution of urea synthesis in vertebrates: the piscine connection. Science 243:72–75

Morel A, O'Carroll A-M, Brownstein MJ, Lolait SJ (1992) Molecular cloning and expression of a rat V_{1a} arginine vasopressin receptor. Nature (Lond) 356:523–526

Morel F, Jard S (1963) Inhibition of frog (*Rana esculenta*) antidiuretic action of vasotocin by some analogues. Am J Physiol 204:227–232

Morel F, Jard S (1968) Actions and functions of the neurohypophysial hormones and related peptides in lower vertebrates. In: Berde B (ed) Neurohypophysial hormones and similar peptides. Springer, Berlin Heidelberg New York, pp 655–716

Morgan IJ, Potts WTW (1995) The effects of adrenoreceptor agonists phenylephrine and isoproterenol on the intracellular ion concentrations of branchial epthelial cells of brown trout (*Salmo trutta* L.). J Comp Physiol B 165:458–463

Morley M, Scanes CG, Chadwick A (1981) The effect of ovine prolactin on sodium and water transport across the intestine of the fowl (*Gallus domesticus*). Comp Biochem Physiol 68A:61–66

Morris R (1960) General problems of osmoregulation with special reference to cyclostomes. Symp Zool Soc Lond 1:1–16

Morris R (1965) Studies on salt and water balance in *Myxine glutinosa* L. J Exp Biol 42:359–371

Morris R, Pickering AD (1976) Changes in the ultrastructure of the gills of the river lamprey, *Lampetra fluviatilis* (L.), during the anadromous spawning migration. Cell Tissue Res 173:271–277

Mosimann R, Imboden H, Felix D (1996) The neuronal role of angiotensin II in thirst, sodium appetite, cognition and memory. Biol Rev 71:545–559

Motais R (1970) Effects of actinomycin D on the branchial Na-K-dependent ATPase activity in relation to sodium balance in the eel. Comp Biochem Physiol 34:497–501

Motais R, Maetz J (1964) Action des hormones neurohypophysaire sur les échanges de sodium (mesurés a l'aide du radio-sodium Na24) chez un téléostéen euryhalin: *Platichthys flesus*. Gen Comp Endocrinol 4:210–224

Motais R, Maetz J (1965) Comparaison des échanges de sodium chez un téléostéen euryhalin (le flet) et un téléostéen sténolhalin (le serran) en eaux de mer. Importance relative du tube digestif et de la branchie dans ces échanges. CR Acad Sci (Paris) 261:532–535

Motais R, Maetz J (1967) Arginine vasotocine et évolution de la perméabilité branchiale au sodium cours du passage d'eau douce en eau de mer chez le Flet. J Physiol (Paris) 59: 271

Motais R, Garcia Romeu F, Maetz J (1966) Exchange diffusion effect and euryhalinity in teleosts. J Gen Physiol 50:391–422

Motais R, Isaia J, Rankin JC, Maetz J (1969) Adaptive changes of the water permeability of the teleostean gill epithelium in relation to external salinity. J Exp Biol 51:529–546

Mount D, Delpire E, Gamba G, Hall AE, Poch E, Hoover RS, Hebert SC (1998) The electroneutral cation-chloride cotransporters. J Exp Biol 201:2091–2102

Mullins LJ (1950) Osmotic regulation in fish as studied with radioisotopes. Acta Physiol Scand 21:303–314

Munsick RA (1964) Neurohypophysial hormones of chicken and turkeys. Endocrinology 75: 104–112

Munsick RA (1966) Chromatographic and pharmacologic characterization of the neurohypophysial hormones of an amphibian and a reptile. Endocrinology 78:591–599

Munsick RA, Sawyer WH, Van Dyke HB (1960) Avian neurohypophysial hormones: pharmacological properties and tentative identification. Endocrinology 66:860–871

Munz FW, McFarland WN (1964) Regulatory function of a primitive vertebrate kidney. Comp Biochem Physiol 13:381–400

Murray BR, Hume ID, Dickman CR (1995) Digestive tract characteristics of the spinifex hopping-mouse, *Notomys alexis*, and the sandy inland mouse, *Pseudomys hermannsburgensis*, in relation to diet. Aust Mammal 18:93–97

Murray RW, Potts WTW (1961) The composition of the endolymph, perilymph and other body fluids in elasmobranchs. Comp Biochem Physiol 2:65–75

Murrish DE, Schmidt-Nielsen K (1970a) Exhaled air temperature and water conservation in lizards. Respir Physiol 10:151–158

Murrish DE, Schmidt-Nielsen K (1970b) Water transport in the cloaca of lizards: active or passive? Science 170:324–326

Naftalin RJ, Pedley KC (1999) Regional crypt function in rat large intestine in relation to fluid absorption and growth of the pericryptal sheath. J Physiol (Lond) 514:211–227

Naftalin RJ, Zammit PS, Pedley KC (1999) Regional differences in rat large intestinal crypt function in relation to dehydrating capacity in vitro. J Physiol (Lond) 514:201–210

Nagamatsu S, Chan SJ, Falkmer S, Steiner DF (1991) Evolution of the insulin gene superfamily. Sequence of a preproinsulin-like growth factor cDNA from the Atlantic hagfish. J Biol Chem 266:2397–2402

Nagashima K, Ando M (1994) Characterization of esophageal desalination in the seawater eel, *Anguilla japonica*. J Comp Physiol B 164:47–54

Nagy KA (1972) Water and electrolyte budgets of a free-living desert lizard, *Sauromalus obsesus*. J Comp Physiol 79:39–62

Nagy KA (1988) Seasonal patterns of water and energy balance in desert vertebrates. J Arid Environ 14:201–210

Nagy KA, Gruchacz MJ (1994) Seasonal water and energy metabolism of the desert-dwelling kangaroo rat (*Dipodomys merriami*). Physiol Zool 67:1461–1478

Nakagawa Y, Kosugi H, Miyajima A, Arai K-I, Yokota S (1994) Structure of the gene encoding the α subunit of the human granulocyte-macrophage colony stimulating factor. Implications for the evolution of the cytokine receptor superfamily. J Biol Chem 269:10905–10912

Naray-Féjes-Toth A, Canessa CM, Cleaveland ES, Aldrich G, Féjes-Toth G (1999) sgk is an aldosterone-induced kinase in the renal collecting duct. Effects on epithelial Na^+ channels. J Biol Chem 274:16973–16978

Nechay B, Boyarsky S, Catacutan-Labay P (1968) Rapid migration of urine into the intestine of chickens. Comp Biochem Physiol 26:369–370

265

Newsome AE (1965a) The abundance of red kangaroos, *Megaleia rufa* (Demarest) in central Australia. Aust J Zool 13:269–287

Newsome AE (1965b) The distribution of red kangaroos, *Megaleia rufa* (Demarest), about sources of persistent food and water in central Australia. Aust J Zool 13:289–299

Newsome AE (1971) The ecology of red kangaroos. Aust Zool 16:32–50

Nicoll CS, Mayer GL, Russell SM (1986) Structural features of prolactins and growth hormones that can be related to their biological activities. Endocr Rev 7:169–203

Nielsen S, Terris J, Smith CP, Hediger MA, Ecelbarger CA, Knepper MA (1996) Cellular and subcellular localization of the vasopressin-regulated urea transporter in rat kidney. Proc Natl Acad Sci USA 93:5495–5500

Nilsson S (1984) Innervation and pharmacology of the gills. In: Hoar WS, Randall DJ (eds) Fish physiology, Vol X, Gills Part A. Academic Press, Orlando, pp 185–227

Nilsson S, Sundin L (1998) Gill blood flow control. Comp Biochem Physiol 119A:137–147

Nishimura H (1978) Physiological evolution of the renin-angiotensin system. Jpn Heart J 19: 806–822

Nishimura H (1987) Role of the renin-angiotensin system in osmoregulation. In: Pang PKT, Schreibman MP, Sawyer WH (eds) Vertebrate endocrinology: fundamentals and biomedical implications. Vol 2 Regulation of water and electrolytes. Academic Press, San Diego, pp 157–187

Nishimura H, Imai M (1982) Control of renal function in freshwater and marine teleosts. Fed Proc 41:2355–2360

Nishimura H, Sawyer WH (1976) Vasopressor, diuretic, and natriuretic responses to angiotensins by the American eel, *Anguilla rostrata*. Gen Comp Endocrinol 29:337–348

Nishimura H, Koseki C, Patel TB (1996) Water transport in collecting ducts of Japanese quail. Am J Physiol 271:R1535–R1543

Nishimura H, Qin Z, Shimada T (1997) Evolution of angiotensin receptors and signaling. In: Kawashima S, Kikuyama S (eds) XIIIth International Congress of Comparative Endocrinology vol 2. Yokohama (Japan), Monduzzi Editore, Bologna, pp 1299–1305

Nolly HL, Fasciola JC (1973) The specificity of the renin-angiotensin system through the phylogenetic scale. Comp Biochem Physiol 44A:639–645

Norbury G, Norbury D (1992) The impact of red kangaroos on the rangelands. West Australian (WA) J Agric 33:57–61

Noso T, Nicoll CS, Kawauchi H (1993a) Lungfish prolactin exhibits close tetrapod relationships. Biochim Biophys Acta 1164:159–165

Noso T, Nicoll CS, Polenov AL, Kawauchi H (1993b) The primary structure of sturgeon prolactin: phylogenetic implications. Gen Comp Endocrinol 91:90–95

Novelli A (1936) Lobulo posterior de hipofisis e imbibicion de los batrachios. II. Mecanismo de su accion. Rev Soc Argent Biol 12:163–164

O'Connor WJ, Potts DJ (1969) The external water exchanges of normal laboratory dogs. Q J Exp Physiol 54:244–265

Oddie CJ, Blaine EH, Bradshaw SD, Coghlan JP, Denton DA, Nelson JF, Scoggins BA (1976) Blood corticosteroids in Australian marsupial and placental mammals and one monotreme. J Endocrinol 69:341–348

O'Grady SM, Field M, Nash NT, Rao MC (1985) Atrial natriuretic factor inhibits Na-K-Cl cotransport in teleost intestine. Am J Physiol 249:C531–C534

Oguri M (1964) Rectal glands of marine and fresh-water sharks: comparative histology. Science 144:1151–1152

Ohmart RD (1972) Physiological and ecological observations concerning the salt-secreting nasal glands of the roadrunner. Comp Biochem Physiol 43A:311–316

Oide M (1967) Effects of inhibitors on transport of water and ions in isolated intestine and Na-K ATPase in intestinal mucosa of the eel. Ann Zool Jpn 40:130–135

Oide M, Utida S (1968) Changes in intestinal absorption and renal excretion of water during adaptation to sea-water in Japanese eels. Mar Biol 1:172–177

Okawara Y, Karakida T, Yamaguchi K, Kobayashi H (1985) Diurnal rhythm of water intake and plasma angiotensin II in the Japanese quail (*Coturnix coturnix japonica*). Gen Comp Endocrinol 58:89–92

Olson KR, Duff DW (1992) Cardiovascular and renal effects of eel and rat atrial natriuretic peptide in rainbow trout, *Salmo gairdneri.* J Comp Physiol B 162:408–425

Olson KR, Duff DW (1993) Single-pass gill extraction and tissue distribution of atrial natriuretic peptide in trout. Am J Physiol 265:R124–R131

Ono M, Takayama Y, Rand-Weaver M, Sakata J, Yasunaga Y, Noso T, Kawauchi H (1990) cDNA cloning of somatolactin, a pituitary protein related to growth hormone and prolactin. Proc Natl Acad Sci USA 87:4330–4334

Onstott D, Elde R (1986) Immunohistochemical localization of urotensin I. corticotropin-releasing factor, urotensin II, and serotonin immunoreactivities in the caudal spinal cord of non-teleost fishes. J Comp Neurol 249:205–225

Ophir E, Arieli Y, Raber P, Marder J (2000) The role of β-adrenergic receptors in the cutaneous water evaporation mechanism in the heat-acclimated pigeon (*Columba livia*). Comp Biochem Physiol 125A:63–74

Ortiz RM, Worthy GAJ (2000) Effects of capture on adrenal steroid and vasopressin concentrations in free-ranging bottlenose dolphins (*Tursiops truncatus*). Comp Biochem Physiol 125A:317–324

Ortiz RM, Adams S, Costa DP, Ortiz CL (1996) Plasma vasopressin levels and water conservation in postweaned northern elephant seal pups (*Mirounga angustirostris*). Mar Mammal Sci 12:99–106

Ortiz RM, Worthy GAJ, MacKenzie DS (1998) Osmoregulation in wild and captive West Indian manatees (*Trichechus manatus*). Physiol Zool 71:449–457

Ortiz RM, Worthy GAJ, Byers FM (1999) Estimation of water turnover rates of captive West Indian manatees (*Trichechus manatus*) held in fresh water and salt water. J Exp Biol 202: 33–38

Ortiz RM, Wade CE, Ortiz CL (2000) Prolonged fasting increases the response of the renin-angiotensin-aldosterone system, but not vasopressin levels, in postweaned northern elephant seal pups. Gen Comp Endocrinol 119:217–223

Osawa R, Woodall PF (1992) A comparative study of macroscopic and microscopic dimensions of the intestine of five macropods (Marsupialia: Macropodidae). II. Relationship with feeding habits and fibre content of the diet. Aust J Zool 40, part 1:99–113

O'Shea JE, Bradshaw SD, Stewart T (1993) Renal vasculature and excretory system of the agamid lizard, *Ctenophorus ornatus.* J Morphol 217:287–299

Osmolska H (1979) Nasal salt gland in dinosaurs. Acta Palaeontol Pol 24:99–113

Osono E, Nishimura H (1994) Control of sodium and chloride transport in the thick ascending limb of the avian nephron. Am J Physiol 2:R455–R462

Owens A, Wigham T, Doneen B, Bern HA (1977) Effects of environmental salinity and of hormones on urinary bladder function in the euryhaline teleost, *Gillichthys mirabilis.* Gen Comp Endocrinol 33:526–530

Packard GC (1999) Water relations of chelonian eggs and embryos: is wetter better? Am Zool 39:289–303

Pang PKT, Sawyer WH (1974) Effects of prolactin on hypophysectomized mudpuppies, *Necturus maculosus.* Am J Physiol 226:458–462

Pang PKT, Sawyer WH (1978) Renal and vascular responses of the bullfrog (*Rana catesbeiana*) to mesotocin. Am J Physiol 235:F151–F155

Pang PKT, Griffith RW, Atz JW (1977) Osmoregulation in elasmobranchs. Am Zool 17:365–377

Pang PKT, Uchiyama M, Sawyer WH (1982) Endocrine and neural control of amphibian renal function. Fed Proc 41:2365–2370

Pang PKT, Furspan PB, Sawyer WH (1983) Evolution of neurohypophyseal actions in vertebrates. Am Zool 23:655–662

Parkins WM (1931) An experimental study of bilateral adrenalectomy in the fowl. Anat Rec 51: Suppl 39

Parmalee JT, Renfro JL (1983) Esophageal desalination of seawater in flounder: role of active sodium transport. Am J Physiol 245:R888–R893

Parry G (1961) Osmotic and ionic changes in blood and muscle ofmigrating salmonids. J Exp Biol 38:411–427

Parry G (1966) Osmotic adaptation in fishes. Biol Rev 41:392–444

Parsons RH, Schwartz R (1991) Role of circulation in maintaining Na$^+$ and K$^+$ concentration in pelvic patch in *Rana catesbeiana*. Am J Physiol 261:R686–R689

Pärt P, Wright PA, Wood CM (1998) Urea and water permeability in dogfish (*Squalus acanthias*) gills. Comp Biochem Physiol 119A:117–123

Pärt P, Wood CM, Gilmour KM, Perry SF, Laurent P, Zadunaisky J, Walsh P (1999) Urea and water permeability in the ureotelic gulf toadfish (*Opsanus beta*). J Exp Zool 283:1–12

Payan P, Maetz J (1970) Balance hydrique et minérale chez les élasmobranchs: arguments en faveur d'un contrôle endocrinien. Bull Inf Sci Tech Commiss Energ Atom (France) 146:77–96

Payan P, Maetz J (1971) Balance hydrique chez les elasmobranches: arguments en faveur d'un controle endocrinien. Gen Comp Endocrinol 16:535–554

Payan P, Goldstein L, Forster RP (1973) Gills and kidneys in ureosmotic regulation in euryhaline skates. Am J Physiol 224:367–372

Peaker M (1978) Excretion of potassium from the orbital region in *Testudo carbonaria*: a salt gland in terrestrial tortoises? J Zool (Lond) 184:421–422

Peek WD, Youson JH (1979) Transformation of the interlamellar epthelium of the gills of the anadromous sea lamprey, *Petromyzon marinus* L., during metamorphosis. Can J Zool 57: 1318–1332

Peltonen L, Arieli Y, Pyörnilä A, Marder J (1998) Adaptive changes in the epidermal structure of the heat-acclimated rock pigeon (*Columba livia*): a comparative electron microscopy study. J Morphol 235:17–29

Perlman DF, Goldstein L (1999) Organic osmolyte channels in cell volume regulation in vertebrates. J Exp Zool 283:725–733

Perrott MN, Carrick S, Balment RJ (1991) Pituitary and plasma arginine vasotocin levels in teleost fish. Gen Comp Endocrinol 83:68–74

Perrott MN, Sainsbury RJ, Balment RJ (1993) Peptide hormone-stimulated second messenger production in the teleostean nephron. Gen Comp Endocrinol 89:387–395

Perry SF (1997) The chloride cell: structure and function in the gills of freshwater fishes. Annu Rev Physiol 59:325–347

Perry SF, Payan P, Girard JP (1984) Adrenergic control of branchial chloride transport in the isolated perfused head of the freshwater trout (*Salmo gairdneri*). J Comp Physiol B 154:269–274

Perry SF, Gilmour KM, Wood CM, Pärt P, Laurent P, Walsh PJ (1998) The effects of arginine vasotocin and catecholamines on nitrogen excretion and cardio-respiratory physiology of the gulf toadfish, *Opsanus beta*. J Comp Physiol B 168:461–472

Peter RE, Chang JP (1997) Growth hormone secretion in fish. In: Kawashima S, Kikuyama S (eds) XIIIth International congress of comparative endocrinology vol 1. Yokohama (Japan), Monduzzi Editore, Bologna, pp 915–919

Peterson CC (1996) Anhomeostasis: seasonal water and solute relations in two populations of the desert tortoise (*Gopherus agassizii*) during chronic drought. Physiol Zool 69:1324–1358

Petriella S, Reboreda JC, Otero M, Segura ET (1989) Antidiuretic responses to osmotic cutaneous stimulation in the toad, *Bufo arenarum*. J Comp Physiol B 159:91–95

Pettus D (1958) Water relationships in *Natrix sipedon*. Copeia 1958:207–211

Peyrot A, Ferreri F, Mazzi V, Socino M (1963) Il metabolismo del sodio e del potassio nel *Tretone crestato*. II. Effetti dell'ipofisectomia. Boll Soc Ital Biol Sper 40:220–222

Phillips JG, Chester Jones I (1957) The identity of adrenocortical secretions in lower vertebrates. J Endocrinol 16:iii

Phillips JG, Chester Jones I, Bellamy D (1962a) Biosynthesis of adrenocortical hormones by adrenal glands of lizards and snakes. J Endocrinol 25:233–237

Phillips JG, Chester Jones I, Bellamy D, Greep RO, Day LR, Holmes WN (1962b) Corticosteroids in the blood of *Myxine glutinosa* L. (Atlantic hagfish). Endocrinology 71:329–331

Pic P, Mayer-Gostan N, Maetz J (1974) Branchial effects of epinephrine in the seawater-adapted mullet. I. Water permeability. Am J Physiol 226:698–702

Pickering AD, Heller H (1959) Chromatographic and biological characteristics of fish and frog neurohypophysial extracts. Nature (Lond) 184:1463–1464

Pickering BT (1967) The neurohyophysial hormones of a reptile species, the cobra (*Naja naja*). J Endocrinol 39:285–294

Pickford GE, Grant FB (1967) Serum osmolaity in the coelacanth *Latimeria chalumnae*: urea retention and ion regulation. Science 155:568–570

Pickford GE, Philllips JG (1959) Prolactin, a factor promoting the survival of hypophysectomized killifish in fresh water. Science 130:454–455

Pickford GE, Griffith RW, Torretti J, Hendlez E, Epstein FH (1970a) Branchial reduction and renal stimulation of (Na^+, K^+)-ATPase by prolactin in hypophysectomized killifish in fresh water. Nature (Lond) 228:378–379

Pickford GE, Pang PKT, Weinstein E, Torretti J, Hendler E, Epstein FH (1970b) The response of the hypophysectomized cyprinodont, *Fundulus heteroclitus*, to replacement therapy with cortisol: effects on blood serum and sodium-potassium-activated adenosinetriphosphatase in the gills, kidney and intestinal mucosa. Gen Comp Endocrinol 14:524–534

Pierson PM, Guibbolini ME, Lahlou B (1996) A V_1-type receptor for mediating the neurohypophysial-induced ACTH release in trout pituitary. J Endocrinol 149:109–115

Pilley CM, Wright PA (2000) The mechanisms of urea transport by early life stages of rainbow trout (*Oncorhynchus mykiss*). J Exp Biol 203:3199–3207

Pisam M, Caroff A, Rambourg A (1987) Two types of chloride cells in the gill epithelium of a freshwater-adapted euryhaline fish: *Lebistes reticulatus*; their modification during adaptation to saltwater. Am J Anat 179:40–50

Platzack B, Schaffert C, Hazon N, Conlon JM (1998) Cardiovascular actions of dogfish urotensin I in the dogfish, *Scyliorhinus canicula*. Gen Comp Endocrinol 109:269–275

Potts WTM, Evans DH (1967) Sodium and chloride balance in the killifish *Fundulus heteroclitus*. Biol Bull (Woods Hole) 133:411–425

Potts WTM, Forster MA, Rudy PP, Howells GP (1967) Sodium and water balance in the cichlid teleost *Tilapia mossambiica*. J Exp Biol 47:461–470

Poulos JE, Gower WR, Friedl FE, Vesely DL (1995) Atrial natriuretic peptide gene expression within invertebrate hearts. Gen Comp Endocrinol 100:61–68

Poulsen TL (1965) Countercurrent multipliers in avain kidneys. Science 148:389–391

Poulsen TL, Bartholomew GA (1962) Salt balance in the savannah sparrow. Physiol Zool 35:109–119

Powell DW (1999) Water transport revisited. J Physiol (Lond) 514:1

Prager D, Melmed S (1993) Editorial: insulin and insulin-like growth factor receptors: are there functional distinctions. Endocrinology 132:1419–1420

Prange HD, Schmidt-Nielsen K (1969) Evaporative water loss in snakes. Comp Biochem Physiol 28:973–975

Preest MR, Beuchat CA (1997) Ammonia excretion in humming birds. Nature (Lond) 386:561–562

Preston GM, Carroll TP, Guggino WB, Agre P (1992) Appearance of water channels in *Xenopus* oocytes expressing red cell CHIP28 protein. Science 256:385–387

Price KS (1967) Fluctuations in two osmoregulatory components, urea and sodium chloride, of the clearnose skate, *Raja eglanteria*, Bosc1802. II. Natural variation in the salinity of the external medium. Comp Biochem Physiol 23:77–82

Prinzinger R, Prebmar A, Schleucher E (1991) Body temperature in birds. Comp Biochem Physiol 99A:499–506

Propper C, Hillyard SD, Johnson WE (1995) Central angiotensin II induces thirst related responses in an amphibian. Horm Behav 29:41–52

Pruett SJ, Hoyt DF, Stiffler DF (1991) The allometry of osmotic and ionic regulation in Amphibia with emphasis in intraspecific scaling in larval *Ambystoma tigrinum*. Physiol Zool 64:1173–1199

Prunet P, Boeuf G, Bolton JP, Young G (1989) Smoltification and seawater adaptation in Atlantic salmon (*Salmo salar*): plasma prolactin, growth hormone and thyroid hormones. Gen Comp Endocrinol 74:355–364

Prunet P, Pisam M, Claireaux JP, Boeuf G, Rambourg A (1994) Effects of growth hormone on gill chloride cells in juvenile Atlantic salmon (*Salmo salar*). Am J Physiol 266:R850–R857

Race GJ, Wu HM (1961) Adrenal cortex functional zonation in the whale, *Physeter catadon*. Endocrinology 68:156–158

Rand-Weaver M, Pottinger M, Sumpter JP (1992) Somatolactin, a novel pituitary protein: purification and plasma levels during reproductive maturation in coho salmon. J Endocrinol 133: 393–403

Rand-Weaver M, Pottinger TG, Sumpter JP (1993) Plasma somatolactin concentrations in salmonid fish are elevated by stress. J Endocrinol 138:509–515

Randall DJ, Perry SF (1992) Catecholamines. In: Hoar WS, Randall DJ, Farrell AP (eds) Fish physiology, Vol XII Part B The cardiovascular system. Academic Press, San Diego, pp 255–300

Randall DJ, Wood CM, Perry SF, Bergman HL, Maloiy GMO, Mommsen TP, Wright PA (1989) Urea excretion as a strategy for survival in a fish living in a very alkaline environment. Nature (Lond) 337:165–166

Rankin JC, Maetz J (1971) A perfused telostean gill preparation: vascular actions of neurohypophysial hormones and catecholamines. J Endocrinol 51:621–635

Redei E, Hilderbrand H, Aird F (1995) Corticotropin-release-inhibiting factor is preprothyrotropin-releasing hormone-(178–199). Endocrinology 136:3557–3563

Reid IA (1971) Renin secretion in a monotreme (*Tachyglossus aculeatus*). Comp Biochem Physiol 40A:249–255

Reid IA, McDonald IA (1969) The renin-angiotensin system in a marsupial (*Trichosurus vulpecula*). J Endocrinol 44:231–240

Reid IA, McDonald IR (1968) Bilateral adrenalectomy and steroid replacement in the marsupial *Trichosurus vulpecula*. Comp Biochem Physiol 26:613–625

Reid SG, Perry SF (1994) Storage and differential release of catecholamines in rainbow trout (*Oncorhynchus mykiss*) and American eel (*Anguilla rostrata*). Physiol Zool 67:216–237

Reilly RF, Ellison DH (2000) Mammalian distal tubule: physiology, pathophysiology, and molecular biology. Physiol Rev 80:277–313

Reina RD, Cooper PD (2000) Control of salt gland activity in the hatchling green sea turtle, *Chelonia mydas*. J Comp Physiol B 170:27–35

Reinecke M (1989) Atrial natriuretic peptide – localization, structure, function, and phylogeny. In: Holmgren S (ed) The comparative physiology of regulatory peptides. Chapman and Hall, London, pp 3–33

Reinke EE, Chadwick CS (1939) Inducing land stage of *Triturus viridescens* to assume water habitat by pituitary implantations. Proc Soc Exp Biol Med (NY) 40:691–693

Renfro JL (1975) Water and ion transport by the urinary bladder of the teleost, *Pseudopleuronectes americanus*. Am J Physiol 228:52–61

Renkin EM, Tucker VT (1996) Atrial natriuretic peptide as a regulator of transvascular fluid balance. News Physiol Sci 11:138–143

Rey P (1937) Recherches expérimentales sur l'économie de l'eau chez les batrachiens. Physiol Chem Biol 13:1081–1144

Rice GE (1982) Plasma arginine vasotocin concentrations in the lizard *Varanus gouldii* (Gray) following water loading, salt loading, and dehydration. Gen Comp Endocrinol 47:1–6

Rice GE, Skadhauge E (1982) Caecal water and electrolyte absorption and the effects of acetate and glucose, in dehydrated, low-NaCl diet hens. J Comp Physiol B 147:61–64

Rice GE, Bradshaw SD, Prendergast FJ (1982) The effects of bilateral adrenalectomy on renal function in the lizard, *Varanus gouldii* (Gray). Gen Comp Endocrinol 47:182–189

Rice GE, Árnason SS, Arad Z, Skadhauge E (1985) Plasma concentrations of arginine vasotocin, prolactin, aldosterone and corticosterone in relation to oviposition and dietary NaCl in the domestic fowl. Comp Biochem Physiol 81A:769–777

Richards AN, Schmidt CF (1924) A description of the glomerular circulation in the frogs kidney and observations concerning the action of adrenalin and various other substances on it. Am J Physiol 71:178–189

Riegel JA (1998) Analysis of fluid dynamics in perfused glomeruli of the hagfish *Eptatretus stouti* (Lockington). J Exp Biol 201:3097–3104

Riegel JA (1999) Secretion of primary urine by glomeruli of hagfish kidney. J Exp Biol 202: 947–955

Roberts J (1991a) Effects of water deprivation on renal function and plasma arginine vasotocin in the feral chicken, *Gallus gallus* (Phasianidae). Aust J Zool 39:439–446

Roberts J, Dantzler WH (1989) Glomerular filtration rate in concious unrestrained starlings under dehydration. Am J Physiol 256:R836–R839

Roberts JD (1981) Terrestrial breeding in the Australian leptodactylid frog *Myobatrachus gouldii* (Gray). Aust Wildl Res 8:451–462

Roberts JD (1984) Terrestrial egg deposition and direct development in *Arenophryne rotunda* Tyler, a myobatrachid frog from coastal sand dunes at Shark Bay, W.A. Aust Wildl Res 11: 191–200

Roberts JR (1991b) Renal function and plasma arginine vasotocin during water deprivation in an Australian parrot, the galah (*Cacatua roseicapilla*). J Comp Physiol B 161:620–625

Roberts JS, Schmidt-Nielsen B (1966) Renal ultrastructure and excretion of salt and water by three terrestrial lizards. Am J Physiol 211:476–486

Robertshaw D, Taylor CR (1969) A comparison of sweat gland activity in eight species of East African bovids. J Physiol (Lond) 203:135–143

Robertson JD (1954) The chemical composition of the blood of some aquatic chordates, including members of the Tunicata, Cyclostomata and Osteichthyes. J Exp Biol 31:424–442

Robertson JD (1957) The habitat of early vertebrates. Biol Rev 32:156–187

Robertson JD (1965) Studies on the chemical compostion of muscle tissue. III. The mantle muscle of cephalopod molluscs. J Exp Biol 42:153–175

Robertson JD (1975) Osmotic constituents of the blood plasma and parietal muscle of *Squalus acanthias* L. Biol Bull (Woods Hole) 148:303–319

Robertson JD (1976) Chemical composition of the body fluids and muscle of the hagfish *Myxine glutinosa* and the rabbit-fish *Chimaera monstrosa*. J Zool (Lond) 178:261–277

Robertson SI, Smith EN (1982) Evaporative water loss in the spiny soft-shelled turtle *Trionyx spiniferus*. Physiol Zool 55:124–129

Robinson JD (1995) Steps in the Na^+-K^+ pump and Na^+-K^+-ATPase (1939–1962). News Physiol Sci 10:184–188

Robinson KW, MacFarlane WV (1957) Plasma antidiuretic activity in marsupials during exposure to heat. Endocrinology 60:679–680

Robinson B, Kare MR, Beauchamp GK (1980) Comparative aspects of salt preference and intake in birds. In: Fregly MJ, Bernard RA (eds) Biological and behavioral aspects of salt intake. Academic Press, New York, pp 69–81

Romer AS (1955) Vertebrate paleontology. Chicago University Press, Chicago

Romero MF, Boron WF (1999) Electrogenic Na^+/HCO_3^- cotransporters: cloning and physiology. Annu Rev Physiol 61:699–723

Rooke IJ, Bradshaw SD, Langworthy RA (1983) Aspects of the water, electrolyte and carbohydrate physiology of the silvereye, *Zosterops lateralis* (Aves). Aust J Zool 31:695–704

Rooke IJ, Bradshaw SD, Langworthy RA, Tom JA (1986) Annual cycle of physiological stress and condition of the silvereye, *Zosterops lateralis* (Aves). Aust J Zool 34:493–501

Rosenberg J, Hurwitz S (1987) Concentration of adrenocortical hormones in relation to cation homeostasis in birds. Am J Physiol 253:R20–R24

Rosenberg J, Pines M, Hurwitz S (1988) Regulation of aldosterone secretion by avian adrenocortical cells. J Endocrinol 118:447–453

Rosenbloom AA, Fisher DA (1974) Radioimmunoassay of arginine vasotocin. Endocrinology 95: 1726–1732

Roth JJ (1973) Vascular supply to the ventral pelvic regions of anurans as related to water balance. J Morphol 140:443–460

Rouillé Y, Chauvet M-T, Chauvet J, Acher J (1988) Dual duplication of neurohypophysial hormones in an Australian marsupial: mesotocin, oxytocin, lysine vasopressin and arginine vasopressin in a single gland of the Northern bandicoot (*Isoodon macrourus*). Biochem Biophys Res Commun 154:346–350

Rubal A, Haim A, Chosniak I (1995) Resting metabolic rates and daily energy intake in desert and non-desert murid rodents. Comp Biochem Physiol 112A:511–515

Rudy PP, Wagner RC (1970) Water permeability of the Pacific haghish *Polistotrema stouti* and the staghorn sculpin *Leptocottus armatus*. Comp Biochem Physiol 34:399–403

Ruibal R (1962) The adaptive value of bladder water in the toad, Bufo cognatus. Physiol Zool 35:218–223

Russell JM (2000) Sodium-potassium-cotransport. Physiol Rev 80:211–276

Sakaguchi H, Takei Y (1998) Characterization of C-type natriuretic peptide receptors in the gill of dogfish, *Triakis scyllia*. J Endocrinol 156:127–134

Sakaguchi H, Suzuki H, Hagiwara H, Kaiya H, Takei Y, Ito M, Shibabe S, Hirose S (1996) Whole-body autoradiography and microautoradiography in eels after intra-arterial administration of 125 I-labeled eel ANP. Am J Physiol 271:R926–R935

Sakamoto T, Hirano T (1993) Expression of insulin-like growth factor I gene in osmoregulatory organs during seawater adaptation in the salmonid fish: possible mode of osmoregulatory action of growth hormone. Proc Natl Acad Sci USA 90:1912–1916

Sakamoto T, Ogasawara T, Hirano T (1990) Growth hormone kinetics during adaptation to hyperosmotic environment in rainbow trout. J Comp Physiol B 160:1–6

Sakamoto T, Shepherd BS, Madsen SS, Nishioka RS, Siharath K, Rickman NH, Bern HA, Grau EG (1997) Osmoregulatory actions of growth hormone and prolactin in an advanced teleost. Gen Comp Endocrinol 106:95–101

Sakata J-I, Kangawa K, Matsuo H (1988) Identification of new atrial natriuretic peptides in frog heart. Biochem Biophys Res Commun 155:1388–1345

Saltiel AR (1996) Diverse signalling pathaways in the cellular actions of insulin. Am J Physiol 270:E375–E385

Samson WK (1998) Proadrenomeduliin-derived peptides. Front Neuroendocrinol 19:100–127

Samson WK (1999) Adrenomedullin and the control of fluid and electrolyte homeostasis. Annu Rev Physiol 61:363–389

Samson WK, Resch ZT, Murphy TC, Vargas TT, Schell DA (1999) Adrenomedullin: is there physiological relevance in the pathology and pharmacology? News Physiol Sci 14:255–259

Sandor T, Lamoureux J, Lanthier A (1964) Adrenocortical function in reptiles. The in vitro biosynthesis of adrenal cortical steroids by adrenal slices of two common North American turtles, the slider turtle (*Pseudemys scripta elegans*) and the painted turtle (*Chrysemys picta picta*). Steroids 4:213–227

Sandor T, DiBattista JA, Mehdi AZ (1984) Glucocorticoid receptors in the gill tissue of fish. Gen Comp Endocrinol 53:353–364

Sandra O, Le Rouzic P, Cauty C, Edery M, Prunet P (2000) Expression of the prolactin receptor (tiPRL-R) gene in tilapia *Oreochromis niloticus*: tissue distribution and cellular localization in osmoregulatory organs. J Mol Endocrinol 24:215–224

Sands JM, Timmer RT, Gunn RB (1997) Urea transporters in kidney and erythrocytes. Am J Physiol 273:F321–F339

Sardet C, Pisam M, Maetz J (1979) The surface epithelium of teleostean fish gills. Cellular and junctional adaptation of the chloride cell in relation to salt adaptation. J Cell Biol 80:96–117

Sardet C, Franchi A, Pouyssegur J (1989) Molecular cloning, primary structure, and expression of the human growth factor-activatable Na$^+$/H$^+$ antiporter. Cell 56:271–280

Saunders RL, McCormick SD, Henderson EB, Eales JG, Johnson CE (1985) The effect of orally administered 3, 5, 3^1-triiodo-thyronine on growth and salinity tolerance in Atlantic salmon (*Salmo salar*). Aquaculture 45:143–156

Sawyer WH (1951) Effects of posterior pituitary extracts on permeability of frog skin to water. Am J Physiol 164:44–48

Sawyer WH (1957) Increased renal reabsorption of osmotically free water by the toad (*Bufo marinus*) in response to neurohypophysial hormones. Am J Physiol 189:564–568

Sawyer WH (1960) Increased water permeability of the bullfrog (*Rana catesbeiana*) bladder in vitro in response to synthetic oxytocin and arginine vasotocin and to neurohypophysial extracts from non-mammalian vertebrates. Endocrinology 66:112–120

Sawyer WH (1966) Diuretic and natriuretic responses of lungfish (*Protopterus aethiopicus*) to arginine vasotocin. Am J Physiol 210:191–197

Sawyer WH (1968) Phylogenetic aspects of neurohypophysial hormones. In: Berde B (ed) Neurohypophysial hormones and similar polypeptides. Springer, Berlin Heidelberg New York, pp 717–747

Sawyer WH, Sawyer MK (1952) Adaptive responses to neurohypophysial fractions in vertebrates. Physiol Zool 25:84–98

Sawyer WH, Schisgall RM (1956) Increased permeability of the frog bladder to water in response to dehydration and neurohypophysial extracts. Am J Physiol 187:312–314

Sawyer WH, Munsick RA, Van Dyke HB (1959) Pharmacological evidence for the presence of arginine vasotocin and oxytocin in neurohypophysial extracts from cold blooded vertebrates. Nature (Lond) 184:1464–1465

Sawyer WH, Munsick RA, Van Dyke HB (1960) Pharmacological characteristics of neurohypophysial hormones from a marsupial (*Didelphys virginiana*) and a montreme (*Tachyglossus (Echidna) aculeatus*). Endocrinology 67:137–138

Sawyer WH, Munsick RA, Van Dyke HB (1961) Evidence for the presence of arginine vasotocin (8-arginine oxytocin) and oxytocin in neurohypophysial extracts from amphibians and reptiles. Gen Comp Endocrinol 1:30–36

Sawyer WH, Valtin H, Sokol HW (1964) Neurohypophysial principles in rats with familial diabetes insipidus (Brattleboro strain). Endocrinology 74:153–155

Sawyer WH, Blair-West JR, Simpson PA, Sawyer MK (1976) Renal responses of Australian lungfish to vasotocin, angiotensin II, and NaCl infusion. Am J Physiol 231:593–602

Sawyer WH, Uchiyama M, Pang PKT (1982) Control of renal function in lungfishes. Fed Proc 41:2361–2364

Scheide JJ, Zadunaisky JA (1988) Effect of atriopeptin II on isolated opercular epithelium of *Fundulus heteroclitus*. Am J Physiol 254:R27–R32

Schlessinger J (2000) Cell signaling by receptor tyrosine kinases. Cell 103:211–225

Schmid HA, Simon E (1996) Vasotocin acts as a dipsogen in ducks at concentrations stimulating subfornical organ neurons in vitro. J Comp Physiol B 165:615–621

Schmid WD (1965) Some aspects of the water economies of nine species of amphibians. Ecology 46:261–269

Schmid WD, Barden RE (1965) Water permeability and lipid content of amphibian skin. Comp Biochem Physiol 15:423–427

Schmidt-Nielsen B (1964) Organ systems in adaptation: the excretory system. In: Handbook of physiology. Sect 4. Adaptation to the environment. American Physiological Society, Washington, pp 215–243

Schmidt-Nielsen B, Skadhauge E (1967) Function of the excretory system of the crocodile (*Crocodylus acutus*). Am J Physiol 212:973–980

Schmidt-Nielsen K (1960) The salt secreting glands of marine birds. Circulation 21:955–967

Schmidt-Nielsen K (1964a) Desert animals. Physiological problems of heat and water. Clarendon Press, Oxford

Schmidt-Nielsen K (1964b) Terrestrial animals in dry heat: desert rodents: In: Dill DB, Adolph EF, Wilbur CG (eds) Handbook of physiology Sect 4. Adaptation to the environment. American Physiological Society, Washington, pp 493–507

Schmidt-Nielsen K, Bentley PJ (1966) Desert tortoise *Gopherus agassizii*: cutaneous water loss. Science 154:911

Schmidt-Nielsen K, Fänge R (1958) Salt glands in marine reptiles. Nature (Lond) 182:783–785

Schmidt-Nielsen K, Kim YT (1964) The effects of salt intake on the size and function of the salt gland of ducks. Auk 81:160–172

Schmidt-Nielsen K, Lee P (1962) Kidney function in the crab-eating frog (*Rana cancrivora*). J Exp Biol 39:167–177

Schmidt-Nielsen K, Schmidt-Nielsen B, Jarnum SA, Houpt TR (1957) Body temperature of the camel and its relation to water economy. Am J Physiol 188:103–112

Schmidt-Nielsen K, Jørgensen CB, Osaki H (1958) Extrarenal salt excretion in birds. Am J Physiol 193:101–107

Schmidt-Nielsen K, Borut A, Lee P, Crawford EC (1963) Nasal salt excretion and the possible function of the cloaca in water conservation. Science 142:1300–1301

Schmidt-Nielsen K, Crawford EC, Bentley PJ (1966a) Discontinuous respiration in the lizard *Sauromalus obesus*. Fed Proc 25:506

Schmidt-Nielsen K, Dawson TJ, Crawford EC (1966b) Temperature regulation in the echidna (*Tachyglossus aculeatus*). J Cell Comp Physiol 67:63–71

Schmidt-Nielsen K, Crawford EC, Newsome AE, Rawson KS, Hamel HT (1967) Metabolic rate of camels: effect of body temperature and dehydration. Am J Physiol 212:341–346

Schmidt-Nielsen K, Hainsworth FR, Murrish DE (1970) Counter-current heat exchange in the respiratory passages: effect on water and heat balance. Respir Physiol 9:263–276

Schmück R, Kobelt F, Linsenmair KE (1988) Adaptations of the reed frog *Hyperolius viridiflavus* (Amphibia, Anura, Hyperoliidae) to its arid environment. V. Iridophores and nitrogen metabolism. J Comp Physiol B 158:537–546

Schnermann J (1998) Juxtaglomerular cell complex in the regulation of renal salt excretion. Am J Physiol 274:R263–R279

Schofield JP, Jones DSC, Forrest JN (1991) Identification of C-type natriuretic peptide in heart of spiny dogfish shark (*Squalus acanthias*). Am J Physiol 261:F734–F739

Schreiber AM, Specker JL (1999) Metamorphosis in the summer flounder, *Paralichthys dentatus*: changes in gill mitochondria-rich cells. J Exp Biol 202:2475–2494

Schreiber AM, Specker JL (2000) Metamorphosis in the summer flounder, *Paralichthys dentatus*; thyroidal status influences gill mitochondria-rich cells. Gen Comp Endocrinol 117: 238–250

Schreibman MP, Kallman KD (1969) The effect of hypophysectomy on freshwater survival in teleosts of the order Antherinformes. Gen Comp Endocrinol 13:27–38

Schultheiss H (1977) The hormonal regulation of urea excretion in the Mexican axolotl (*Ambystoma mexicanum* Cope). Gen Comp Endocrinol 31:45–52

Schultz SG (1997) Pump-leak parallelism in sodium-absorbing epithelia: The role of ATP-regulated potassium channels. J Exp Zool 279:476–483

Schütz H, Gerstberger R (1990) Atrial natriuretic factor controls salt gland secretion in the Pekin duck (*Anas platyrhynchos*) through interaction with high affinity receptors. Endocrinology 127:1718–1726

Schwabl H, Bairlein F, Gwinner E (1991) Basal and stress-induced corticosterone levels in garden warblers, *Sylvia borin*, during migration. J Comp Physiol B 161:576–580

Seidel ME (1978) Terrestrial dormancy in the turtle *Kinosternon flavescens*; respiratory metabolism and dehydration. Comp Biochem Physiol 61A:1–4

Seidelin M, Madsen SS (1999) Endocrine control of Na$^+$,K$^+$-ATPase and chloride cell development in brown trout (*Salmo trutta*): interaction of insulin-like growth factor-I with prolactin and growth hormone. J Endocrinol 162:127–135

Seidelin M, Madsen SS, Byrialsen A, Kristiansen K (1999) Effects of insulin-like growth factor-I and cortisol on Na$^+$,K$^+$-ATPase expression in osmoregulatory tissues of brown trout (*Salmo trutta*). Gen Comp Endocrinol 113:331–342

Sernia C (1980) Physiology of the adrenal cortex in monotremes. In: Schmidt-Nielsen K, Bolis L, Taylor CR (eds) Comparative physiology: primitive mammals. Cambridge University Press, Cambridge, pp 303–315

Serventy DL (1971) Biology of desert birds. In: Farner DS, King JR, Parkes KC (eds) Avian biology, vol 1. Academic Press, London, pp 287–339

Sharp PJ (1997) Comparative regulation of prolactin secretion. In: Kawashima S, Kikuyama S (eds) XIIIth International congress of comparative endocrinology, vol 2. Yokohama (Japan), Monduzzi Editore, Bologna, pp 951–955

Sharratt BM, Chester Jones I, Bellamy D (1964a) Water and electrolyte composition of the body and renal function of the eel (*Anguilla anguilla* L.). Comp Biochem Physiol 11:9–18

Sharratt BM, Chester Jones I, Bellamy D (1964b) Adaptation of the silver eel (*Anguilla anguilla* L.) to seawater and to artificial media together with observations on the role of the gut. Comp Biochem Physiol 11:19–30

Sheppard DN, Welsh MJ (1999) Structure and function of the CFTR chloride channel. Physiol Rev 79 Suppl:S23–S45

Shigaev A, Asher C, Latter H, Garty H, Reuveny E (2000) Regulation of sgk by aldosterone and its effects in the epithelial Na$^+$ channel. Am J Physiol 278:F613–F619

Shimkets RA, Lifton R, Canessa CM (1998) In vivo phosphorylation of the epithelial sodium channel. Proc Natl Acad Sci USA 95:3301–3305

Shirley HV, Nalbandov AV (1956) Effects of neurohypophysectomy in domestic chickens. Endocrinology 58:477–483

Shoemaker VH (1964) The effects of dehydration on electrolyte concentrations in a toad, *Bufo marinus*. Comp Biochem Physiol 13:261–271

Shoemaker VH (1988) Physiological ecology of amphibians in arid environments. J Arid Environ 14:145–153

Shoemaker VH, Waring H (1968) Effect of hypothalamic lesions on the water-balance response of a toad (*Bufo marinus*). Comp Biochem Physiol 24:47–54

Shoemaker VH, Licht P, Dawson WR (1966) Effects of temperature on kidney function in the lizard *Tiliqua rugosa*. Physiol Zool 39:244–252

Shoemaker VH, Balding D, Ruibal R, McClanahan LL (1972a) Uricotelism and low evaporative water loss in a South American frog. Science 175:1018–1020

Shoemaker VH, Nagy KA, Bradshaw SD (1972b) Studies on the control of electrolyte excretion by the nasal gland of the lizard *Dipsosaurus dorsalis*. Comp Biochem Physiol 42A:749–757

Shpun S, Katz U (1995) Renal function at steady state in a toad (*Bufo viridis*) acclimated in hyperosmoric NaCl and urea solutions. J Comp Physiol B 164:646–652

Shpun S, Katz U (1999) Renal response of euryhaline toad (*Bufo viridis*) to acute immersion in tap water, NaCl, or urea solutions. Physiol Biochem Zool 72:227–237

Shrimpton JM, McCormick SD (1999) Responsiveness of gill Na$^+$/K$^+$-ATPase to cortisol is related to gill corticosteroid receptor concentration in juvenile rainbow trout. J Exp Biol 202:987–995

Shrimpton JM, McCormick SD (1999) Regulation of gill cytosolic corticosteroid receptors in juvenile Atlantic salmon: interaction effects of growth hormone with prolactin and triiodothyronine. Gen Comp Endocrinol 112:262–274

Shuttleworth TJ, Hildebrandt J-P (1999) Vertebrate salt glands: short- and long-term regulation of function. J Exp Zool 283:689–701

Shuttleworth TJ, Thompson JL, Dantzler WH (1987) Potassium secretion by the nasal salt glands of the desert lizard, *Sauromalus obesus*. Am J Physiol 253:R83–R90

Silva P, Solomon R, Spokes K, Epstein FH (1977) Ouabain inhibition of gill Na-K-ATPase: relationship to active chloride transport. J Morphol 199:419–426

Silva P, Solomon RJ, Epstein FH (1999a) Transport mechanisms that mediate the secretion of chloride by the rectal gland of *Squalus acanthias*. J Exp Zool 279:504–508

Silva P, Solomon RJ, Epstein FN (1999b) Mode of activation of salt secretion by C-type natriuretic peptide in the shark rectal gland. Am J Physiol 277:R1725–R1732

Silveira PF, Schiripa LN, Carmona E, Picarelli ZP (1992) Circulating vasotocin in the snake *Bothrops jararaca*. Comp Biochem Physiol 103A:59–64

Silver RB, Soleimani M (1999) H$^+$-K$^+$-ATPases: regulation and role in pathological states. Am J Physiol 276:F799–F811

Simon E, Gray DA (1991) Control of renal handling of potassium loads in ducks with active salt glands. Am J Physiol 261:R231–R238

Simpson P, Blair-West JR (1971) Renin levels in the kangaroo, the wombat and other marsupial species. J Endocrinol 51:79–90

Singer SJ, Nicolson GL (1972) The fluid mosaic model of the structure of cell membranes. Science 175:720–731

Skadhauge K (1964) Effects of unilateral infusion of arginine-vasotocin into the portal circulation of the avian kidney. Acta Endocrinol 47:321–330

Skadhauge E (1967) In vivo perfusion studies of the cloacal water and electrolyte resorption in the fowl (*Gallus domesticus*). Comp Biochem Physiol 24:7–18

Skadhauge E (1968) The cloacal storage of water in the rooster. Comp Biochem Physiol 24:7–18

Skadhauge E (1969) The mechanism of salt and water absorption in the intestine of the eel (*Anguilla anguilla*) adapted to waters of various salinities. J Physiol (Lond) 204:135–158

Skadhauge E (1974) Renal concentrating ability in selected West Australian birds. J Exp Biol 61:269–276

Skadhauge E (1981) Osmoregulation in birds. Springer, Berlin Heidelberg New York

Skadhauge E, Duvdevani I (1977) Cloacal absorption of NaCl and water in the lizard *Agama stellio*. Comp Biochem Physiol 56A:275–279

Skadhauge E, Maetz J (1967) Etude in vivo de l'absorption intestinale d'eau et d'électrolytes chez *Anguilla anguilla* adapté à des milieux de salinité diverses. CR Acad Sci (Paris) 265:347–350

Skadhauge E, Maloney SK, Dawson TJ (1991) Osmotic adaptation of the emu (*Dromaius novaehollandiae*). J Comp Physiol B 161:173–178

Skalstad I, Norday ES (2000) Experimental evidence of seawater drinking in juvenile hooded (*Cystophora cristata*) and harp seals (*Phoca groenlandica*). J Comp Physiol B 170:393–401

Skog EB, Folkow LP (1994) Nasal heat and water exchange is not an effector mechanism for water balance regulation in grey seals. Acta Physiol Scand 151:233–240

Smith CP, Wright PA (1999) Molecular characterization of an elasmobranch urea transporter. Am J Physiol 276:R622–R626

Smith DC (1956) The role of endocrine organs in the salinity tolerance of trout. Mem Soc Endocrinol 5:83–98

Smith HW (1930a) Metabolism of the lungfish *Protopterus aethiopicus*. J Biol Chem 88:97–130

Smith HW (1930b) The absorption and secretion of water and salts by marine teleosts. Am J Physiol 93:480–505

Smith HW (1931) The absorption and secretion of salts by the elasmobranch fishes. II. Marine elasmobranchs. Am J Physiol 98:296–310

Smith HW (1936) The retention and physiological role of urea in the Elasmobranchii. Biol Rev 11:49–82

Smith HW (1951) The kidney. Oxford University Press, New York

Smith MW (1964) The in vitro absorption of water and solutes from the intestine of the goldfish, Carassius auratus. J Physiol (Lond) 175:38–49

Smith MW (1967) Influences of temperature acclimatization on the temperature-dependence and ouabain-sensitivity of goldfish intestinal adenosine triphosphatase. Biochem J 105:65–71

Smith NF, Eddy FB, Struthers AD, Talbot C (1991) Renin, atrial natriuretic peptide and blood plasma ions in parr and smolts of Atlantic salmon, *Salmo salar* L., and rainbow trout, *Oncorhynchus mykiss* (Walbaum), in fresh water and after short-term exposure to sea water. J Exp Biol 157:63–74

Sokabe H, Mizogami S, Murase T, Sakai F (1966) Renin and euryhalinity in the Japanese eel, *Anguilla japonica*. Nature (Lond) 212:952–953

Solomon R, Taylor M, Dorsey D, Silva P, Epstein FH (1985) Atriopeptin stimulation of rectal gland function in *Squalus acanthias*. Am J Physiol 249:R348–R354

Solomon R, Protter A, McEnroe G, Porter JG, Silva P (1992) C-type natriuretic peptides stimulate chloride secretion in the rectal gland of *Squalus acanthias*. Am J Physiol 262:R707–R711

Specker JL, Kishida M, Huang L, King DS, Nagahama Y, Ueda H, Anderson TR (1993) Immunocytochemical and immunogold localization of two prolactin isoforms in the same pituitary cells in the same granules in tilapia (*Oreochromis mossambicus*). Gen Comp Endocrinol 89:29–38

Sperber I (1944) Studies on the mammalian kidney. Zool Bidr Upp 22:249–431

Spight TM (1967a) The water economy of salamanders: exchange of water with the soil. Biol Bull (Woods Hole) 132:126–132

Spight TM (1967b) The water economy of salamanders; water uptake after dehydration. Comp Biochem Physiol 20:767–771

Spight TM (1968) The water economy of salamanders: evaporative water loss. Physiol Zool 41:195–203

Spindler B, Verrey F (1999) Aldosterone action: induction of p21ras and fra-2 and transcription-independent decrease in myc, jun, and fos. Am J Physiol 276:C1154–C1161

Spotila JR, Berman EN (1976) Determination of skin resistance and the role of the skin in controlling water loss in amphibians and reptiles. Comp Biochem Physiol 55A:407–411

Spring KR (1999) Epithelial fluid transport – a century of investigation. News Physiol Sci 14:92–98

Stallone JN, Braun EJ (1985) Contribution of glomerular and tubular mechanisms to antidiuresis in concious domestic fowl. Am J Physiol 249:F842–F850

Stallone JN, Braun EJ (1986a) Osmotic and volemic regulation of plasma arginine vasotocin in concious domestic fowl. Am J Physiol 250:R644–R657

Stallone JN, Braun EJ (1986b) Regulation of plasma arginine vasotocin in concious water-deprived domestic fowl. Am J Physiol 250:R658–R664

Stallone JN, Braun EJ (1988) Regulation of plasma antidiuretic hormone in the dehydrated kangaroo rat (*Dipodomys spectabilis* M.). Gen Comp Endocrinol 69:119–127

Steggerda FR (1937) Comparative study of water metabolism in amphibians injected with pituitrin. Proc Soc Exp Biol Med (NY) 36:103–106

Stevens CE, Hume ID (1995) Comparative physiology of the vertebrate digestive system. 2nd edn. Cambridge University Press, Cambridge

Stewart AD (1968) Genetic variation in the neurohypophysial hormones of the mouse (*Mus musculus*). J Endocrinol 41:xix–xx

Stiffler DF, Roach SC, Pruett SJ (1984) A comparison of the responses of the amphibian kidney to mesotocin, isotocin, and oxytocin. Physiol Zool 57:63–69

Stiffler DF, De Ruyter ML, Hanson PB, Marshall M (1986) Interrenal function in larval *Ambystoma tigrinum*. 1. Responses to alterations in external electrolye concentrations. Gen Comp Endocrinol 62:290–297

Stiffler DF, DeRuyter ML, Talbot CR (1990) Osmotic and ionic regulation in the aquatic caecilian *Typhlonectes compressicauda* and the terrestrial caecilian *Icthyophis kohtaoensis*. Physiol Zool 63:649–668

Stille WT (1952) The nocturnal amphibian fauna of the southern Lake Michigan beach. Ecology 33:149–162

Stoicovici F, Pora EA (1951) Comportarea la variatiuni de salinitate Rome: Acad Repub Pop Rom pp Fil Cluj 2:159–219

Stolte H, Schmidt-Nielsen B, Davis L (1977) Single nephron function in the kidney of the lizard *Sceloporus cyanogenys*. Zool Jahrb Physiol 81:S219–244

Storey KB, Storey JM (1990) Metabolic rate depression and biochemical adaptation to anaerobiosis, hibernation and estivation. Q Rev Biol 65:145–141

Strahan R (1959) Pituitary hormones of *Myxine* and *Lampetra*. In: Trans Asia and Oceania Reg. Congress Endocrin, vol 1, No 24 Kyoto (Japan)

Subash Peter MC, Lock RAC, Wendelaar Bonga SE (2000) Evidence for an osmoregulatory role of thyroid hormones in the freshwater Mozambique tilapia, *Oreochromis mossambicus*. Gen Comp Endocrinol 120:157–167

Suzuki H, Takahashi A, Takei Y (1992) Different molecular forms of C-type natriuretic peptide from the brain and heart of an elasmobranch, *Triakis scyllia*. J Endocrinol 135:317–323

Suzuki H, Kubokawa K, Nagasawa H, Urano A (1995) Sequence analysis of vasotocin cDNAs of the lamprey, *Lampetra japonica*, and the hagfish *Eptatretus burgeri*: evolution of cyclostome vasotocin precursors. J Mol Endocrinol 14:67–77

Suzuki R, Togashi K, Ando M, Takei Y (1994) Distribution and molecular forms of C-type natriuretic peptide in plasma and tissue of a dogfish, *Triakis scyllia*. Gen Comp Endocrinol 96:378–384

Sweeting RM, McKeown BA (1987) Growth hormone and seawater adaptation in coho salmon, *Oncorhynchus kisutch*. Comp Biochem Physiol 88A:147–151

Szarski H (1962) The origin of the Amphibia. Q Rev Biol 35:189–241

Takada M, Shomazaki S (1988) Effect of prolactin on transcutaneous Na transport in the Japanese newt, *Cynops pyrrogaster*. Gen Comp Endocrinol 69:141–145

Takei Y (1994) Structure and function of natriuretic peptides in vertebrates. In: Davey KG, Peter RE, Tobe SS (eds) Perspectives in comparative endocrinology. National Research Council of Canada, Ottawa, pp 155–165

Takei Y, Balment RJ (1993) Natriuretic factors in non-mammalian vertebrates. In: Brown A, Rankin C, Balment RJ (eds) New insights into vertebrate kidney function. Cambridge University Press, Cambridge, pp 351–385

Takei Y, Kaiya H (1998) Antidiuretic effect of eel ANP infused at physiological doses in seawater-adapted eels, *Anguilla japonica*. Zool Sci 15:399–404

Takei Y, Hasegawa Y, Watanabe TX, Nakajima K, Hazon N (1993) A novel angiotensin I isolated from an elasmobranch fish. J Endocrinol 139:281–285

Takei Y, Takahashi A, Watanabe TX, Nakajima K, Ando M (1994a) Eel ventricular natriuretic peptide: isolation of low molecular size form and characterization of plasma form by homologous radioimmunoassay. J Endocrinol 141:81–89

Takei Y, Takano M, Itahara Y, Watanabe TX, Nakajima K, Conklin DJ, Duff DW, Olsen KR (1994b) Rainbow trout ventricular natriuretic peptide: isolation, sequencing, and determination of biological activity. Gen Comp Endocrinol 96:420–426

Takei Y, Ueki M, Nishizawa T (1994c) Eel ventricular natriuretic peptide: cDNA cloning and mRNA expression. J Mol Endocrinol 13:339–345

Takei Y, Butler DG, Watanabe TX, Oudit GY (1998) Tetrapod-type [Asp1] angiotensin I is present in a holostean fish, *Amia calva*. Gen Comp Endocrinol 110:140–146

Taplin LE (1984) Drinking of fresh water but not salt water by the estuarine crocodile (*Crocodylus porosus*). Comp Biochem Physiol 77A:763–767

Taplin LE (1985) Sodium and water budgets of the fasted estuarine crocodile, *Crocodylus porosus*, in sea water. J Comp Physiol B 155:501–513

Taplin LE (1988) Osmoregulation in crocodilians. Biol Rev 63:333–377

Taplin LE, Grigg GC (1981) Salt glands in the tongue of the estuarine crocodile *Crocodylus porosus*. Science 212:1045–1047

Taubenhaus M, Fritz IB, Morton JV (1956) In vitro effects of steroids upon electrolyte transfer through frog skin. Endocrinology 59:458–462

Taylor CR (1969) The eland and the oryx. Sci Am 220:88–95

Taylor CR, Spinage CA, Lyman CP (1969) Water relations of the water buck and East African antelope. Am J Physiol 217:630–634

Taylor GC, Franklin CE, Grigg GC (1995) Salt loading stimulates secretion by the lingual salt glands in unrestrained *Crocodylus porosus*. J Exp Zool 272:490–495

Taylor RE, Barker SB (1965) Transepidermal potential difference: development in anuran larvae. Science 148:1612–1613

Technau G (1936) Die Nasendrüse der Vögel. J Ornithol 84:511–617

Templeton JR (1964) Nasal salt excretion in terrestrial lizards. Comp Biochem Physiol 11: 223–229

Templeton JR (1966) Responses of the lizard nasal gland to chronic hypersalemia. Comp Biochem Physiol 18:563–572

Templeton JR, Murrish DE, Randall E, Mugaas J (1968) The effect of aldosterone and adrenalectomy on nasal salt excretion of the desert iguana, *Dipsosaurus dorsalis*. Am Zool 8:818–819

Templeton JR, Murrish DE, Randall EM, Mugaas JN (1972a) Salt and water balance in the desert iguana *Dipsosaurus dorsalis*. II. The effect of aldosterone and adrenalectomy. Z Vergl Physiol 76:255–269

Templeton JR, Murrish DE, Randall EM, Mugaas JN (1972b) Salt and water balance in the desert iguana, *Dipsosaurus dorsalis*. I. The effect of dehydration, rehydration, and full hydration. Z Vergl Physiol 76:245–254

Tervonen V, Arjamaa O, Kokkonen K, Rushoaho H, Vuolteenaha O (1998) A novel cardiac hormone related to A-, B- and C-type natriuretic peptides. Endocrinology 139:4021–4025

Tervonen V, Ruskoaho H, Vuolteenaho O (2000) Novel cardiac peptide hormone in several teleosts. J Endocrinol 166:407–418

Thaysen JH (1960) Handling of alkali metals by exocrine glands other than the kidney. In: Ussing HH, Kruhøffer P, Thaysen JH, Thorn NA (eds) Handbuch der experimentellen Pharmakologie, vol 13. Springer, Berlin Heidelberg New York, pp 424–507

Therien AG, Blostein R (2000) Mechanisms of sodium pump regulation. Am J Physiol 279: C541–C566

Thibault G, Amiri F, Garcia R (1999) Regulation of natriuretic peptide secretion by the heart. Annu Rev Physiol 61:193–217

Thomas DH (1982) Salt and water excretion by birds: the lower intestine as an integrator of renal and intestinal excretion. Comp Biochem Physiol 71A:527–535

Thomas DH (1997) The ecophysiological role of the avian lower gastrointestinal tract. Comp Biochem Physiol 118A:247–255

Thomas DH, Phillips JG (1975) Studies in avian adrenal function. I. Survival and mineral balance, following adrenalectomy in domestic ducks (*Anas platyrhynchos*). Gen Comp Endocrinol 26:394–403

Thomas DH, Robin AP (1977) Comparative studies of thermoregulatory and osmoregulatory behaviour and physiology of five species of sandgrouse (Aves: Pterocliidae) in Morocco. J Zool (Lond) 183:229–249

Thomas DH, Skadhauge E (1979) Chronic aldosterone therapy and the control of transepithelial transport of ions and water by the colon and coprodeum of the domestic fowl (*Gallus domesticus*) in vivo. J Endocrinol 83:239–250

Thomas DH, Skadhauge E (1988) Transport function and control in bird caeca. Comp Biochem Physiol 90A:591–596

Thomas DH, Skadhauge E (1989) Function and regulation of the avian caecal bulb: influence of dietary NaCl and aldosterone on water and electrolyte fluxes in the hen (*Gallus domesticus*) perfused in vivo. J Comp Physiol B 159:51–60

Thomas DH, Jallageas M, Munck BG, Skadhauge E (1980) Aldosterone effects on electrolyte transport of the lower intestine (coprodeum and colon) of the fowl (*Gallus domesticus*) in vitro. Gen Comp Endocrinol 40:44–51

Thompson AJ, Sargent JR (1977) Changes in the levels of chloride cells and Na-K-dependent ATPase in the gills of yellow and silver eels adapting to sea water. J Exp Biol 200:33–40

Thompson GG, Withers PC (1997) Patterns of gas exchange and extended non-ventilatory periods in small goannas (Squamata: Varanidae). Comp Biochem Physiol 118A:1411–1417

Thomson KS (1980) The ecology of Devonian lobe-finned fishes. In: Panchen AL (ed) The terrestrial environment and the origins of land vertebrates. Academic Press, London, pp 187–222

Thorn N (1960) The total body contents of sodium and potassium. Total exchangeable sodium and potassium. In: Ussing HH, Kruhøffer P, Thaysen JH, Thorn NA (eds) Handbuch der experimentellen Pharmakologie, vol 13. Springer Berlin Göttingen Heidelberg, pp 226–232

Thorson TB (1955) The relationship of water economy to terrestrialism in amphibians. Ecology 36:100–116

Thorson TB (1961) The partitioning of body water in Osteichthyes: phylogenetic and ecological implications of aquatic vertebrates. Biol Bull (Woods Hole) 120:238–254

Thorson TB, Svihla A (1943) Correlation of the habitats of amphibians with their ability to survive the loss of body water. Ecology 24:374–381

Thorson TB, Watson DE (1975) Reassigment of the African freshwater stingray, *Potamotrygon garouaensis*, to the genus *Dasyatis*, on physiologic and morphologic grounds. Copeia 1975: 701–712

Tieleman BI, Williams JB (1999) The role of hyperthermia in the water economy of desert birds. Physiol Biochem Zool 72:87–100

Tieleman BI, Williams JB (2000) The adjustment of avian metabolic rates and water fluxes to desert enviroments. Physiol Biochem Zool 73:461–479

Tierney ML, Luke G, Cramb G, Hazon N (1995) The role of the renin-angiotensin system in the control of blood pressure and drinking in the European eel, *Anguilla anguilla*. Gen Comp Endocrinol 100:39–48

Tierney ML, Takei Y, Hazon N (1997) The presence of angiotensin II receptors in elasmobranchs. Gen Comp Endocrinol 105:9–17

Tipsmark CK, Madsen SS (2001) Rapid modulation of Na$^+$/K$^+$-ATPase activity in osmoregulatory tissues of salmonid fish. J Exp Biol 204:701–709

Toledo RC, Jared C (1993) Cutaneous adaptations to water balance in amphibians. Comp Biochem Physiol 105A:593–608

Toop T, Donald JA, Evans DH (1995a) Natriuretic peptide receptors in the kidney and the ventral and dorsal aortae of the Atlantic hagfish, *Myxine glutinosa* (Agnatha). J Exp Biol 198: 1875–1882

Toop T, Donald JA, Evans DH (1995b) Localization and the characteristics of the natriuretic peptide receptors in the gills of the Atlantic hagfish, *Myxine glutinosa* (Agnatha). J Exp Biol 198:117–126

Toop T, Donald JA, Evans DH (1995c) Natriuretic peptide receptors in the kidney and the ventral and dorsal aortae of the Atlantic hagfish *Myxine glutinosa* (Agnatha). J Exp Biol 198: 1875–1882

Torre-Bueno JR (1978) Evaporative cooling and water balance during flight in birds. J Exp Biol 75:231–236

Towle DW, Gilman ME, Hempel JD (1977) Rapid modulation of gill Na$^+$ + K$^+$-dependent ATPase activity during acclimation of the killifish *Fundulus heteroclitus* to salinity change. J Exp Zool 202:179–186

Townson R (1799) Tracts and observations in natural history and physiology. Printed for the author by J White, London

Toyoda F, Matsuda K, Yamamoto K, Kikuyama S (1996) Involvement of endogenous prolactin in the expression of courtship behavior in the newt, *Cynops pyrrhogaster*. Gen Comp Endocrinol 102:191–196

Tsipoura N, Scanes CG, Burger J (1999) Corticosterone and growth hormone levels in shorebirds during spring and fall migration stopover. J Exp Zool 284:645–651

Tsuchida T, Takei Y (1998) Effects of homologous atrial natriuretic peptide on drinking and plasma ANG II levels in eels. Am J Physiol 275:R1605–R1610

Tuchmann-Duplessis H (1948) Development des caractères sexuels du *Triton* traité par des hormones hypophysaires gonadotrope et lactogènes. CR Soc Biol 142:629–630

Tucker VA (1968) Respiratory exchange and water loss in the flying budgerygah. J Exp Biol 48: 67–87

Tyrrell CL, Cree A (1998) Relationships between corticosterone concentration and season, time of day and confinement in a wild reptile (tuatara *Sphenodon punctatus*). Gen Comp Endocrinol 110:97–108

Uchida S, Kaneko T, Tagawa M, Hirano T (1998) Localization of cortisol receptor in branchial chloride cells in chum salmon fry. Gen Comp Endocrinol 109:175–185

Uchiyama M, Murakami T (1994) Effects of AVT and vascular antagonists on kidney function and smooth muscle contraction in the river lamprey, *Lampetra japonica*. Comp Biochem Physiol 107A:493–499

Uchiyama M, Saito N, Shimada T, Murakami T (1994) Pituitary and plasma vasotocin levels in the lamprey, *Lampetra japonica*. Comp Biochem Physiol 107A:23–36

Uchiyama M, Takeuchi T, Matsuda K (1997) Physiological and pharmacological effects of homologous natriuretic peptides in the bullfrog, *Rana catesbeiana*. In: Kawashima S, Kikuyama S (eds) XIIIth international congress of comparative endocrinology, vol 2. Yokohama (Japan), Monduzzi Editore, Bologna, pp 1243–1246

Uranga J, Sawyer WH (1960) Renal responses of the bullfrog to oxytocin, arginine-vasotocin and frog neurohypophysial extract. Am J Physiol 198:1287–1290

Ussing HH (1947) Interpretation of the exchange of radiosodium in isolated muscle. Nature (Lond) 160:262–263

Ussing HH (1960) The alkali metal ions in isolated systems and tissues. In: Ussing HH, Kruhøffer P, Thaysen JH, Thorn NA (eds) Handbuch der experimentellen Pharmakologie. Springer, Berlin Heidelberg New York, pp 1–195

Ussing HH, Zerhan K (1951) Active transport of sodium as a source of electric current in the short-circuited isolated frog skin. Acta Physiol Scand 23:110–117

Utida S, Hirano T, Oide M, Kamiya M, Saisyu S, Oide M (1966) Na^+-K^+ activated adenosinetriphosphatase in gills and Cl-activated alkaline phosphatase in intestinal mucosa with special reference to salt adaptation of eels. In: Proc XI Pacific science congress, Tokyo (Japan), 7, 5

Uva B, Vallarino M, Mandich A, Isola G (1982) Plasma aldosterone levels in the female tortoise *Testudo hermanni* Gmelin in different experimental conditions. Gen Comp Endocrinol 46: 116–123

Vaandrager AB, Tilly BC, Smolenski A, Schneider-Rasp S, Bott AGM, Edixhoven M, Scholte BJ, Jarchau T, Walter U, Lohmann SM, Poller WC, de Jonge HR (1997) cGMP stimulation of cystic fibrosis transmembrane conductance regulator Cl⁻ channels co-expressed with cGMP-dependent protein kinase type II but not type Iβ. J Biol Chem 272:4195–4200

Valentich JD, Karnaky KJ, Moran WM (1995) Phenotypic expression and natriuretic peptide-activated chloride secretion in cultured shark (*Squalus acanthias*) rectal gland epithelial cells. In: Wood CM, Shuttleworth TJ (eds) Cellular and molecular approaches to fish ionic regulation. Academic Press, San Diego, pp 173–205

Valtin H, Schroeder HA, Benirschke K, Sokol HW (1962) Familial hypothalamic diabetes insipidus in rats. Nature (Lond) 196:1109–1110

Vander AJ (1995) Renal physiology. McGraw-Hill, Health Professions Division, New York

Van Kesteren RE, Smit AB, de With ND, Geraerts WPM, Joose J (1992a) Evolution of the vasopressin/oxytocin superfamily: characterization of a cDNA encoding a vasopressin-related precursor, preproconopressin, from the mollusc *Lymnaea stagnalis*. Proc Natl Acad Sci USA 89:4593–4597

Van Kesteren RE, Smit AB, de With ND, Van Minnen J, Dirks RW, Van der Schors R, Joose J (1992b) A vasopressin-related peptide in the mollusc *Lymnaea stagnalis*: peptide structure, prohormone organization, evolutionary and functional aspects of *Lymnaea* conopressin. Prog Brain Res 92:47–57

Vaughan J, Donaldson C, Bittencourt J, Perrin MH, Lewis K, Sutton S, Chan R, Turnbull AV, Lovejoy D, Rivier C, Rivier J, Sawchenko PE, Vale W (1995) Urocortin, a mammalian neuropeptide related to urotensin I and to corticotropin-releasing hormone. Nature (Lond) 378:287–292

Verrey F (1999) Early aldosterone action toward filling the gap between transcription and transport. Am J Physiol 277:F319–F327

Verrey F, Schaerer E, Zoerkler P, Paccolat MP, Geering K, Kraehenbuhl J-P, Rossier BC (1987) Regulation by aldosterone of Na⁺,K⁺-ATPase mRNAs, protein synthesis, and sodium transport in cultured kidney cells. J Cell Biol 104:1231–1237

Vinson GP, Whitehouse BJ, Goddard C, Sibley CP (1979) Comparative and evolutionary aspects of aldosterone secretion and zona glomerulosa function. J Endocrinol 81:5P–24P

Vleck CM (1993) Homones, reproduction, and behaviour in birds of the Sonoran Desert. In: Sharp PJ (ed) Avian endocrinology. Journal of Endocrinology Ltd., Bristol, pp 73–86

Vleck CM, Priedkalns J (1985) Reproduction in zebra finches: hormone levels and effect of dehydration. Condor 87:37–46

von Sechendorff Hoff K, Hillyard SD (1991) Angiotensin II stimulates cutaneous drinking in the toad *Bufo punctatus*. Physiol Zool 64:1165–1172

von Sechendorff Hoff K, Hillyard SD (1993) Toads taste sodium with their skin: sensory function in a transporting epithelium. J Exp Biol 183:347–351

Wade JB, Stetson DL, Lewis SA (1981) ADH action: evidence for a membrane shuttle mechanism. Ann NY Acad Sci 372:106–117

Wakabayashi S, Shigekawa M, Pouyssegur J (1997) Molecular physiology of vertebrate Na⁺/H⁺ exchangers. Physiol Rev 77:51–74

Wake MH (1993) Evolution of oviductal gestation in amphibians. J Exp Zool 266:394–413

Walker AM, Hudson CL, Findley TJ, Richards AN (1937) The total molecular concentration and the chloride concentration of fluid from different segments of the renal tubule of amphibia. The site of chloride resorption. Am J Physiol 118:121–129

Walker EP (1964) Mammals of the World. The Johns Hopkins Press, Baltimore

Wallis M (1992) The expanding growth hormone/prolactin family. J Mol Endocrinol 9:185–188

Walsh PJ (1997) Evolution and regulation of urea synthesis and ureotely in (Batrachoidid) fishes. Annu Rev Physiol 59:299–323

Walsh PJ (1998) Nitrogen excretion and metabolism. In: Evans DH (ed) The physiology of fishes, 2nd edn. CRC Press, Boca Raton, pp 199–214

Walsh PJ, Heitz MJ, Campbell CE, Cooper GJ, Medina M, Wang YS, Goss GG, Vincek V, Wood CM, Smith CP (2000) Molecular characterization of a urea transporter in the gill of the gulf toadfish (*Opsanus beta*). J Exp Biol 203:2357–2364

Wang J, Barbry P, Maiyar AC, Rozansky DJ, Bhargava A, Leong M, Firestone GL, Pearce D (2001) SGK integrates insulin and mineralocorticoid regulation of epithelial sodium transport. Am J Physiol 280:F303–F313

Warburg MR (1965a) The influence of ambient temperature and humidity on the body temperature and water loss from two Australian lizards, *Tiliqua rugosa* (Gray) and *Amphibolurus* barbatus (Cuvier). Aust J Zool 13:331–350

Warburg MR (1965b) Studies on the water economy of some Australian frogs. Aust J Zool 13:317–330

Warburg MR (1967) On thermal and water balance of three central Australian frogs. Comp Biochem Physiol 20:27–43

Warburg MR (1971) The water economy of Israel amphibians: the urodeles *Triturus vittatus* (Jenyns) and *Salamandra salamandra* (L.). Comp Biochem Physiol 40A:1055–1063

Warburg MR (1997) Ecophysiology of amphibians inhabiting xeric environments. Springer, Berlin Heidelberg New York

Warburg MR, Goldenberg S (1978) The effect of oxytocin and vasotocin on water balance in two urodeles followed throughout their life cycle. Comp Biochem Physiol 60A:113–114

Ward DT, Hammond TG, Harris HW (1999) Modulation of vasopressin-elicited water transport by trafficking of aquaporin2-containing vesicles. Annu Rev Physiol 61:683–697

Waring H (1963) Color change mechanisms in cold blooded vertebrates. Academic Press, New York

Waring H, Moir RJ, Tyndale-Biscoe CH (1966) Comparative physiology of marsupials. In: Advances in comparative physiology and biochemistry, vol 2. Academic Press, New York, pp 237–376

Warne JM, Balment RJ (1995) Effect of acute manipulation of blood volume and osmolality on plasma [AVT] in seawater flounder. Am J Physiol 269:R1107–R1112

281

Warne JM, Hazon N, Rankin JC, Balment RJ (1994) A radioimmunoassay for the determination of arginine vasotocin (AVT): plasma and pituitary concentrations in fresh- and seawater fish. Gen Comp Endocrinol 96:438–444

Watlington CO (1968) Effect of catecholamines and adrenergic blockade on sodium transport of isolated frog skin. Am J Physiol 214:1001–1007

Watlington CO (1998) Focus on "Aldosterone responsiveness of A6 cells is restored by cloned rat mineralocorticoid receptor". Am J Physiol 274:C38

Waugh D, Conlon JM (1993) Purification and characterization of urotensin II from the brain of a teleost (trout, Oncorhynchus mykiss) and an elasmobranch (skate, Raja rhina). Gen Comp Endocrinol 92:419–427

Waugh D, Youson J, Mims SD, Sower S, Conlon JM (1995) Urotensin II from the river lamprey (Lampetra fluviatilis), the sea lamprey (Petromyzon marinus) and the paddlefish (Polyodon spathula). Gen Comp Endocrinol 99:323–332

Weaver D, Walker L, Alcorn D, Skinner S (1994) The contributions of renin and vasopressin to the adaptation of the Australian spinifex hopping mouse (Notomys alexis) to free water deprivation. Com Biochem Physiol 108A:107–116

Webster MK, Goya L, Ge Y, Maiyar AC, Firestone GL (1993) Characterization of sgk, a novel member of the serine/threonine protein kinase gene family which is transcriptionally induced by glucocorticoids and serum. Mol Cell Biol 13:2031–2040

Weisinger RS, Coghlan JP, Denton DA, Fan DA, Hatzikostas S, McKinley MJ, Nelson JF, Scoggins BA (1980) ACTH-elicited sodium appetite in sheep. Am J Physiol 239:E45–E50

Weisinger RS, Denton DA, Di Nicolantonio R, McKinley MJ, Muller AF, Tarjan E (1987) Role of angiotensin in sodium appetite of sodium-deplete sheep. Am J Physiol 253:R482–R488

Weisinger RS, Denton DA, McKinley MJ, Miselis RR, Park RG, Simpson JB (1993) Forebrain lesions that disrupt water homeostasis do not eliminate the sodium appetite of sodium deficiency in sheep. Brain Res 628:166–178

Weiss M, McDonald IR (1965) Corticosteroid secretion in a monotreme Tachyglossus aculeatus. J Endocrinol 33:203–210

Weiss M, McDonald IR (1967) Corticosteroid secretion in kangaroos (Macropus cangurus Major and M.(Megaleia) rufus). J Endocrinol 39:251–261

Westenfelder C, Birch FM, Baranowski RL, Rosenfeld MJ, Shiozawa DK, Kablitz G (1988) Atrial natriuretic factor and salt adaptation in the teleost Gila atraria. Am J Physiol 255: F1281–F1286

Whybron PJ (1981) Evidence for the presence of nasal salt glands in the Hadrosauridae (Ornithischia). J Arid Environ 4:43–57

Wikgren B (1953) Osmotic regulation in some animals with special reference to temperature. Acta Zool Fenn 71:1–102

Wilkie MP (1997) Mechanisms of ammonia excretion across fish gills. Comp Biochem Physiol 118A:39–50

Williams JB, Pacelli MM, Braun EJ (1991a) The effect of water deprivation on renal function in concious unrestrained Gambel's quail (Callipepla gambelii). Physiol Zool 64:1200–1216

Williams JB, Withers PC, Bradshaw SD, Nagy KA (1991b) Metabolism and water flux of captive and free-living Australian parrots. Aust J Zool 39:131–142

Wilson JM, Randall DJ, Vogl AW, Iwama GK (1997) Immmunolocalization of proton-ATPase in the gills of the elasmobranch, Squalus acanthias. J Exp Zool 278:78–86

Wilson JX, Van Pham D, Tan-Wilson HI (1985) Angiotensin and converting enzyme regulate extrarenal salt excretion in ducks. Endocrinology 117:135–140

Wimalawansa SJ (1996) Calcitonin-gene related peptide and its receptors: molecular genetics, physiology, pathophysiology, and therapeutic potentials. Endocrine Rev 17:533–585

Wingfield JC, Schabe H, Mattocks PW (1990) Endocrine mechanisms in migration. In: Gwinner E (ed) Bird migration. Springer, Berlin Heidelberg New York, pp 232–256

Wingfield JC, Vleck CM, Moore MC (1992) Seasonal changes of the adrenocortical response to stress in birds of the Sonoran Desert. J Exp Zool 264:419–428

Wingo CS, Smolka AJ (1995) Function and structure of H-K-ATPase in the kidney. Am J Physiol 269:F1–F16

Wingstrand KG (1966) Comparative anatomy and evolution of the hypophysis. In: Harris GW, Donovan BT (eds) The pituitary gland, vol I. University of California Press, Berkeley, pp 58–126

Winter MJ, Hubbard PC, McCrohan CR, Balment RJ (1999) A homologous radioimmunoassay for the measurement of urotensin II in the euryhaline flounder, *Platichthys flesus*. Gen Comp Endocrinol 114:249–256

Winter MJ, McCrohan CR, Balment RJ (2000) Plasma urotensin (UII) levels fall in response to acute hypotonic challenge in euryhaline flounder, *Platichthys flesus*. J Endocrinol 167: Suppl C7

Withers PC (1983) Energy, water and solute balance in the ostrich *Struthio camelus*. Physiol Zool 56:568–579

Withers PC (1993) Metabolic depression during aestivation in Australian frogs, *Neobatrachus* and *Cyclorana*. Aust Zool 41:467–473

Withers PC (1995) Cocoon formation and structure in the aestivating Australian desert frogs, *Neobatrachus* and *Cyclorana*. Aust J Zool 43:429–441

Withers PC (1998a) Evaporative water loss and the role of cocoon formation in Australian frogs. Aust J Zool 46:405–418

Withers PC (1998b) Urea: diverse functions of a "waste product". Clin Exp Pharmacol Physiol 25:722–727

Withers PC, Guppy M (1996) Do Australian frogs co-accumulate counteracting solutes with urea during aestivation? J Exp Biol 199:1809–1816

Wong CKC, Chan DKO (2001) Effects of cortisol on chloride cells in the gill epithelium of Japanese eel, *Anguilla japonica*. J Endocrinol 168:185–192

Wood CM, Hopkins TE, Hogstrand C, Walsh PJ (1995) Pulsatile urea excretion in the ureogenic toadfish *Opsanus beta*: an analysis of rates and routes. J Exp Biol 198:1729–1741

Wood CM, Hopkins TE, Walsh PJ (1997) Pulsatile urea excretion in the toadfish (*Opsanus beta*) is due to a pulsatile excretion mechanism, not a pulsatile production mechanism. J Exp Biol 200:1039–1046

Wood CM, Gilmour KM, Perry SF, Pärt P, Laurent P, Walsh PJ (1998) Pulsatile urea excretion in gulf toadfish (*Opsanus beta*): evidence for activation of a specific facilitated diffusion transport system. J Exp Biol 201:805–817

Wood CM, Gilmour KM, Perry SF, Laurent P, Zadunaisky J, Walsh PJ (1999) Urea and water permeability in the ureotelic gulf toadfish (*Opsanus beta*). J Exp Zool 283:1–12

Woodall PF, Skinner JD (1993) Dimensions of the intestine, diet and faecal water loss in some African antelope. Zoologist (Lond) 229:457–471

Woolley P (1959) The effects of posterior lobe pituitary extracts on blood pressure in several vertebrate classes. J Exp Biol 36:453–458

Wright A, Chester Jones I, Phillips JG (1957) The histology of the adrenal in prototheria. J Endocrinol 15:100–107

Wright G (1984) Immunocytochemical study of growth hormone, prolactin, and thyroid-stimulating hormone in the adenohypophysis of the sea lamprey, *Petromyzon marinus*. Gen Comp Endocrinol 55:269–274

Wright JW, Harding JW (1980) Body dehydration in xeric adapted rodents: does the renin-angiotensin system play a role? Comp Biochem Physiol 66A:181–188

Wright JW, Krebs LT, Stobb JW, Harding JW (1995) The angiotensin IV system: functional implications. Front Neuroendocrinol 16:23–52

Wright P (1995) Nitrogen excretion: three end products, many physiological roles. J Exp Biol 198:273–281

Wright PA, Land MD (1998) Urea production and transport in teleost fishes. Comp Biochem Physiol 119A:47–54

Wyndham E (1980) Total body lipids in the budgerygah, *Melopsittacus undulatus* (Psittaciformes: Playcercidae) in inland mid-eastern Australia. Aust J Zool 28:239–247

Xavier F (1974) La pseudogestation chez *Nectophrynoides occidentalis* Angel. Gen Comp Endocrinol 22:98–115

Xavier F, Ozon R (1971) Recherches sur l'activité endocrine de l'ovaire de *Nectophrynoides occidentalis* ANGEL (amphibien anoure vivipare). II. Synthèse in vitro de stéroids. Gen Comp Endocrinol 16:30–40

Xu J-C, Lytle C, Zhu TT, Payne JA, Benz JE, Forbush III B (1994) Molecular cloning and functional expression of the bumetanide-sensitice Na-K-Cl cotransporter. Proc Natl Acad Sci USA 91:2201–2205

Yada T, Hirano T, Grau EG (1994) Changes in plasma levels of two prolactins and growth hormone during adaptation to different salinities in the euryhaline tilapia, *Oreochromis mossambicus*. Gen Comp Endocrinol 93:214–223

Yamamoto T, Sasaki S (1998) Aquaporins in the kidney: emerging new aspects. Kidney Int 54:1041–1051

Yamashita T, Matsuda K, Hayashi H, Hanaoka Y, Tanaka S, Yamamoto K, Kikuyama S (1993) Isolation and characterization of two forms of Xenopus prolactin. Gen Comp Endocrinol 91:307–317

Yancey PH, Clark ME, Hand SC, Bowlus RD, Somero GN (1982) Living with water stress: evolution of osmolyte systems. Science 217:1214–1222

Yapp WB (1956) Two physiological considerations in bird migration. Wilson Bull 68:312–319

Yokota S, Iwata K, Fujii Y, Ando M (1997) Ion transport across the skin of the mudskipper *Periophthalmus modestus*. Comp Biochem Physiol 118A:903–910

Yokota SD, Hillman SS (1984) Adrenergic control of the anuran cutaneous hydroosmotic response. Gen Comp Endocrinol 53:309–314

Yorio T, Bentley PJ (1977) Asymmetrical permeability of the integument of tree frogs (HYLIDAE). J Exp Biol 67:197–204

Yorio T, Bentley PJ (1978) The permeability of the skin of the aquatic anuran *Xenopus laevis* (Pipidae). J Exp Biol 72:285–289

Yorio T, Bentley PJ (1979) Evaporative water loss in anuran amphibia: a comparative study. Comp Biochem Physiol 62A:1005–1009

Yoshihara A, Kozawa H, Minamino N, Kangawa K, Matsuo H (1990) Isolation and sequence determination of frog C-type natriuretic peptide. Biochem Biophys Res Commun 173:591–598

You G, Smith CP, Kanai Y, Lee W-S, Stelzner M, Hediger MA (1993) Cloning and characterization of the vasopressin regulated urea transporter. Nature (Lond) 365:844–847

Young G, Björnsson BT, Prunet P, Lin RJ, Bern HA (1989) Smoltification and seawater adaptation in coho salmon (*Oncorhynchus kisutch*): plasma prolactin, growth hormone, thyroid hormones, and cortisol. Gen Comp Endocrinol 74:335–345

Zadunaisky J, Candia OA (1962) Active transport of sodium and chloride by the isolated skin of the South American frog, *Leptodactylus ocellatus*. Nature (Lond) 195:1004

Zadunaisky J, Degnan KJ (1980) Chloride active transport and osmoregulation. In: Lahlou B (ed) Epithelial transport in lower vertebrates. Cambridge University Press, Cambridge, pp 185–196

Zeuthen T (1991) Secondary active transport of water across ventricular cell membrane of choroid plexus epithelium of *Necturus maculosus*. J Physiol (Lond) 444:153–173

Zoeller TR, Moore FL (1988) Brain arginine vasotocin concentrations related to sexual behaviors and hydromineral balance in an amphibian. Horm Behav 22:66–75

Index